VII International Conference: Topical Problems of Continuum Mechanics (TPCM 2021)

Journal of Physics: Conference Series Volume 2231

Online
4 – 8 October 2021

ISBN: 978-1-7138-5487-6
ISSN: 1742-6588

Printed from e-media with permission by:

Curran Associates, Inc.
57 Morehouse Lane
Red Hook, NY 12571

Some format issues inherent in the e-media version may also appear in this print version.

Institute of Physics
Dirac House, Temple Back
Bristol BS1 6BE UK

Phone: 44 1 17 929 7481
Fax: 44 1 17 920 0979

techtracking@iop.org

Additional copies of this publication are available from:

Curran Associates, Inc.
57 Morehouse Lane
Red Hook, NY 12571 USA
Phone: 845-758-0400
Fax: 845-758-2633
Email: curran@proceedings.com
Web: www.proceedings.com

TABLE OF CONTENTS

VII International Conference on Topical Problems of Continuum Mechanics October 4-8, 2021, Tsaghkadzor, Armenia

The VII International Conference was held 4-8 October 2021 hosted by the Institute of Mechanics of the National Academy of Sciences of the Republic of Armenia. The conference aims to bring together leading academic scientists, researchers, and engineers to exchange and share their experience and results on all aspects of continuum mechanics. It also aims to provide the premier interdisciplinary forum for researchers, practitioners, and educators to present and discuss the most recent innovations, trends, and concerns, practical challenges encountered, and solutions adopted in the field of continuum mechanics.

The conference program covers such topics as:

- General Problems
- Contact Mechanics
- Mechanics of Growing Solids
- Mechanics of Technological Processes
- Mechanics of Additive Manufacturing
- Thin-walled Structures
- Elasticity and Viscoelasticity
- Plasticity
- Tribology
- Fracture Mechanics
- Coupled Fields in Solids
- Composites
- Soil Mechanics
- Fluid Mechanics
- Optimal Design and Control Theory
- Mechanics of Nano-systems
- Structural Mechanics
- Biomechanics
- Hydraulics and Hydraulic Facilities
- Experimental Mechanics

During the conference, a joint discussion of fundamental, computational, and applied problems of the new scientific area by leading mechanicians, mathematicians, physicists, and technologists, exchange of the latest ideas and achievements, as well as working out the main directions of its further development were carried out. Representatives from Russia (30), Armenia (17), and Ukraine (1) were participated in the conference.

The geography of participants from Russia is quite wide: Moscow (14 from 5 Institutions), Rostov-on-Don (7 from 2 Institutions), Saint Petersburg (3 from 2 Institutions), Ufa (2), Perm, Saratov, Ekaterinburg, Yaroslavl. The keynote lectures were delivered by professors: S. M. Aizikovich (Russia), A. S. Avetisyan (Armenia), V. N. Hakobyan (Armenia), V. E. Nazaikinsky (Russia), G. N. Kuvyrkin (Russia), V. N. Zimin (Russia).

48 oral and 33 poster addresses have been presented at the conference. The materials of the conference consisting of 73 brief papers have been published at the beginning of the conference.

Figure 1. Conference session

Figure 2. Conference session

V. N. Hakobyan
A. O. Vatulyan
A. V. Sahakyan
M. A. Sumbatyan

Peer Review Statement

All papers published in this volume have been reviewed through processes administered by the Editors. Reviews were conducted by expert referees to the professional and scientific standards expected of a proceedings journal published by IOP Publishing.

- **Type of peer review:** Undefined
- **Conference submission management system:** Morressier
- **Number of submissions received:** 32
- **Number of submissions sent for review:** 0
- **Number of submissions accepted:** 30
- **Acceptance Rate (Submissions Accepted / Submissions Received × 100):** 93.8
- **Average number of reviews per paper:** 0
- **Total number of reviewers involved:** 0
- **Contact person for queries:**
  Name: Avetik Sahakyan
  Email: avsahakyan@gmail.com
  Affiliation: National Academy of Sciences of Armenia - Institute of Mechanics

Exact and approximate solutions of the system of interrelated equations of the theory of creep and long-term strength

A R Arutyunyan* **and R R Saitova****
St. Petersburg State University, St. Petersburg, Russia
E-mail: *a.arutyunyan@spbu.ru, **rigastr@yandex.ru

Abstract. The problem of creep and long-term strength of metallic materials and alloys is considered. Under the long action of high temperatures and relatively small stresses, many metallic alloys and pure metals loose plasticity and undergo brittle fracture. This problem in mechanics of materials is known as the phenomenon of thermal brittleness. To solve this problem Kachanov and Rabotnov introduced the concept of damage into the mechanics of materials. Also, they suggested a system of interrelated kinetic equations for creep deformation and damage parameter. In this work, a modified system of interrelated kinetic equations for creep deformation and damage parameter are formulated. In these equations, the mass conservation law is taking into account and the relative changes of material density are considered as a damage parameter. The results of numerous studies on the changes of porosity and density of various metals and alloys due to the formation and development of micropores and microcracks under conditions of high-temperature creep make it possible to consider density as an integral measure of the accumulation of structural microdefects. Different analytical and numerical solutions of the considered system of kinetic equations are obtained. It is shown, that the character of the curves according to different solutions is identical and agrees with the corresponding experimental results. It was shown, that the numerical solution completely coincides with the case of purely brittle fracture and small deformations. The long-term strength criterion is formulated. Comparison of theoretical creep and long-term strength curves with experimental results for 2.25Cr-1Mo steel is given. A good agreement between the corresponding theoretical and experimental results is observed.

1. Introduction

The problem of creep and long-term strength of metallic materials and alloys is considered. Under the long action of high temperatures and relatively small stresses, many metallic alloys and pure metals loose plasticity and undergo brittle fracture. This problem is known as the phenomenon of thermal brittleness. These effects are observed in elements of many important engineering objects, so the problem of brittle fractures became a subject of numerous theoretical and experimental researches.

To solve this problem, L. M. Kachanov [1], Y. N. Rabotnov [2] introduced the concept of damage in mechanics of materials. To describe the brittle region of the experimental long-term strength curve, a system of simple kinetic equations for the damage parameter and creep deformation was proposed and a long-term strength criterion was formulated.

In the works of R. A. Arutyunyan [3] a modified version of the system of kinetic equations of Kachanov-Rabotnov was proposed. The mass conservation law was taking into account and the damage parameter was specified in the form of relative changes of the material density.

In the scientific literature there are numerous experimental studies on the changes of the porosity and density of various metals and alloys due to the formation and development of micropores and microcracks under conditions of high-temperature creep [4–6]. The results of these studies make it possible to consider density as an integral measure of the accumulation of structural micro-defects, and specified the damage parameter as relative changes of density.

2. A modified version of the Kachanov-Rabotnov system of kinetic equations

Let us consider the system of interrelated kinetic equations of R. A. Arutyunyan for the creep rate and damage parameter:

$$\frac{d\varepsilon}{dt} = B\sigma_0^m \psi^{m-\beta} e^{m\varepsilon}, \tag{1}$$

$$\frac{d\psi}{dt} = -A\sigma_0^n \psi^{n-\alpha} e^{n\varepsilon}, \tag{2}$$

where $\psi = 1 - \omega = \rho/\rho_0$ is continuity parameter, ρ_0 is initial, ρ is current specimen density, A, B, m, n, α, and β are constants.

The exact analytical solution of the proposed system of kinetic equations (1), (2) in the form $\psi(t)$ and $\varepsilon(t)$ is not possible. In this regard, the exact solution of system (1), (2) was obtained in the form $\psi(\varepsilon)$:

$$\psi = \left[1 + \frac{A\sigma_0^{n-m}(1-n+\alpha+m-\beta)}{B(n-m)}(1-e^{(n-m)\varepsilon})\right]^{\frac{1}{1-n+\alpha+m-\beta}}. \tag{3}$$

Various cases of approximate solutions of system (1), (2) were analyzed in [7, 8]. The case of purely brittle fracture and small deformations was considered, when the following approximations $e^{m\varepsilon} \approx 1$ and $e^{n\varepsilon} \approx 1$ were taken into account. In these case

$$\frac{d\varepsilon}{dt} = B\sigma_0^m \psi^{m-\beta}, \tag{4}$$

$$\frac{d\psi}{dt} = -A\sigma_0^n \psi^{n-\alpha}. \tag{5}$$

The solution of system (4), (5) for the creep deformation and the damage parameter under the initial conditions $t = 0$, $\varepsilon = 0$, $\psi = 1$ has the following form:

$$\psi = [1 - (\alpha - n + 1)A\sigma_0^n t]^{\frac{1}{\alpha-n+1}}, \tag{6}$$

$$\varepsilon = \frac{B\sigma_0^{m-n}}{A(m-\beta+\alpha-n+1)}\left\{1 - [1-(\alpha-n+1)A\sigma_0^n t]^{\frac{m-\beta}{\alpha-n+1}+1}\right\}, \tag{7}$$

The solution of the system (4), (5) is also obtained in the form $\psi(\varepsilon)$:

$$\psi = \left[-\frac{A}{B}(-n+m+\alpha-\beta+1)\sigma_0^{n-m}\varepsilon + 1\right]^{\frac{1}{-n+m+\alpha-\beta+1}}. \tag{8}$$

The case of purely brittle fracture and small deformations was considered when $e^{m\varepsilon} \approx 1 + m\varepsilon$, $e^{n\varepsilon} \approx 1 + n\varepsilon$, so the system (1), (2) can be written as:

$$\frac{d\varepsilon}{dt} = B\sigma_0^m \psi^{m-\beta}(1 + m\varepsilon), \tag{9}$$

$$\frac{d\psi}{dt} = -A\sigma_0^n \psi^{n-\alpha}(1 + n\varepsilon). \tag{10}$$

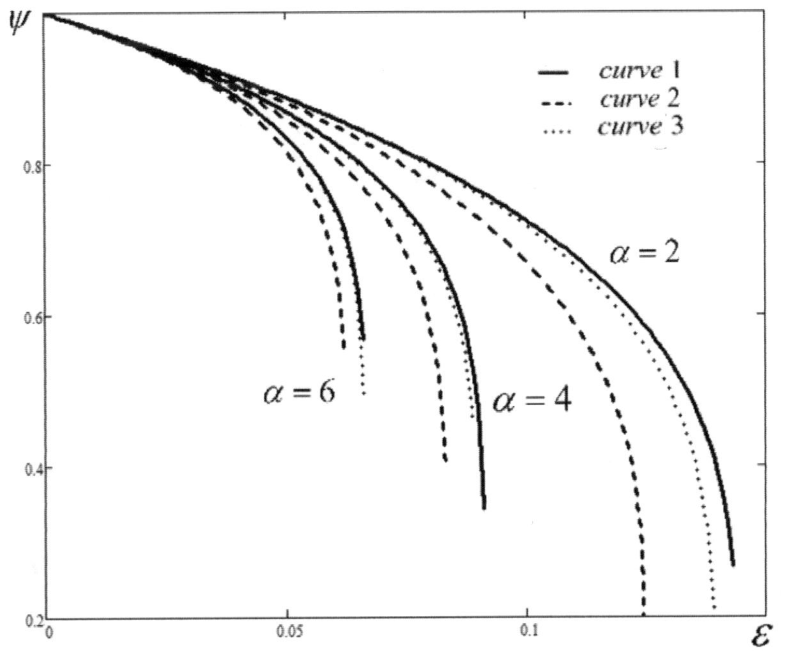

Figure 1. Theoretical curves $\psi(\varepsilon)$ according to solutions (3) (curve *1*), (8) (curve *2*), and (11) (curve *3*) for different values of α.

The solution of the system (9), (10) in the form $\psi(\varepsilon)$ has the following form

$$\psi = \left\{1 - \frac{A}{B} A\sigma_0^{n-m}(1 - n + \alpha + m - \beta)\left[\frac{(m-n)\ln(\varepsilon m + 1)}{m^2} + \frac{\varepsilon n}{m}\right]\right\}^{\frac{1}{1-n+\alpha+m-\beta}}. \quad (11)$$

On the conditions $m = \beta$, $n = 1 + \alpha$ the system (9), (10) were reduced to second-order nonlinear ordinary differential equation [8]. So the solution in the form $\psi(t)$ and $\varepsilon(t)$ was received as follows

$$\psi = \exp\left\{\frac{A\sigma_0^{n-m}(m-n)}{Bm^2}(e^{mB\sigma_0^m t} - 1) - \frac{A\sigma_0^n(m-n)}{m}t\right\}, \quad (12)$$

$$\varepsilon = \frac{e^{Bm\sigma_0^m t} - 1}{m}. \quad (13)$$

The theoretical curves of the changes of the continuity parameter according to the solutions (3) (curve *1*), (8) (curve *2*), and (11) (curve *3*) for different values of the parameter $\alpha = 2, 4, 6$ are shown on the figure 1. In the calculations, the following values of the coefficients were used: $A = 10^{-12}\,\text{MPa}^{-2}\text{h}^{-1}$, $B = 5 \times 10^{-17}\,\text{MPa}^{-4}\text{h}^{-1}$, $\sigma_0 = 100\,\text{MPa}$, $n = 2$, $m = 4$, $\beta = 1$.

Figure 1 shows that the solutions at approximations $e^{m\varepsilon} \approx 1 + m\varepsilon$, $e^{n\varepsilon} \approx 1 + n\varepsilon$ (curve *3*) are located closer to the exact solution (curve *1*) and with an increase of the parameter α these solutions practically coincide.

In addition, an exact solution of the system of kinetic equations (1), (2) was obtained using numerical methods (Runge-Kutte method of the fourth order). For this case, the capabilities of the Mathcad mathematical package were used.

The theoretical curves of the changes of the continuity parameter according to the equation (6) (curve *1*) and according to the numerical solution (curve *2*) are shown on the figure 2. In

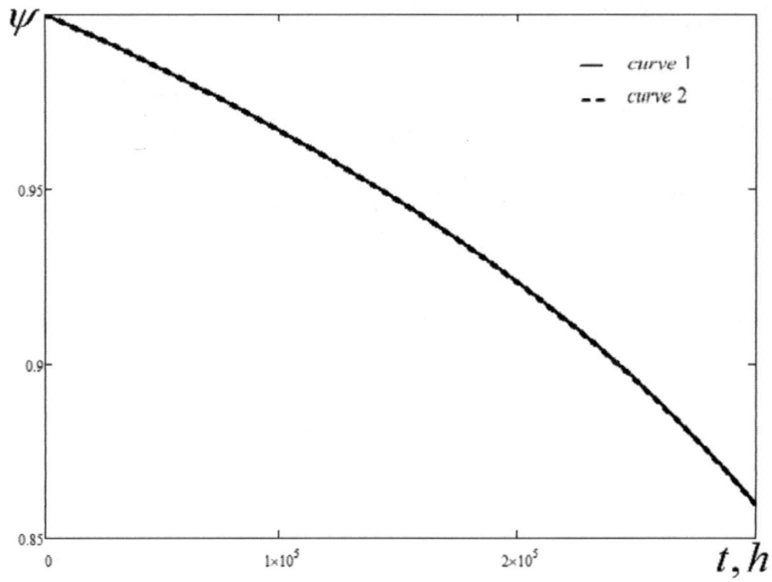

Figure 2. Theoretical curves $\psi(t)$ according to the numerical solution of the system (1), (2) (curve *2*) and for the case of brittle fracture and small deformations according to the equation (6) (curve *1*).

the calculations the following values of the coefficients were used: $A = 2.99 \times 10^{-11}\,\mathrm{MPa^{-2}h^{-1}}$, $n = 2$, $\sigma = 100\,\mathrm{MPa}$, $\alpha = 8.634$. From figure 2 can be seen that these curves and solutions coincide.

Taking the fracture condition at the form $t = t_f$, $\psi = 0$ from the equation (6) we can obtain the following long-term strength criterion

$$t_f = \frac{1}{(\alpha - n + 1)A\sigma_0^n}. \tag{14}$$

For $\alpha = 2n$ criterion (8) coincides with the Kachanov-Rabotnov long-term strength criterion [1, 2].

3. Comparison of the obtained solutions with experimental results

The comparison of theoretical creep curves (relations (7) and (13)) and long-term strength curves (relation (14)) with experimental results for 2.25Cr-1Mo steel at 500°C according [9] were given. The corresponding curves are shown on the figures 3 and 4. In the calculations the following values of the coefficients were used: $A = 5.1 \times 10^{-14}\,\mathrm{MPa^{-2}h^{-1}}$, $B = 1.8 \times 10^{-9}\,\mathrm{MPa^{-4}h^{-2}}$, $\sigma = 98\,\mathrm{MPa}$, $n = 4$, $m = 2$, $\alpha = 8$, $\beta = 2$.

A good agreement between theoretical and experimental creep curves is observed.

On figure 5 the continuity curves according to the relations (6) and (12) are shown for the same values of the coefficients which are used for figures 3 and 4.

4. Conclusions

Under the action of relatively small stresses and high temperatures, metallic materials and alloys become brittle and fracture with a small value of residual deformations. This problem is known as the problem of thermal brittleness of metals. In this paper to solve this problem, the interrelated kinetic equations for creep deformation and the damage parameter are formulated,

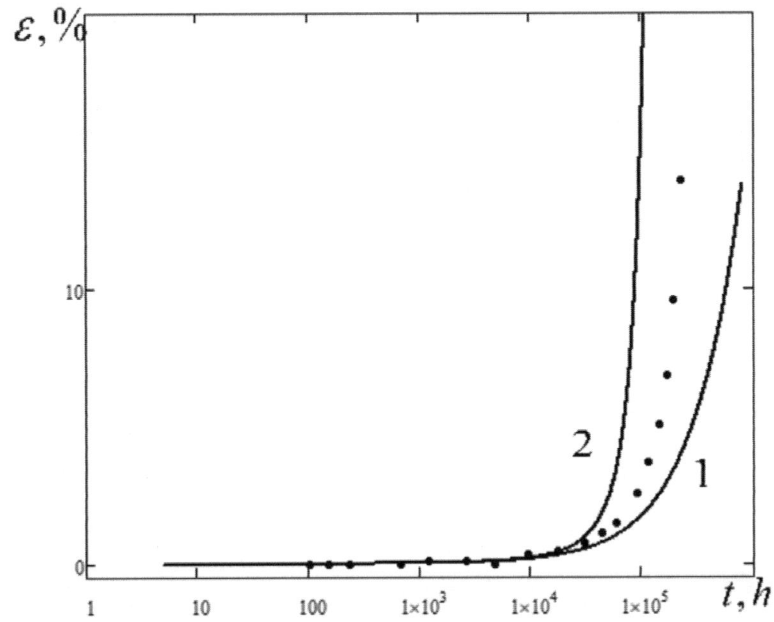

Figure 3. The comparison of theoretical creep curves by relation (7) (curve *1*) and relation (13) (curve *2*) with experimental results for 2.25Cr-1Mo steel at 500°C according [9].

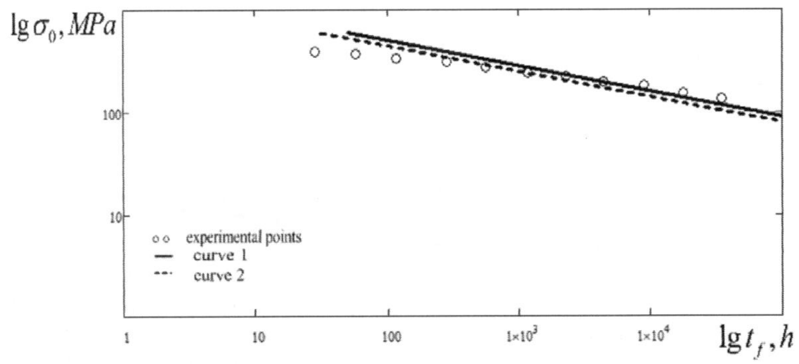

Figure 4. Long-term strength curves according to criterion (14) (curve *1*), the Kachanov-Rabotnov criterion (curve *2*) and experimental points for 2.25Cr-1Mo steel at 500°C by [9].

taking into account the mass conservation law. The relative changes of the material density are considered as a damage parameter. The results of numerous studies on changes of the porosity and density of various metals and alloys due to the formation and development of micropores and microcracks under conditions of high-temperature creep allow us to consider the density as an integral measure of the accumulation of structural microdefects. Different analytical and numerical solutions of the considered system of kinetic equations are obtained and their comparison is given. The character of the curves according to different solutions is identical and agrees with the corresponding experimental results. With increase of the value of the parameter alfa, these solutions become closer to each other and the numerical solution completely coincides with the case of purely brittle fracture and small deformations (when $e^{m\varepsilon} \approx 1+m\varepsilon$, $e^{n\varepsilon} \approx 1+n\varepsilon$). The criterion of long-term strength is formulated. The theoretical creep and long-term strength

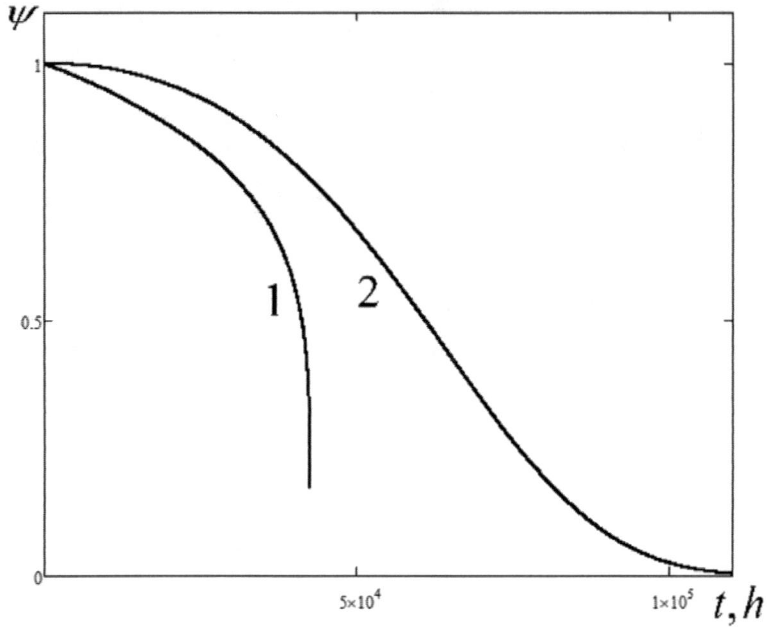

Figure 5. Continuity curves according to relations (6) (curve *1*) and (12) (curve *2*) at the same values of the coefficients.

curves are compared with experimental results for 2.25Cr-1Mo steel. A good agreement of the corresponding theoretical and experimental curves is observed.

References

[1] Kachanov L M 1958 Time of the rupture process under creep conditions *Izvestiya AN SSSR. Otdelenie Tekhnicheskikh Nauk* No. 8 26–31

[2] Rabotnov Y N 1959 *Problems of Strength of Materials and Structures* (Moscow: Izdat. AN SSSR) [in Russian]

[3] Arutyunyan R A 2017 The problem of high-temperature creep and long-term strength in the mechanics of materials *Doklady Physics* **62** (8) 384–6 doi: 10.1134/S1028335817080018

[4] Novozhilov V V 1965 On plastic loosening *Prikladnaya Matematika i Mekhanika* **29** (4) 681–9

[5] Betekhtin V I 1997 Porosity of Solids *St. Petersburg Academy of Sciences for Strength Problems. Transactions* **1** 202–10

[6] Lokoshchenko A M 2015 *Creep and Long-Term Strength of Metals* (Moscow: Fizmatlit) [in Russian]

[7] Arutyunyan A, Arutyunyan R, and Saitova R 2019 The criterion of high-temperature creep of metals based on relative changes of density *WSEAS Transactions on Applied and Theoretical Mechanics* **14** 140–4

[8] Polyanin A D and Zaitsev V F 2003 *Handbook of Exact Solutions for Ordinary Differential Equations.* 2nd Edition (Boca Raton: Chapman & Hall/CRC)

[9] Hore S and Ghosh R N 2011 Computer simulation of the high temperature creep behaviour of Cr-Mo steels *Materials Science and Engineering: A* **528** (19-20) 6095–102 doi: 10.1016/j.msea.2011.04.050

Modelling and investigation of the stress state in a finite cylinder with a circular crack under sudden torsional loading

O V Demydov* **and V G Popov**
National University "Odessa Maritime Academy", Odessa, Ukraine
E-mail: *alexandr.v.demidov@gmail.com

Abstract. This paper is devoted to the analysis of the dynamic stress state of a finite cylinder with an internal circular crack. One of the cylinder's bases is fixed, and a time-dependent torque is applied to the other base. For solving this problem, a modified finite difference method was applied. This method consists in the difference approximation of only the time derivative. As a result, the original problem is reduced to a set of sequentially solved homogeneous boundary value problems. Their solutions contain an unknown displacement jump on the crack. The integral equation for the unknown jump is reduced to the Fredholm integral equation of the second kind, which was solved numerically. This numerical solution made it possible to obtain an approximate formula for calculating the stress intensity factor (SIF).

1. Introduction

The presence of a large number of cylinder-shaped elements among structural parts, machine and mechanisms parts is one of the reasons for the popularity of elasticity theory axisymmetric problems for cylindrical bodies. Up to the present time, their static stress state is well studied [1–10], but there are much fewer works with the analysis of the stress state under dynamic loading. Basically infinite bodies [11–16] or finite cylinders under harmonic loading [17, 18] were considered.

A possible explanation is the complexity and incorrectness of the Laplace transform numerical inversion. Mixed numerical-experimental methods make it possible to avoid these difficulties [19–22]. However, the need for numerous experiments and the preparation of samples makes these methods more complicated and more expensive than numerical-analytical ones, and practically inapplicable for internal defects cases.

A modified method of finite difference in time [23] was used in this paper. The problem of determining the SIF in the region of a circular crack in a finite cylinder under torsion was solved. Until now, such a problem has been considered for the harmonic moment [18]. The authors of this paper also solved problems with a similar statement for a cylinder with a ring-shaped crack [24] and a cylinder with different types of delamination [25].

2. Problem formulation

The study object is a finite-length solid circular cylinder (radius – r_0, length – a) made of a linear elastic isotropic material, weakened by an axisymmetric internal circular crack (radius –

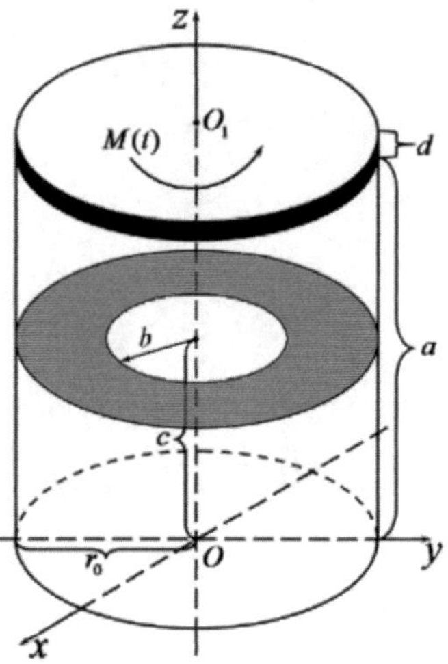

Figure 1. A sketch of a cracked finite-length cylinder subjected to torsional loading.

r_0) located parallel to the rigidly fixed bottom end of the cylinder at a height c from it. The top end of the cylinder is coupled with a rigid plate (radius – r_0, thickness – d), which is subjected to a time-dependent torque $M(t)$. The lateral surface of cylinder and, naturally, crack surface are considered to be free of stresses (figure 1). It is expedient to consider the cracked cylinder torsion problem in a cylindrical coordinate system, assuming that the origin and the Oz axis coincide with the centre of the bottom end and the cylinder axis respectively.

Due to the axisymmetric nature of the torsion problem, the radial and axial displacement components are equal to zero in the entire cylinder with a crack, and the only nonzero component is the angular displacement $\bar{w}(r,z,t)$, which is independent of the azimuth.

In order to generalize the study results, the following dimensionless quantities are introduced:

$$\bar{w}(r,z,t) = r_0 w(\eta,\zeta,\tau), \quad \gamma = \frac{a}{r_0}, \quad l = \frac{c}{a}, \quad \beta = \frac{b}{r_0},$$
$$r = r_0\eta, \quad 0 \le \eta \le 1, \quad z = a\zeta, \quad 0 \le \zeta \le 1, \quad t = \frac{r_0\tau}{c_2}, \quad \tau \in (0,+\infty), \quad c_2^2 = \frac{G}{\rho},$$

where ρ, G are the mass density and shear modulus for the cylinder material.

Non-demensional angular displacement satisfies the equation

$$D_{\eta\zeta} w = \frac{\partial^2 w}{\partial \tau^2}, \quad D_{\eta\zeta} = \frac{\partial^2}{\partial \eta^2} + \frac{1}{\eta}\frac{\partial}{\partial \eta} - \frac{1}{\eta^2} + \frac{1}{\gamma^2}\frac{\partial^2}{\partial \zeta^2} \tag{1}$$

Assuming that the cylinder is initially stationary, equation (1) should be considered with zero initial conditions.

According to the above statements, the mixed boundary conditions in dimensionless quantities can be formulated as follows.

Shear stress on the lateral surface of the cylinder and angular displacement at the bottom end of it:

$$\tau_{\varphi\eta}(1,\zeta,\tau) = 0, \quad 0 \le \zeta \le 1, \quad w(\eta,0,\tau) = 0, \quad 0 \le \eta \le 1 \quad \tau \in [0,+\infty). \tag{2}$$

At the top cylinder end, the angular displacement must satisfy the equality:

$$w(\eta,1,\tau) = \eta\alpha(\tau), \quad 0 \le \eta \le 1, \quad \tau \in [0,+\infty), \tag{3}$$

where $\alpha(\tau)$ — the unknown angle of the plate rotation, determined from the equation of its motion

$$\frac{\pi\gamma m_0}{2}\ddot{\alpha}(\tau) = M_0(\tau) - M_R(\tau), \quad \alpha(0) = 0, \quad \dot{\alpha}(0) = 0,$$

where m_0 is ratio of the plate mass to cylinder mass, M_0 and M_R are dimensionless moments.

The shear stress absence and the displacement jump on the crack surface mean that the following equalities must be satisfied:

$$\tau_{\varphi\zeta}(\eta,l,\tau)=0, \quad 0\leq\eta\leq\beta, \quad w(\eta,l+0,\tau)-w(\eta,l-0,\tau)=\chi(\eta,\tau), \quad 0\leq\eta\leq\beta, \quad \tau\in[0,+\infty). \quad (4)$$

where $\chi(\eta,\tau)$ is an unknown non-dimensional displacement jump in the plane of the crack, and $\chi(\eta,\tau)\equiv 0$, $\eta\geq\beta$.

To solve the formulated initial-boundary value problems (1)–(4) a method based on the difference approximation of time derivatives, detailed in [23] was applied. For this purpose, we create a time grid:

$$\tau_k = \sum_{\nu=1}^{k} h_\nu, \quad h_\nu = \tau_k - \tau_{k-1} \quad (\tau_0 = 0), \ k = 1,2,3,\dots, \quad h_i \neq h_j, \quad i \neq j.$$

We introduce the designation $w(\eta,\zeta,\tau_k) = w_k(\eta,\zeta)$ and use the left difference for approximation time derivatives.

Then, the following differential equations were found from initial conditions and equation (1)

$$w_0 = 0, \quad D_{\eta\zeta}w_1 = \frac{w_1}{h_1^2}, \quad D_{\eta\zeta}w_k - \frac{w_k}{h_k^2} = \frac{w_{k-2}}{h_k h_{k-1}} - \frac{w_{k-1}}{h_k}\left(\frac{1}{h_k} + \frac{1}{h_{k-1}}\right), \quad k = 2,3,\dots$$

The analytical solution to these equations is considerably complicated by the fact that on the right side there are displacements values in the two previous moments of time. To avoid these difficulties, according to [23], we write angular displacement, angle of rotation of the plate and stress in the form of a linear combination of new unknown functions:

$$w_k = \sum_{\nu=1}^{k} C_{k\nu}U_\nu, \quad \alpha_k = \sum_{\nu=1}^{k} C_{k\nu}A_\nu, \quad \tau_{\varphi\eta k} = \sum_{\nu=1}^{k} C_{k\nu}\tau_{\varphi\eta\nu}, \quad \tau_{\varphi\zeta k} = \sum_{\nu=1}^{k} C_{k\nu}\tau_{\varphi\zeta\nu}. \quad (5)$$

It was shown in [23], that if we choose the coefficients in formulas (5) according to the formulas

$$C_{kk} = 1, \quad k = 1,2,3,\dots, \quad C_{k,k-1} = \frac{h_{k-1}}{h_{k-1} - h_k}, \quad k = 2,3,\dots,$$

$$C_{k,\nu} = \frac{h_\nu^2}{h_k^2 - h_\nu^2}\left[\frac{h_k}{h_{k-1}}C_{k-2,\nu} - \left(1 + \frac{h_k}{h_{k-1}}\right)C_{k-1,\nu}\right], \quad k = 3,4,\dots, \quad \nu = 1,2,\dots,k-2,$$

then the functions U_ν satisfy the homogeneous Helmholtz equations:

$$D_{\eta\zeta}U_\nu - \kappa_\nu^2 U_\nu = 0, \quad \nu = 1,2,\dots, \quad \kappa_\nu = h_\nu^{-1}. \quad (6)$$

The boundary conditions on the lateral surface and at the cylinder ends with respect to these functions can be written as follows:

$$\tau_{\varphi\eta\nu}(1,\zeta) = 0, \quad 0 < \zeta < 1, \quad U_\nu(\eta,0) = 0, \quad U_\nu(\eta,1) = \eta A_\nu, \quad 0 < \eta < 1. \quad (7)$$

A_ν is determined from the equation:

$$\pi\gamma m_0\kappa_\nu^2 A_\nu = 2(\mu_{0\nu} - \mu_{R\nu}), \quad \mu_{R\nu}(\tau) = 2\pi\int_0^1 \eta^2 \tau_{\varphi\zeta\nu}(\eta, 1)\, d\eta, \tag{8}$$

where $\mu_{0\nu}$ can be found from the recurrence relation.

And the boundary conditions for the crack will take the form:

$$\tau_{\varphi\zeta\nu}(\eta, 0) = \begin{cases} \chi_{2\nu}(\eta), & 0 \le \eta \le \beta, \\ 0, & \beta < \eta \le 1, \end{cases} \quad \chi_2(\eta, \tau_k) = \sum_{\nu=1}^{k} C_{k\nu}\chi_{2\nu}(\eta). \tag{9}$$

3. Solution to the problem

According to the superposition method, the solution to the boundary-value problem (6)–(9) can be expressed as:

$$U_\nu(\eta, \zeta) = U_\nu^0(\eta, \zeta) + U_\nu^*(\eta, \zeta).$$

The first term is the solution to the uncracked cylinder problem (6)–(8) and it is given by the formula:

$$U_\nu^0(\eta, \zeta) = A_\nu \eta \frac{\sinh(\gamma\zeta\kappa_\nu)}{\sinh(\gamma\kappa_\nu)}.$$

The second term is the solution that satisfies the zero conditions on the ends and lateral surface of the cylinder. The condition on the crack (9) is held. Hence equality should be satisfied:

$$\tau_{\varphi\zeta\nu}^*(\eta, l) = -\tau_{\varphi\zeta\nu}^0(\eta, l), \quad 0 < \eta < \beta. \tag{10}$$

The solution to this boundary value problem is constructed by the integral transform method, analogous to papers [24, 25] and it has the form:

$$U_\nu^*(\eta, \zeta) = \int_0^\beta \xi\chi_\nu(\xi)[S(\xi, \eta, \zeta) + D(\xi, \eta, \zeta)]\, d\xi, \tag{11}$$

where

$$S(\xi, \eta, \zeta) = \frac{1}{2}\int_0^\infty \lambda J_1(\eta\lambda) J_1(\xi\lambda)[F(\lambda, \zeta - l) + F(\lambda, \zeta + l)]\, d\lambda,$$

$$D(\xi, \eta, \zeta) = 2\sum_{j=1}^{\infty} \frac{\lambda_j}{\gamma^2}\cos(\lambda_j l)\sin(\lambda_j \zeta)\frac{K_2(q_{j\nu})}{I_2(q_{j\nu})} I_1(q_{j\nu}\xi) I_1(q_{j\nu}\eta),$$

$$F(\lambda, \zeta \pm l) = \operatorname{sgn}(\zeta \pm l)\frac{\sinh[\alpha_\nu(1 - |\zeta \pm l|)]}{\sinh \alpha_\nu}, \quad q_{j\nu} = \left(\frac{\lambda_j^2}{\gamma^2} + \kappa_\nu^2\right)^{1/2},$$

$$\lambda_j = \pi j, \quad \alpha_\nu = \gamma\kappa_\nu(\lambda^2 + \kappa_\nu^2)^{1/2}.$$

The solution (11) contains the unknown function $\chi_\nu(\xi)$. In order to determine it, the following integral equation was obtained by using the condition on the crack (10) and integration by parts:

$$\int_0^\beta \xi\psi_\nu(\xi)[F_1(\xi, \eta) + D_1(\xi, \eta)]\, d\xi = A_\nu \eta\gamma\kappa_\nu \frac{\cosh(\gamma\kappa_\nu l)}{\sinh(\gamma\kappa_\nu)}, \tag{12}$$

where

$$\psi_\nu(\xi) = \frac{1}{\xi}\frac{d}{d\xi}[\xi\chi(\xi)], \quad F_1(\xi, \eta) = \left.\frac{\partial S(\xi, \eta, \zeta)}{\partial \zeta}\right|_{\zeta = l}, \quad D_1(\xi, \eta) = \left.\frac{\partial D(\xi, \eta, \zeta)}{\partial \zeta}\right|_{\zeta = l}.$$

Next, equation (12) was reduced to the Fredholm equation of the second kind in accordance with the known method used in [25].

The new unknown function was introduced

$$\psi_\nu(\xi) = -\frac{2}{\pi}\frac{1}{\xi}\frac{d}{d\xi}\int_\xi^\beta \frac{\tau\varphi_\nu(\tau)}{\sqrt{\tau^2-\xi^2}}d\tau.$$

And the Weber-Sonin operator was applied

$$D_2[f] = \frac{d}{dx}\int_0^x \frac{y\,dy}{\sqrt{x^2-y^2}}\int_0^y f(\eta)\,d\eta.$$

Due to these transformations, the introduction of designations

$$\lambda = u\kappa_\nu, \quad \sqrt{u^2+1} = p, \quad \tau = \beta y, \quad \varphi_\nu(\tau) = \varphi_\nu(\beta y) = \beta g_\nu(y), \quad y \in [0,1], \quad x = \beta s, \quad s \in [0,1]$$

and even extension of the function $g_\nu(y)$ on the interval $[-1,1]$, equation (12) is reduced to the Fredholm integral equation of the second kind:

$$g_\nu(s) = \frac{2\beta}{\pi\gamma}\int_{-1}^1 g_\nu(y)[B(y,s)+Q(y)-Q(y-s)]\,dy - g_\nu(0) = -\frac{2\beta\kappa_\nu A_\nu}{\sinh(\gamma\kappa_\nu)}s^2, \tag{13}$$

where $B(Y)$ and $Q(Y)$ are represented as uniformly convergent series and proper integrals.

Equation (13) contains unknown A_ν, equation (8) should be used to determine it.

An approximate solution to equation (13), as in [24, 25], is sought in the form of an interpolation polynomial.

Integrals were approximated according to the quadrature Gauss-Legendre formula and a system of linear algebraic equations with respect to the unknown function values in the interpolation nodes was obtained. After solving the system, the unknown function is approximated by the interpolation polynomial

$$g_\nu(y) \approx g_\nu^n(y) = \sum_{m=1}^n g_{\nu m}\frac{P_n(y)}{(y-y_m)P_n'(y_m)}, \quad g_{\nu m} = g_\nu(y_m), \quad m = 1,2,3,\ldots,n,$$

where $P_n(y)$ is the nth Legendre polynomial, and y_m is the polynomial root.

4. Results of numerical analysis

The SIF is one of the criteria of destruction and is used in fracture mechanics to describe the stress state in the region of defects. It can be determined by the formula:

$$\bar{K}_k = \lim_{r\to b+0}\sqrt{r-b}\,\bar{\tau}_{\varphi zk}(r,0).$$

The non-dimensional SIF value can be calculated by the formulas:

$$K(\tau_k) = \frac{\bar{K}(t_k)}{G\sqrt{b}}, \quad K(\tau_k) = \sum_{\nu=1}^k C_{k\nu}K_\nu, \quad K_\nu = -\frac{\sqrt{\beta}}{\sqrt{2\pi}\gamma}g_\nu. \tag{14}$$

In order to investigate the influence of various factors on the dependence of the SIF on the dimensionless time $\tau = c_2t/r_0$, numerical experiments were carried out using the formula (14). To study the SIF behaviour immediately after loading, the time grid nodes were condensed near

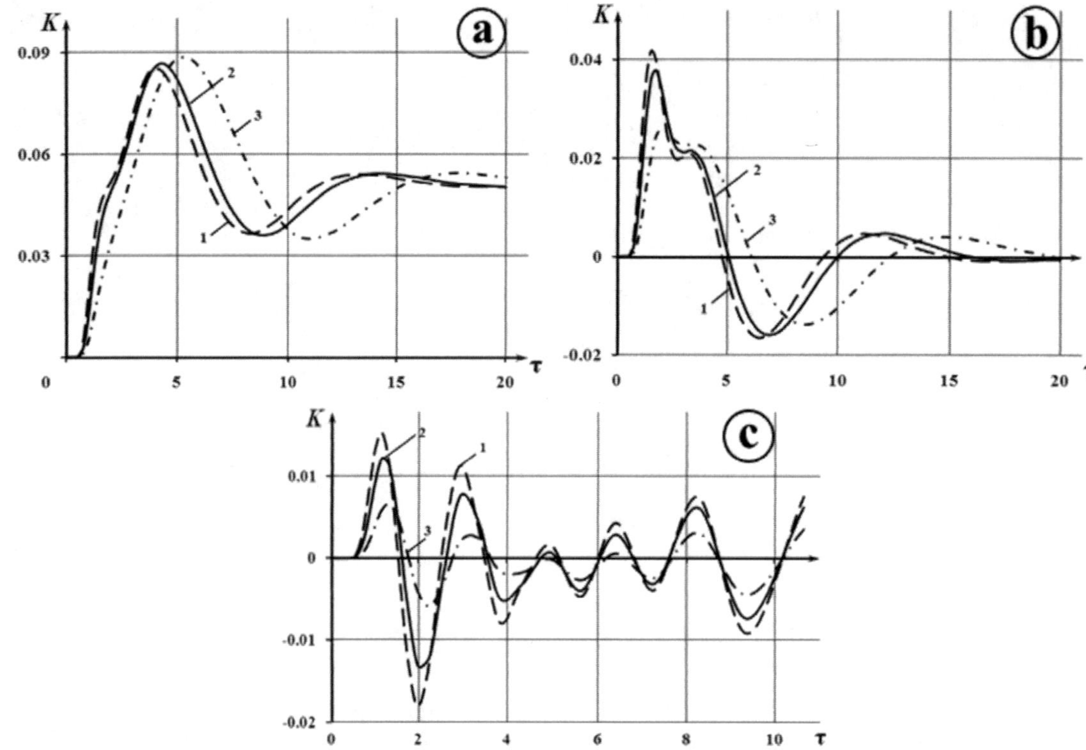

Figure 2. Non-dimensional SIFs time dependence for different types of load: a) $M_0(\tau) = H(\tau)$; b) $M_0(\tau) = H(\tau) - H(\tau - 1)$; c) $M_0(\tau) = H(\tau)\cos(3\tau)$.

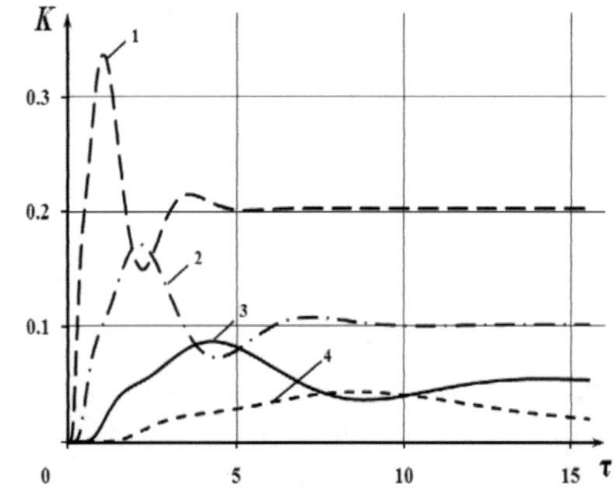

Figure 3. The relative cylinder length influence on non-dimensional SIF value.

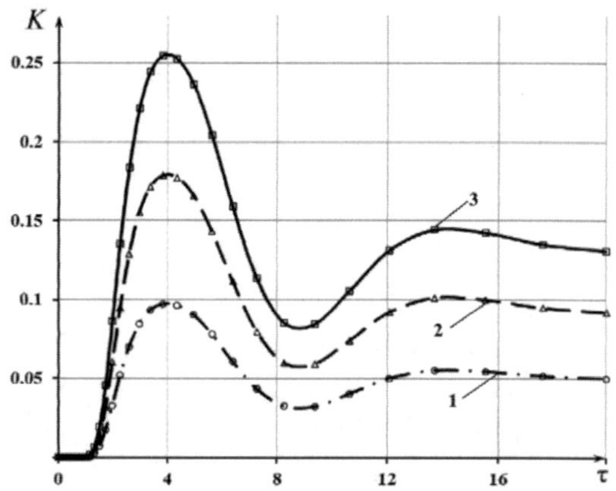

Figure 4. The relative crack radius influence on non-dimensional SIF value.

the point $\tau = 0$. During these calculations, it was considered that dimensionless thickness of the plate is $\delta = d/a = 0.1$, the dimensionless radius of the circular crack is $\beta = b/r_0 = 0.5$.

The influence of factors such as the type of load, mass of the plate, geometrical parameters of the cylinder and cracks on the time dependence of the SIFs was studied. According to the results of numerical experiments, graphs of the dimensionless SIFs time dependences were plotted.

Figure 2 demonstrates the loading type influence. The influence of a suddenly applied torsional load (figure 2*a*), the torsional load by a suddenly applied moment of the unit length (figure 2*b*), and harmonic torque $M_0(\tau) = H(\tau)\cos(3\tau)$ (figure 2*c*) was evaluated. The dimensionless cylinder length is $\gamma = a/r_0 = 2$. Curves *1–3* correspond to different values of the relative plate mass density: $\bar{\rho} = \rho_{\text{plate}}/\rho_{\text{cyl}}$: 0.25, 1, 4.

In the case of constant and oscillating loading (figures 2*a*, 2*c*), a transition to a stationary regime is clearly observed. The maximum values of SIFs are observed during the transient period, and they are 2–2.5 times greater than the values of static SIFs and SIFs at steady harmonic oscillations. It shows that the growth of a crack and destruction of the cylinder are possible precisely during the transient period. Under pulsed loading (figure 2*b*), the SIF increases to a maximum value, and then its value tends to zero, making damped oscillations.

Figures 3 and 4 show the results of studying the influence of the relative cylinder length and the relative crack radius, respectively, on the values of the SIF under the suddenly applied constant torque at the relative plate mass density $\bar{\rho} = \rho_{\text{plate}}/\rho_{\text{cyl}} = 1$. Curves *1–3* in the figure 3 correspond to the values of the relative cylinder length $\gamma = a/r_0$: 0.5, 1, 2, 4 with the relative crack radius $\beta = b/r_0 = 0.5$. Curves *1–3* in the figure 4 correspond to the values of the relative crack radius $\beta = b/r_0$: 0.5, 0.75, 0.95 with the relative cylinder length $\gamma = a/r_0 = 2$

It is shown that with an increase in the cylinder relative height, the maximum value of the SIF decreases and the duration of the transient process increases. An increase in the relative crack radius leads to an increase in the maximum SIF and does not affect the duration of the transient process.

Conclusion

Using the example of solving the problem of determining the stress state in a finite cylinder with an internal circular crack under torsional loading, the efficiency of applying the modified method of finite differences in time to solving non-stationary problems of elasticity theory for finite bodies with cracks is shown. This method makes it possible to avoid the difficulties associated with the numerical inversion of the integral Laplace transform and obtain reliable results over a sufficiently long time interval for various types of loading.

However, it should be noted that, in the framework of the above problem statement, no crack can be indefinitely moved nearer to the cylinder ends, since in the case of the crack approaching them the convergence of integrals and series that determine the solution and integral equations kernels deteriorates significantly, and upon reaching the very ends, the integrals, in general, become singular. Consequently, for these borderline cases, it is necessary to solve individual problems. They are given in [25].

There also arise some problems in applying this technique for large amounts of time, due to the step-by-step accumulation of errors.

References

[1] Grinchenko V T 1978 *Equilibrium and Stationary Oscillations of Elastic Bodies of Finite Dimensions* (Kiev: Naukova Dumka) [in Russian]

[2] Chang S S 1985 The general solution of a finite cylinder with a concentric penny-shaped crack under torsion *Engineering Fracture Mechanics* **22** (4) 571-8 doi: 10.1016/0013-7944(85)90120-1

[3] Chau K T and Wei X X 2000 Finite solid circular cylinders subjected to arbitrary surface load. Part I — Analytic solution *International Journal of Solids and Structures* **37** 5707-32 doi: 10.1016/S0020-7683(99)00289-9

[4] Kaman M O and Gecit M R 2006 Cracked semi-infinite cylinder and finite cylinder problems *International Journal of Engineering Science* **44** (20) 1534–55 doi: 10.1016/j.ijengsci.2006.08.009

[5] Kaman M O and Gecit M R 2008 Axisymmetric finite cylinder with one end clamped and the other under uniform tension containing a penny-shaped crack *Engineering Fracture Mechanics* **75** (13) 3909-23 doi: 10.1016/j.engfracmech.2008.03.002

[6] Malits P 2009 Torsion of a cylinder with a shallow external crack *International Journal of Solids and Structures* **46** (16) 3061-7 doi: 10.1016/J.IJSOLSTR.2009.04.009

[7] Meleshko V V and Tokovyy Y V 2013 Equilibrium of an elastic finite cylinder under axisymmetric discontinuous normal loadings *Journal of Engineering Mathematics* **78** (1) 143-66 doi: 10.1007/s10665-011-9524-y

[8] Hassani A R, Faal R T, and Noda N-A 2017 Torsion analysis of finite solid circular cylinders with multiple concentric planar cracks *ZAMM — Journal of Applied Mathematics and Mechanics* **97** (4) 458-72 doi: 10.1002/zamm.201600095

[9] Vaysfeld N and Protcerov Y 2017 Torsion problems of finite cylinders weakened by ring-shaped cracks *Procedia Structural Integrity* **3** 526-44 doi: 10.1016/j.prostr.2017.04.046

[10] Rahnama H and Shokrieh M M 2019 Axisymmetric equilibrium of an isotropic elastic solid circular finite cylinder *Mathematics and Mechanics of Solids* **24** (4) 996-1029 doi: 10.1177/1081286518764013

[11] Martin P A and Wickham G R 1983 Diffraction of elastic waves by a penny-shaped crack: Analytical and numerical results *Proceedigs of the Royal Society A: Mathematical, Physical and Engineering Sciences* **390** (1798) 91–129 doi: 10.1098/rspa.1983.0124

[12] Singh B M, Haddow J B, Vrbik J, and Moodie T B 1980 Dynamic stress intensity factors for penny-shaped crack in twisted plate *Journal of Applied Mechanics* **47** (4) 963-5 doi: 10.1115/1.3153826

[13] Srivastava K N, Palaiya R M and Gupta O P 1982 Interaction of elastic waves with a penny-shaped crack in an infinitely long cylinder *Journal of Elasticity* **12** 143-52 doi: 10.1007/BF00043709

[14] Fu J, Chen Z, Qian L, and Hu K 2014 Transient thermoelastic analysis of a solid cylinder containing a circumferential crack using the CV heat conduction model *Journal of Thermal Stresses* **37** (11) 1324-45 doi: 10.1080/01495739.2014.937214

[15] Pourseifi M and Faal R T 2015 Tension analysis of infinite solid circular cylinders with arbitrary located axisymmetric cracks *Theoretical and Applied Fracture Mechanics* **80** (B) 182-92 doi: 10.1016/j.tafmec.2015.08.003

[16] Kyrylova O I and Mykhas'kiv V V 2019 Harmonic vibration and resonance effects in the case of longitudinal shear of a hollow cylinder with crack *Materials Science* **55** 114-23 doi: 10.1007/s11003-019-00258-3

[17] Ebenezer D D, Ravichandran K, and Padmanabhan C 2015 Free and forced vibrations of hollow elastic cylinders of finite length *The Journal of the Acoustical Society of America* **137** 2927–38 doi: 10.1121/1.4919339

[18] Popov V G 2012 Stress state of a finite elastic cylinder with a circular crack at torsional vibrations *International Applied Mechanics* **48** (4) 430–7 doi: 10.1007/s10778-012-0530-1

[19] Bui H D, Maigre H, and Rittel D 1992 A new approach to the experimental determination of the dynamic stress intensity factor *International Journal of Solids and Structures* **29** (23) 2881-95 doi: 10.1016/0020-7683(92)90146-K

[20] Andreikiv O E, Boiko V M, Kovchyk S E, and Khodan I V 2000 Dynamic tension of a cylindrical specimen with circumferential crack *Materials Science* **36** (3) 382–91 doi: 10.1007/BF02769599

[21] Ivanyts'kyi Y L, Boiko V M, Khodan' I V, and Shtayura S T 2007 Stressed state of a cylinder with external circular crack under dynamic torsion *Materials Science* **43** (2) 203–14 doi: 10.1007/s11003-007-0023-2

[22] Petureau L, Doquet V, and Dieu B 2022 Influence of a static torque on plasticity and asperity-induced closure effects during fatigue crack growth in bainitic steel cylinders loaded in push-pull *International Journal of Fatigue* **155** 106628 doi: 10.1016/j.ijfatigue.2021.106628

[23] Savruk M P 2003 New method for the solution of dynamic problems of the theory of elasticity and fracture mechanics *Materials Science* **39** (4) 465–71 doi: 10.1023/B:MASC.0000010922.84603.8d

[24] Demydov O and Popov V 2019 Stress state in a finite cylinder with outer ring-shaped crack at non-stationary torsion In *Proc. Second Int. Conf. on Theoretical, Applied and Experimental Mechanics. ICTAEM 2019. Structural Integrity* ed. E Gdoutos (Cham: Springer) pp. 215–21 doi: 10.1007/978-3-030-21894-2_41

[25] Demydov O V and Popov V G 2020 Nonstationary torsion problems for the finite cylinder partly coupled with rigid base *Journal of Physics: Conference Series* **1474** 012012 doi: 10.1088/1742-6596/1474/1/012012

Mechanical properties of metal–matrix composites reinforced with carbide structures

N F Morozov, B N Semenov*, V M Smirnov, and E G Zemtsova
St. Petersburg State University, St. Petersburg, Russia
E-mail: *b.semenov@spbu.ru

Abstract. Metal–matrix composites based on an aluminum matrix reinforced with ceramic particles are widely used as structural materials in the aerospace and automotive industries. The main problems in the manufacture are the uniformity of the distribution of particles over the volume, poor adhesion of ceramic particles to the matrix metal, and the formation of aluminum carbide at the interfaces, leading to undesirable embrittlement. We propose a new technique for the manufacture of metal–matrix composites, which consists in creating a composite material, when carbide nanostructures form a framework in the volume of an aluminum matrix, and which allows solving the above problems. A finite element model of deformation of a metal–matrix composite reinforced with carbide structures is constructed.

1. Introduction

Due to their high performance properties, such as high specific strength, wear resistance, metal–matrix composites are now widely used in various industries, such as aerospace, automotive and electronics.

The most common matrix materials of these composites are aluminum and iron, which are reinforced with ceramic particles or fibers, such as silicon carbide (SiC), titanium carbide (TiC), boron carbide (B_4C), and aluminum oxide (Al_2O_3). Of particular interest for applications is the reinforcement of an aluminum matrix with both microparticles and nanoparticles of silicon carbide, which provides a significant strengthening of the metal composite. The magnitude of this strengthening depends on the volume fraction of the reinforcing particles, their geometry and size, adhesion between the matrix and the reinforcing particles, as well as the conditions for introducing the reinforcing particles into the matrix.

The in experiments observed strengthening of metal–matrix composites is the result of the interaction of several strengthening mechanisms: due to the effective transfer of the load between the matrix and the reinforcing elements, due to the formation of internal thermal stresses caused by a mismatch in the coefficients of thermal expansion between the matrix and carbide particles, due to grain refinement matrix (Hall-Petch), Orowan-strengthening caused by the dispersed phase and strengthening due to the mismatch of the elastic moduli of the matrix materials and the particles. Which of these mechanisms are implemented in each specific case is determined both by the geometry, size, concentration of reinforcing particles and the conditions for their introduction into the matrix. It should be noted that along with a noticeable increase in strength in metal–matrix (nano)-composites, there is a significant decrease in the ability to absorb energy (plasticity, deformation to failure).

The influence of volume fraction, size and geometry of reinforcing carbide inclusions on the value of the hardening effect of silicon carbide in a matrix of aluminum alloy was investigated [1–4]. Mechanisms of strengthening for reinforcement with micro-sized particles were proposed and the limits of their application were studied. It was shown that reinforcement with silicon carbide in the form of fibers or plates leads to an increase in the degree of hardening compared to reinforcement with particles at the same volume fraction of inclusions

In recent decades, considerable attention has been paid to the study of the properties of metal–matrix composites reinforced with nano-sized carbide particles, since reinforcement with nanoparticles leads to a higher value of strengthening compared to reinforcement with micro-sized particles at the same volume fraction of particles.

An analysis of the effect of the volume fraction, size of reinforcing particles and grain size of a metal matrix on mechanical properties of a metal matrix nanocomposite based on aluminum, including yield strength, tensile strength and deformation to fracture, was carried out, as well as a quantitative analysis of the strengthening mechanisms [5–8]. It is noted that, upon stirring, the size of the crystallites of the Al–matrix decreases with an increase in the volume fraction of reinforcement and a decrease in the size of nanosized reinforcing particles. In addition, SiC nanoparticles slow down the grain growth of the matrix material during sintering, that is, the ultrafine-grained structure is retained even at a relatively high sintering temperature. With an increase in the volume fraction of SiC nanoparticles, the strength of the metal matrix composite increases. It is shown that there is an optimum volume fraction of reinforcing particles in the matrix, exceeding which a decrease in the strength of the nanocomposite is observed [8]. The contribution of various hardening mechanisms to an increase in the strength of a metal composite with nanosized reinforcing particles of silicon carbide is investigated.

Deformation and failure of metal matrix composites reinforced with silicon carbide particles are modeled by the finite element method [9–11]. The influence of the choice of a representative volume, the distribution of reinforcing particles in it and the interface phase on the deformation curves is estimated and the results of computer modeling are compared with experimental ones. It is noted that without taking into account the actual hardening, finite element modeling of elastic modules and yield strength can be applied only in cases where load transfer is the main mechanism of strengthening [9].

Despite intensive research on metal–matrix (nano)composites, there are still a number of problems that have to be solved in the production of such composites. This is, first of all, ensuring a uniform distribution of reinforcing particles, excluding agglomeration of both reinforcing particles and grains of the matrix material, poor adhesion between the matrix material and the particles.

We have proposed a new method for introducing silicon carbide into a metal matrix, which makes it possible to solve the above problems of manufacturing a metal matrix nanocomposite based on aluminum. A study of the mechanical properties of the obtained composite has been carried out.

2. Description of a new technique for creating a metal–matrix nanocomposite

The method proposed for creating a metal–matrix composite can be divided into the following stages.

At the first stage, silicon carbide (SiC) nanolayers are synthesized by the method of molecular deposition (ML-ALD) [12–13] on the surface of an aluminum powder, the average particle size of which is 5–50 microns. The ML-ALD method is based on alternating chemical reactions of low molecular weight reagents on the surface. As a result of each cycle of surface chemical reactions carried out under strictly controlled conditions on the surface of aluminum particles, there is a build-up of new structural units (carbides) chemically bonded to the original substrate. For one cycle of reactions, the increase is 0.5 nm. By carrying out a certain number of such reactions, it

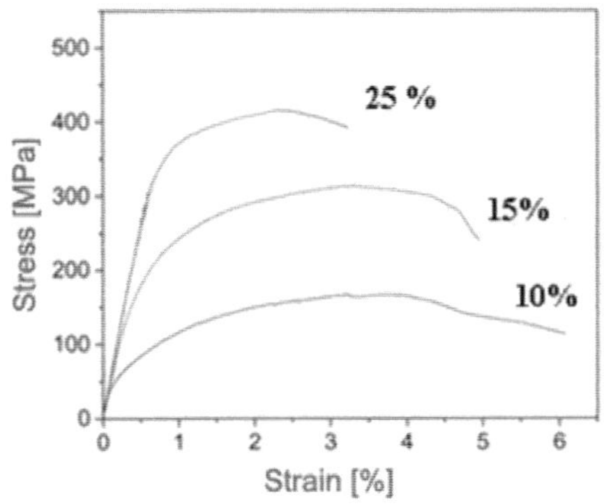

Figure 1. Deformation curves for metal composites with dispersed particles of the following composition: 90% Al + 10% (Al–SiC/Fe); 85% Al + 15% (Al–SiC/Fe); 75% Al + 25% (Al–SiC/Fe).

is possible to synthesize a nanolayer of a substance of a certain thickness. For 100 cycles, the increase was 50 nm.

The synthesis of SiC nanolayers was carried out at a temperature of 200–500°C in a flow-type reactor in a flow of dried inert gas (helium), which simultaneously ensured the removal of gaseous reaction products from the reactor.

The optimal conditions for the synthesis of SiC nanostructures on the surface of an aluminum matrix were found. The optimal temperature is 450°C, gas flow rate is 100 ml/min, synthesis time for one cycle of reactions on the surface is 30 minutes.

The chemical bond between the carbide nanolayers and the substrate ensures a high degree of adhesion of silicon carbide to aluminum. To avoid agglomeration (coarsening) of the obtained dispersed particles (Al–SiC), a Fe layer is applied to their surface.

At the second stage, the synthesized dispersed reinforcing particles (Al–SiC/Fe) were mixed with powder of aluminum particles and then pressed. To reduce porosity at a temperature of 600°C, a sample compacted from dispersed starting materials (aluminum powder and dispersed particles (Al–SiC/Fe)) was sintered into a single structure.

The absence of interphase boundaries between aluminum particles and carbide nanostructures made it possible to reduce the internal porosity of the sample. The presence of an iron nanolayer made it possible to uniformly redistribute the reinforcing carbide particles (Al–SiC/Fe) in the volume of the aluminum matrix, provides a good adhesion between the matrix material and the particles, and limits the agglomeration of both reinforcing particles and grains of the matrix material. The proposed technique makes it possible to solve the above-mentioned problems in the production of metal–matrix composites reinforced with carbide structures and allowed to synthesize aluminum matrix composites with improved mechanical properties.

3. Tensile properties of synthesized aluminum matrix composites

To study the dependence of the mechanical properties of the aluminum matrix composite, samples were prepared with a mass fraction of reinforcing dispersed particles (1%, 3%, 10%, 15%, 25%) with a carbide nanolayer thickness of 50 nm. Deformation curves of samples were obtained on an Shimadzu AG-50kNXD test machine and strength limits were determinated.

It is shown that the reinforcement with dispersed particles (Al–SiC/Fe) leads to a noticeable strengthening of the aluminum matrix composite at a mass concentration of particles of more than 10%. The content of the reinforcing dispersed phase of 1%, 2%, 3% practically does not affect the mechanical properties of the composite. For metal–matrix composites with a mass concentration of dispersion particles (Al–SiC/Fe) of 10%, 15%, and 25%, not only a high degree of strengthening is characteristic, but also a significant plastic deformation before fracture (figure 1). The improvement in the plastic properties of the composite (deformation to fracture) can be explained by the presence of large regions in the matrix, in which there are no silicon carbide particles [14].

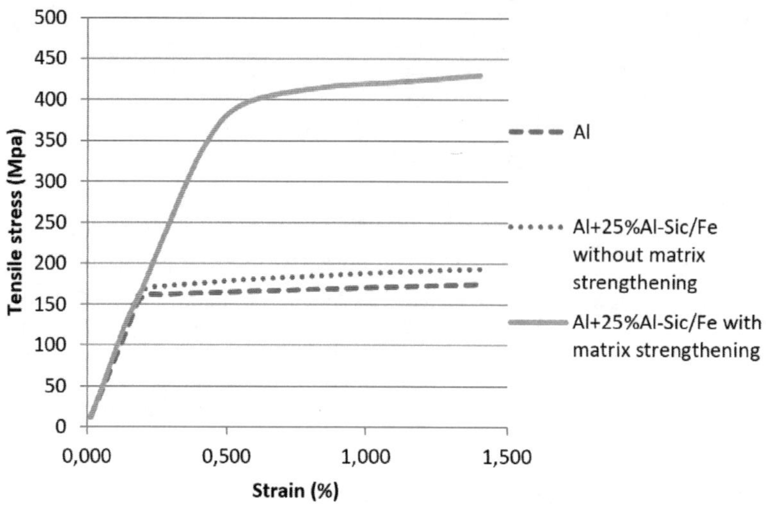

Figure 2. Deformation curves for metal composites with dispersed particles of the following composition: 90% Al + 10% (Al–SiC/Fe); 85% Al + 15% (Al–SiC/Fe); 75% Al + 25% (Al–SiC/Fe).

Table 1. Dependence of the strength properties of the synthesized nanocomposite based on aluminum with a dispersed silicon carbide phase on the concentration of dispersed particles (Al–SiC/Fe).

No	Composites	Ultimate strength, MPa
1	99.7% Al	75
2	99% Al + 1% (Al–SiC/Fe)	80
3	97% Al + 3% (Al–SiC/Fe)	95
4	90% Al + 10% (Al–SiC/Fe)	165
5	85% Al + 15% (Al–SiC/Fe)	310
6	75% Al + 25% (Al–SiC/Fe)	430

The dependence of the ultimate strength on the mass fraction of dispersed particles (Al–SiC/Fe) is shown in table 1. It should be noted that such strengthening is not achieved when reinforcing an aluminum matrix with silicon carbide nanoparticles at the same volume fraction silicon carbide fraction [3–4], which could be explained by weak adhesion between silicon carbide nanoparticles and an aluminum matrix, as well as high porosity at high concentrations of inclusions, leading to embrittlement composite.

Using the finite element method, the mechanical properties of the synthesized metal–matrix composite reinforced with dispersed Al–SiC/Fe particles have been simulated.

In the numerical simulation of the deformation curves, it is necessary to take into account the actual strengthening of the matrix and introduce appropriate changes in the constitutive relations for the matrix material [2, 14]. These values can be determined either from experimental stress-strain curves or from the results of micromechanical modeling, taking into account the contributions of various strengthening mechanisms to the actual strengthening of the metal matrix composite. Disregarding the actual matrix reinforcement in the finite element simulation, load transfer alone is the dominant mechanism for determining modulus and yield strength.

The results of finite element modeling of deformation curves for Al and for a metal matrix composite of the following composition 75% Al + 25% (Al–SiC/Fe) with a carbide nanolayer thickness of 50 nm, taking into account the actual strengthening of the matrix, are shown in figure 2. For comparison, the results of deformation modeling are also presented (figure 2), when only the strengthening mechanism is taken into account due to the effective redistribution of the load between the matrix and reinforcing dispersed particles (Al–SiC/Fe). Considering the low concentration of silicon carbide in the composite, the contribution from this hardening mechanism is insignificant.

Acknowledgments
The research is financially supported by the Russian Science Foundation (Grant No. 20-11-20083).

References
[1] Arsenault R J 1984 The strengthening of aluminum alloy 6061 by fiber and platelet silicon carbide *Materials Science and Engineering* **64** (2) 171–81 doi: 10.1016/0025-5416(84)90101-0
[2] Shi N and Arsenault R J 1994 Plastic flow in SiC/A1 composites-strengthening and ductility *Annual Review of Materials Science* **24** 321–57 doi: 10.1146/annurev.ms.24.080194.001541
[3] Dai L H, Ling Z, and Bai Y L 2001 Size-dependent inelastic behavior of particle-reinforced metal–matrix composites *Composites Science and Technology* **61** (8) 1057–63 doi: 10.1016/S0266-3538(00)00235-9
[4] Kim C-S, Cho K, Manjili M H, and Nezafati M 2017 Mechanical performance of particulate reinforced Al metal–matrix composites (MMCs) and Al metal–matrix nano-composites (MMNCs) *Journal of Materials Science* **52** 13319–49 doi: 10.1007/s10853-017-1378-x
[5] Kamrani S, Riedel R, Seyed Reihani S M, and Kleebe H J 2009 Effect of reinforcement volume fraction on the mechanical properties of Al–SiC nanocomposites produced by mechanical alloying and consolidation *Journal of Composite Materials* **44** 313–26 doi: 10.1177/0021998309347570
[6] El-Daly A A, Abdelhameed M, Hashish M, and Daoush W 2013 Fabrication of silicon carbide reinforced aluminum matrix nanocomposites and characterization of its mechanical properties using non-destructive technique *Materials Science and Engineering: A* **559** 384–93 doi: 10.1016/j.msea.2012.08.114
[7] Casati R and Vedani M 2014 Metal matrix composites reinforced by nano-particles — A review *Metals* **4** 65–83 doi: 10.3390/met4010065
[8] Somashekar H M, Gowthama K, and Suresha B 2020 Microstructure, mechanical properties and sliding wear behaviour of silicon carbide reinforced Al 6026 composites *International Journal of Mechanical and Production Engineering Research and Development* **10** (1) 259–70
[9] Zhang J F, Zhang X X, Wang Q Z, et al. 2018 Simulations of deformation and damage processes of SiCp/Al composites during tension *Journal of Materials Science & Technology* **34** (4) 627–34 doi: 10.1016/j.jmst.2017.09.005
[10] Ghasemi M J, Silani M, Maleki A, and Jamshidian M 2021 Micromechanical simulation and experimental investigation of aluminum-based nanocomposites *Defence Technology* **17** (1) 196–201 doi: 10.1016/j.dt.2020.03.010
[11] Ferguson J B, Thao X, Rohatgi P K, et al. 2014 Computational and analytical prediction of the elastic modulus and yield stress in particulate-reinforced metal matrix composites *Scripta Materialia* **83** 45–8 doi: 10.1016/j.scriptamat.2014.04.004
[12] George S M 2010 Atomic layer deposition: An overview *Chemical Reviews* **110** (1) 111–31 doi: 10.1021/cr900056b
[13] Zemtsova E G, Yurchuk D V, Sidorov Yu V, et al. 2020 Synthesis of metallic composite based on iron matrix and SiC nanostructures and its strength properties *Materials Physics and Mechanics* **46** (1) 122–31 doi: 10.18149/MPM.4612020_12
[14] Tang F, Hagiwara M, and Schoenung J M 2005 Microstructure and tensile properties of bulk nanostructured Al–5083/SiCp composites prepared by cryomilling *Materials Science and Engineering: A* **407** (1-2) 306–14 doi: 10.1016/j.msea.2005.07.056

Journal of Physics: Conference Series 2231 (2022) 012004 doi:10.1088/1742-6596/2231/1/012004

System of cracks under the impact of plane elastic waves

V G Popov

National University "Odessa Maritime Academy", Odessa, Ukraine

E-mail: dr.vg.popov@gmail.com

Abstract. The problem of determining the two-dimensional diffraction field and the stress state arising from the interaction of elastic waves with a system of arbitrarily located cracks is solved. Cracks are located in an unbounded elastic medium which is in a plane deformation state. The method of solution is based on the use of discontinuous solutions of the equations of motion of elastic media and the reduction of the original problem to a system of singular integro-differential equations. The obtained systems are solved numerically the mechanical quadrature method.

1. Introduction

The problem of determining the diffraction field and the stress state in the vicinity of a single crack at interaction with propagating waves can be considered completely solved today. But when in the body contains a system of arbitrarily located cracks the situation becomes much more difficult. Papers containing solutions to similar problems appeared relatively recently, and in the firsts of them the cases of two, most often parallel, cracks were consider [1–7]. Considerable attention was also given to the problems of interaction of elastic waves with systems of periodic and radially located cracks [8–11]. The solution of dynamic problems to determine the fields of displacements and stresses in bodies with arbitrary systems of cracks by the method of boundary integral equations is reflected in works [12–14]. It should be noted that the integral equations to which the original problems are reduced by this method have hipersingularities, which difficulties their numerical solution. In this paper, we propose a solution to the problem of determining the diffraction field and stress state under the impact of plane longitudinal waves or shear waves on a system of cracks. A feature of the proposed solution method is that in order to avoid hypersingular integrals, it is proposed to solve a system of integro-differential equations for jumps of displacements on cracks surface and their derivatives. Previously, this method was used to solve a similar problem in the case of wave of longitudinal shear impacting on a system of defects [15].

2. Statement of the problem and its reduction to a system of singular integro-differential equations

Consider an isotropic unbounded elastic medium under plane deformation conditions, in which Nthrough cracks are located. These cracks in the coordinate plane Oxy (figure 1) are located on segments of length $2d_k$ and centered at points (a_k, b_k), $k = 1, 2, \ldots, N$. Longitudinal waves

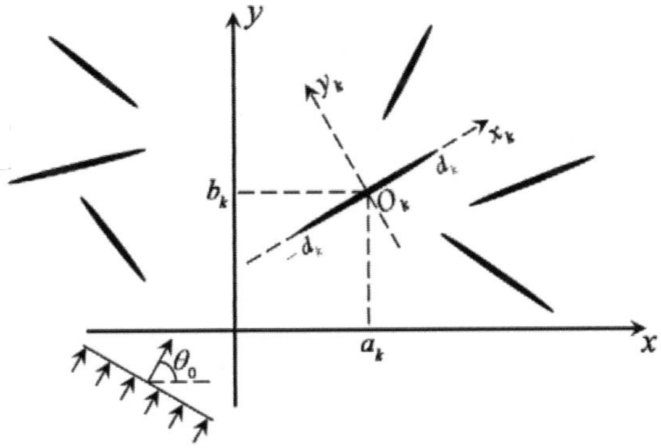

Figure 1. Arbitrary system of through cracks.

or shear waves, which are given by potentials,

$$\begin{aligned}\varphi_0(x,y) &= \frac{A}{\kappa_1}e_1(x,y),\\ \psi_0(x,y) &= \frac{B}{\kappa_2}e_2(x,y),\\ e_j(x,y) &= \exp[i\kappa_j(x\cos\theta_0 + y\sin\theta_0)], \quad j=1,2,\end{aligned} \tag{1}$$

interact with cracks. In formulas (1) took notation: $\kappa_j^2 = \omega^2 c_j^{-2}$, $c_1^2 = \lambda + 2\mu$, $c_2^2 = \mu\rho^{-1}$, ω is oscillations frequency, λ, μ, ρ are Lame coefficients and density of medium, θ_0 is the angle between the direction of wave propagation and the Ox axis. The time dependence is determined by a multiplier $\exp(-i\omega t)$ that is omitted.

Let $u(x,y)$, $v(x,y)$ are the displacement of the wave field scattered by cracks. Then they satisfy the differential equations of motion of an elastic medium for harmonic oscillation under conditions of plane deformation:

$$(\lambda+2\mu)\frac{\partial}{\partial x}\left(\frac{\partial u}{\partial x}+\frac{\partial v}{\partial y}\right)+\mu\Delta u = -\rho\omega^2 u, \quad (\lambda+2\mu)\frac{\partial}{\partial y}\left(\frac{\partial u}{\partial x}+\frac{\partial v}{\partial y}\right)+\mu\Delta v = -\rho\omega^2 v. \tag{2}$$

Here Δ is the two-dimensional Laplace operator. These equations should be considered with boundary conditions on the cracks surfaces, for the formulation of which a local coordinate system is $O_k x_k y_k$ (figure 1) connected with each of them. Let $u^k(x_k,y_k)$, $v^k(x_k,y_k)$, $\tau_{xy}^k(x_k,y_k)$, $\sigma_x^k(x_k,y_k)$, $\sigma_y^k(x_k,y_k)$ are displacements and stresses of the wave field in the coordinate system $O_k x_k y_k$, and, $u^{0k}(x_k,y_k)$, $v^{0k}(x_k,y_k)$, $\tau_{xy}^{0k}(x_k,y_k)$, $\sigma_x^{0k}(x_k,y_k)$, $\sigma_y^{0k}(x_k,y_k)$ are displacements and stresses caused by the acting waves (1). Then, at stress-free surfaces of cracks, the equalities should be realized

$$\sigma_y^k(x_k,0) = -\sigma_y^{0k}(x_k,0), \quad \tau_{yx}^k(x_k,0) = -\tau_{xy}^{0k}(x_k,0), \quad -d_k < x_k < d_k, \quad k=1,2,\ldots,N. \tag{3}$$

Also, displacements are discontinuous on the surface of each crack. The jumps of displacements are denoted by:

$$v^k(x_k,+0) - v^k(x_k,-0) = \chi_{3k}(x_k), \quad u^k(x_k,+0) - u^k(x_k,-0) = \chi_{4k}(x_k), \quad -d_k < x_k < d_k. \tag{4}$$

The problem is to determine the displacements and stresses of the diffraction field in the body under the formulated conditions (3), (4).

For this purpose, for each crack in the local coordinate system, a discontinuous solution of equations (2) with jumps (4) is constructed according to the formulas from [16]:

$$
\begin{aligned}
v^{dl}(x_l,y_l) &= \int_{-d_l}^{d_l} \chi_{3l}(\eta) G_{33}(\eta - x_l, y_l) d\eta + \int_{-d_l}^{d_l} \chi_{4l}(\eta) G_{34}(\eta - x_l, y_l) d\eta, \\
u^{dl}(x_l,y_l) &= \int_{-d_l}^{d_l} \chi_{3l}(\eta) G_{43}(\eta - x_l, y_l) d\eta + \int_{-d_l}^{d_l} \chi_{4l}(\eta) G_{44}(\eta - x_l, y_l) d\eta.
\end{aligned} \tag{5}
$$

Here denoted

$$
G_{33} = \frac{1}{\kappa_2^2}\frac{\partial}{\partial y_l}\left[\left(\kappa_2^2 + 2\frac{\partial^2}{\partial x_l^2}\right) r_1 - 2\frac{\partial^2 r_2}{\partial x_l^2}\right], \quad G_{34} = \frac{1}{\kappa_2^2}\frac{\partial}{\partial y_l}\left[2\left(\kappa_1^2 + 2\frac{\partial^2}{\partial x_l^2}\right) r_1 - \left(\kappa_2^2 + 2\frac{\partial^2}{\partial x_l^2}\right) r_2\right],
$$
$$
G_{43} = \frac{1}{\kappa_2^2}\frac{\partial}{\partial y_l}\left[\left(\kappa_2^2 + 2\frac{\partial^2}{\partial x_l^2}\right) r_1 - 2\left(\kappa_2^2 + \frac{\partial^2}{\partial x_l^2}\right) r_2\right], \quad G_{44} = \frac{1}{\kappa_2^2}\frac{\partial}{\partial y_l}\left[-2\frac{\partial^2 r_1}{\partial x_l^2} + \left(\kappa_2^2 + 2\frac{\partial^2}{\partial x_l^2}\right) r_2\right],
$$
$$
r_j(\eta - x_l, y_l) = -\frac{1}{2\pi}\int_{-\infty}^{+\infty} \frac{\exp[-(\beta^2 - x_j^2)^{1/2}|y_l|]}{2\sqrt{\beta^2 - x_2^2}} \exp[i\beta(\eta - x_l)] d\beta = -\frac{i}{4} H_0^{(1)}\{x_j[(\eta - x_l)^2 + y_l^2]^{1/2}\}.
$$

The corresponding formulas for stresses look like:

$$
\begin{aligned}
\sigma_{xx}^{dl}(x_l,y_l) &= \mu \int_{-d_l}^{d_l} \chi_{3l}'(\eta) E_{03}(\eta - x_l, y_l) d\eta \\
&+ \mu(\kappa_2^2 - 2x_1^2) \int_{-d_l}^{d_l} \chi_{3l}(\eta) r_1(\eta - x_l, y_l) d\eta, + \mu \int_{-d_l}^{d_l} \chi_{4l}'(\eta) E_{04}(\eta - x_l, y_l) d\eta, \\
\sigma_{yy}^{dl}(x_l,y_l) &= \mu \int_{-d_l}^{d_l} \chi_{3l}'(\eta) E_{13}(\eta - x_l, y_l) d\eta \\
&+ \mu \int_{-d_l}^{d_l} \chi_{4l}'(\eta) E_{14}(\eta - x_l, y_l) d\eta - \mu\kappa_2^2 \int_{-d_l}^{d_l} \chi_{3l}(\eta) r_1(\eta - x_l, y_l) d\eta, \\
\tau_{yx}^{dl}(x_l,y_l) &= \mu \int_{-d_l}^{d_l} \chi_{3l}'(\eta) E_{23}(\eta - x_l, y_l) d\eta \\
&+ \mu \int_{-d_l}^{d_l} \chi_{4l}'(\eta - x_l, y_l) d\eta + \mu\kappa_2^2 \int_{-d_l}^{d_l} \chi_{3l}(\eta) r_2(\eta - x_l, y_l) d\eta
\end{aligned} \tag{6}
$$

In formulas (6) for stresses, in purpose to reduce the exponent of the singularity of the integrals, as in [16], integration by parts was completed on account $\chi_{3l}(\pm d_l) = \chi_{4l}(\pm d_l) = 0$. Each discontinuous solution (5), (6) determine the following displacements and stresses in the coordinate system Oxy:

$$
\begin{aligned}
&u^{gl} = u^{dl}\cos\alpha_l - v^{dl}\sin\alpha_l, \quad v^{gl} = u^{dl}\sin\alpha_l + v^{dl}\sin\alpha_l, \\
&\sigma_x^{gl} = \sigma_x^{dl}\cos^2\alpha_l + \sigma_x^{dl}\sin^2\alpha_l - \tau_{xy}^{dl}\sin(2\alpha_l), \quad \sigma_y^{gl} = \sigma_x^{dl}\sin^2\alpha_l + \sigma_x^{dl}\cos^2\alpha_l + \tau_{xy}^{dl}\sin(2\alpha_l), \\
&2\tau_{yx}^{gl} = \sigma_x^{dl}\sin(2\alpha_l) + \sigma_x^{dl}\sin(2\alpha_l) + 2\tau_{xy}^{dl}\cos(2\alpha_l).
\end{aligned} \tag{7}
$$

In formulas (7) and further, α_l is the angle between the axis Ox_l and the axis Ox. Then the displacements of the scattered wave field can be represented as

$$
u(x,y) = \sum_{l=1}^{N} u^{gl}(x,y), \quad v(x,y) = \sum_{l=1}^{N} v^{gl}(x,y) \tag{8}
$$

similar formulas take place and for stresses. Formulas (8) make it possible to find the displacements of the scattered wave field after determining the unknown jumps (4). For this, it is necessary to use the boundary conditions (3). The stresses that enter into these conditions, through discontinuous solutions (5), (6), are expressed by the formulas

$$\begin{aligned}
&\sigma_y^k(x_k, y_k) = \sum_{l=1}^{N} \sigma_y^{kl}(x_k, y_k), \quad \tau_{yx}^k(x_k, y_k) = \sum_{l=1}^{N} \tau_y^{kl}(x_k, y_k), \\
&\sigma_y^{kl} = \sigma_x^{dl} \sin^2 \alpha_{kl} + \sigma_y^{dl} \cos^2 \alpha_{kl} - \tau_{yx}^{dl} \sin(2\alpha_{kl}) \\
&2\tau_{yx}^{kl} = -\sigma_x^{dl} \sin(2\alpha_{kl}) + \sigma_y^{dl} \sin(2\alpha_{kl}) + 2\tau_{yx}^{dl} \cos(2\alpha_{kl}), \quad \alpha_{kl} = \alpha_k - \alpha_l.
\end{aligned} \tag{9}$$

After substitution of expressions (9) into boundary conditions (3) with respect to displacement jumps, a system of $2N$ singular integro-differential equations is obtained. After separating the singular component and some of transformations, this system has the form

$$\begin{aligned}
&\frac{1}{2\pi}\int_{-1}^{1} \varphi_{3k}'(\tau)\left[-\frac{2(1-\xi^2)}{\tau-\varsigma} + R_B^k(\tau-\varsigma)\right]d\tau + \frac{1}{2\pi}\int_{-1}^{1} \varphi_{3k}(\tau)[\kappa_0^2\gamma_k^2 \ln|\tau-\varsigma| + V_{13}^k(\tau-\varsigma)]d\tau \\
&\quad + \sum_{l=1}^{N}\left[\frac{1}{2\pi}\int_{-1}^{1} \varphi_{3l}'(\tau)F_{kl}^{13}(\tau,\varsigma)d\tau + \frac{1}{2\pi}\int_{-1}^{1} \varphi_{4k}'(\tau)F_{kl}^{14}(\tau,\varsigma)d\tau\right] \\
&\quad + \sum_{\substack{l=1 \\ l\neq k}}^{N}\left[\frac{1}{2\pi}\int_{-1}^{1} \varphi_{3l}(\tau)U_{kl}^{13}(\tau,\varsigma)d\tau + \frac{1}{2\pi}\int_{-1}^{1} \varphi_{4k}(\tau)U_{kl}^{13}(\tau,\varsigma)d\tau\right] = f_{1k}(\varsigma), \quad k = 1, 2, \ldots, N, \\
&\frac{1}{2\pi}\int_{-1}^{1} \varphi_{3k}'(\tau)\left[\frac{2(1-\xi^2)}{\tau-\varsigma} + R_{24}^k(\tau-\varsigma)\right]d\tau + \frac{1}{2\pi}\int_{-1}^{1} \varphi_{4k}(\tau)\left[\kappa_0^2\gamma_k^2 \ln|\tau-\varsigma| + V_{24}^k(\tau-\varsigma)\right]d\tau \\
&+ \sum_{\substack{l=1 \\ l\neq k}}^{N}\left[\frac{1}{2\pi}\int_{-1}^{1} \varphi_{3l}'(\tau)F_{kl}^{23}(\tau,\varsigma)d\tau + \frac{1}{2\pi}\int_{-1}^{1} \varphi_{4k}'(\tau)F_{kl}^{24}(\tau,\varsigma)d\tau\right] \\
&+ \sum_{\substack{l=1 \\ l\neq k}}^{N}\left[\frac{1}{2\pi}\int_{-1}^{1} \varphi_{3l}(\tau)U_{kl}^{23}(\tau,\varsigma)d\tau + \frac{1}{2\pi}\int_{-1}^{1} \varphi_{4k}(\tau)U_{kl}^{23}(\tau,\varsigma)d\tau\right] = f_{2k}(\varsigma), \quad k = 1, 2, \ldots, N.
\end{aligned} \tag{10}$$

In this system denoted $\varphi_{sk}(\tau) = d_k^{-1}\chi_{sk}(d_k\tau)$ $(s = 3, 4)$, $\gamma_k = d^{-1}d_k$ (k=1,2,...,N), $d = \max(d_1, d_2, \ldots, d_N)$, $\kappa_0 = \kappa_2 d$, $\xi^2 = \kappa_2^2/\kappa_1^2 = (1-2\nu)/[2(1-\nu)]$, ν is Poisson's ratio of an elastic medium.

The functions R_{13}^k, V_{13}^k, R_{24}^k, V_{24}^k, F_{kl}^{1s}, U_{kl}^{1s}, $s = 3, 4$ are continuous at $-1 \leq \varsigma, \tau \leq 1$, and the functions $f_{1k}(\varsigma)$, $f_{2k}(\varsigma)$ are determined by the potentials of the acted wave (1). Also, it is necessary to add equalities to the system (10),

$$\int_{-1}^{1} \varphi_{sk}'(\tau)d\tau = 0, \quad k = 1, 2, \ldots, N, \quad s = 3, 4, \tag{11}$$

which are the conditions for the closure of the cracks.

3. Numerical solution of the system integro–deferential equation

For an approximate solution of this system, the method described in detail in [16] was applied. For this, the derivatives of unknown functions are represented in the form

$$\varphi_{sk}'(\tau) = (1-\tau^2)^{-0.5}\psi_{sk}(\tau), \quad k = 1, 2, \ldots, N, \quad s = 3, 4, \tag{12}$$

and approximation of functions $\psi_{sk}(\tau)$ by interpolation polynomials

$$\psi_{sk}(\tau) \approx \sum_{m=1}^{n} \psi_{skm} \frac{T_n(\tau)}{(\tau - \tau_m)T_n'(\tau_m)}, \quad k = 1, 2, \ldots, N, \quad s = 3, 4. \tag{13}$$

Here $T_n(\tau)$ are Chebyshev polynomials, τ_m are roots of these polynomials, $\psi_{skm} = \psi_{sk}(\tau_m)$.
As proved in [8], from formulas (12), (13), for the functions themselves follows the approximation

$$\varphi_{sk}(\tau) \approx \sqrt{1-\tau^2} S_{sk}(\tau), \quad S_{sk}(\tau) = 2n^{-1} \sum_{m=1}^{n} \psi_{skm} \sum_{p=1}^{n} p^{-1} T_n(\tau_m) U_{p+1}(\tau). \tag{14}$$

Formulas (12)–(14) make it possible to apply for the numerical solution of system (10) the collocation method which [17] to use as collocation points the roots of the Chebyshev polynomial $U_{n-1}(\zeta)$ $\zeta_j = \cos(\pi j n^{-1})$, $j = 1, 2, \ldots, n-1$. The well-known quadrature formula [17] for the Cauchy integrals, and the Gauss-Chebyshev formulas [18] for regular integrals are used this method. As for the integral with a logarithmic singularity, it is calculated by the quadrature formula derived using the representation (14) [8]:

$$\int_{-1}^{1} \varphi_{sk}(\tau) \ln|\tau - \zeta_j| d\tau = \pi n^{-1} \sum_{m=1}^{n} C_{jm} \psi_{skm},$$

$$C_{jm} = \tau_m [\ln 2 - 0.5\cos(2\sigma_j)] + \sum_{p=2}^{n} p^{-1} \cos(p\beta_m)[\cos((p-1)\sigma_j) - (p+1)^{-1}\cos((p+1)\sigma_j)]$$

$$\sigma_j = j\pi n^{-1}, \quad \beta_m = (2m-1)(2n)^{-1}.$$

As a result, from (10), (11) a system of $2Nn$ linear algebraic equations for the determining values of unknown functions $\psi_{skm} = \psi_{sk}(\tau_m)$ at the interpolation points was obtain.
One of the important characteristics of the stress state caused by the action of waves is the stress intensity factors (SIF) in the vicinity of the ends of the cracks. After the numerical solution of the system of integro-differential equations (10), the approximate values of the SIF are calculated by the formulas:

$$\begin{aligned} &K_{pl}^{\pm} = \mu\sqrt{d} k_{pl}^{\pm}, \quad p = 1, 2, \quad l = 1, 2, \ldots, N, \\ &k_{pl}^{+} = -(1-\xi^2) n^{-1} \sum_{m=1}^{n} (-1)^m \psi_{2+plm} \cot(0.5\beta_m), \\ &k_{pl}^{-} = -(1-\xi^2) n^{-1} (-1)^{p+n} \sum_{m=1}^{n} (-1)^m \psi_{2+plm} \tan(0.5\beta_m). \end{aligned} \tag{15}$$

4. Analysis of numerical results

As an example, the system of 5 cracks, located as shown in figure 2, is considered. In the calculations, the following relationships between geometric sizes were taken:

$$\gamma_1 = 1, \quad \gamma_2 = \gamma_3 = \gamma_4 = \gamma_5 = 1/3, \quad r_0 = r/d = 0.1, \quad 0 < \beta < 180^\circ, \quad \theta_0 = 90^\circ.$$

Graphs in figure 3, figure 4 show the dependence of the dimensionless SIF of the normal stresses of the first crack on the dimensionless wave $\kappa_0 = \kappa_2 d_1$ number at different values of the angle β.

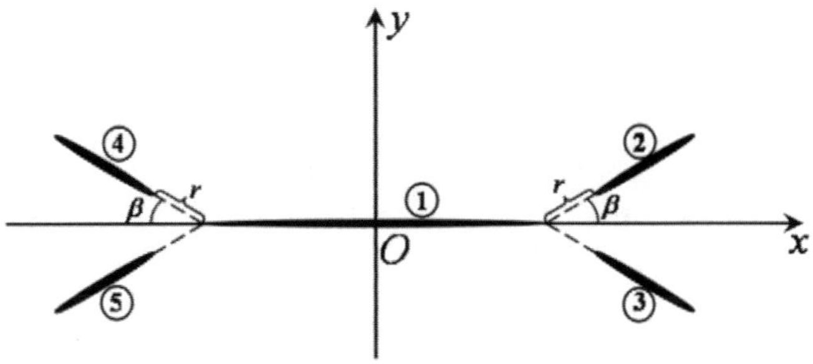

Figure 2. The system of five cracks.

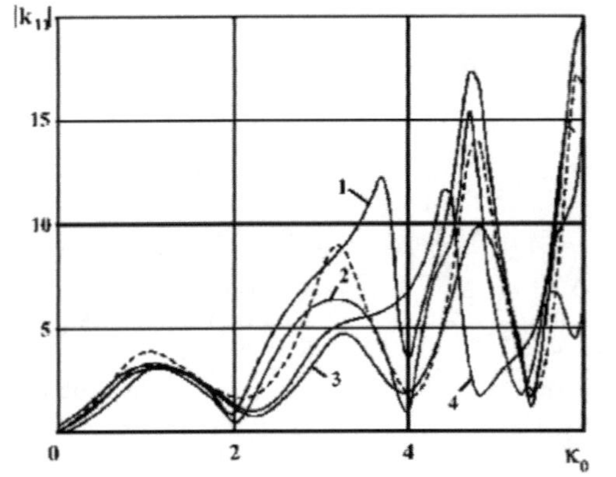

Figure 3. Dimensionless SIF of normall stresses at $\beta = 45°$, $\beta = 60°$, $\beta = 90°$, $\beta = 120°$.

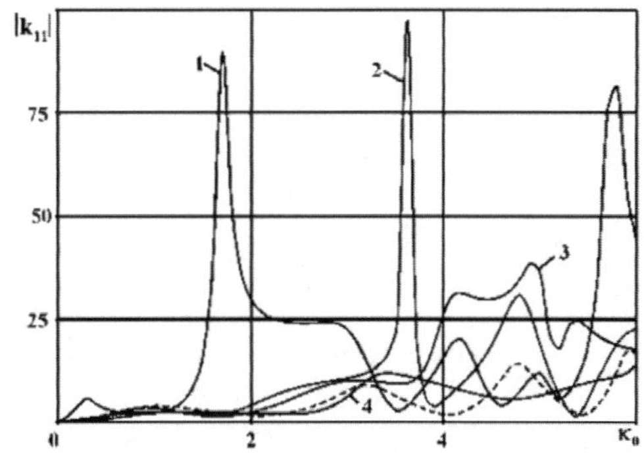

Figure 4. Dimensionless SIF of normal stresses at $\beta = 5°$, $\beta = 30°$, $\beta = 150°$, $\beta = 175°$.

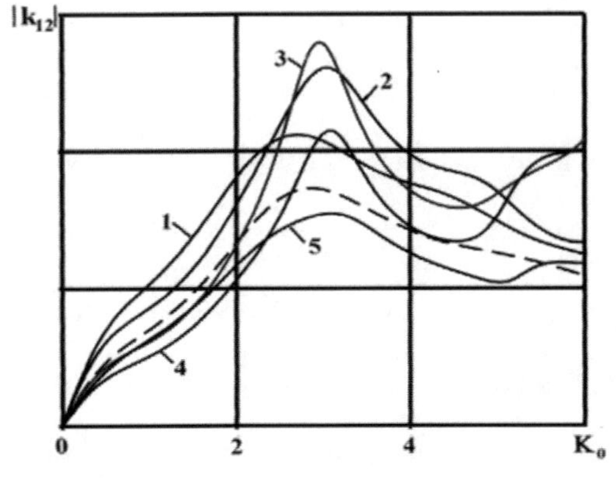

Figure 5. Dimensionless SIF of shear stresses at $\beta = 5°$, $\beta = 30°$, $\beta = 45°$, $\beta = 60°$, $\beta = 75°$.

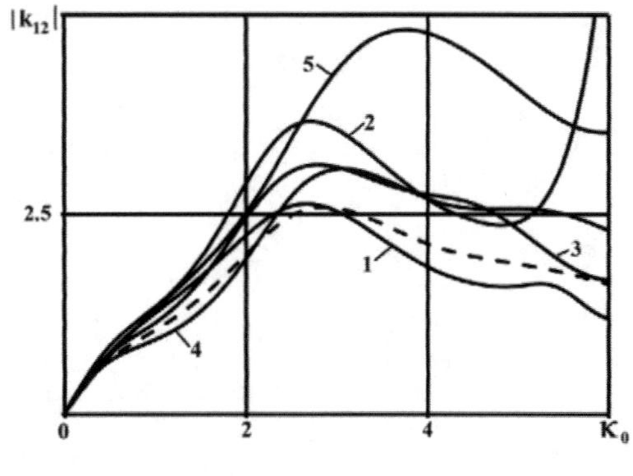

Figure 6. Dimensionless SIF of shear stresses at $\beta = 90°$, $\beta = 120°$, $\beta = 135°$, $\beta = 150°$, $\beta = 175°$.

The dotted line shows the graph corresponding to one crack. Numerical analysis shows that the SIF values are strongly influenced by the angle β formed by cracks. Figure 5 and figure 6 shows similar graphs for the SIF of shear stresses for interaction shear waves the cracks system. The SIF values of both normal and shear stresses remain less than the SIF for a one crack for angles in the range $60° < \beta < 120°$. For other values of the angle β, the reciprocal influence of the cracks leads to a considerable (several times) increase in the SIF values in comparison with its values for a single crack figure 4, figure 6. This effect is most revealed for the SIF of normal stresses. In general, the values of the SIF of normal stresses under the action of a longitudinal wave greatly exceed the values of the SIF of shear stresses under the action of a shear wave.

5. Conclusion

An efficient numerical method is proposed for determining two-dimensional diffraction wave's fields in the interaction of elastic waves with arbitrary systems of through cracks. This method remains effective for systems of cracks of complex geometry and systems of sufficiently nearly located cracks. A considerable effect on the stress state in the body of the configuration of the crack systems was found. The result of this influence is the presence of greatly maxima of the SIF and a complexity dependence of the SIF on frequency.

References

[1] Jain D L and Kanval P P 1972 Diffraction of elastic waves by two coplanar Griffith cracks in an infinite elastic medium *International Journal of Solids and Structures* **8** (7) 961–75 doi: 10.1016/0020-7683(72)90009-1
[2] Srivastava K N, Palaiya R, and Karaulia D S 1983 Interaction of shear waves with two coplanar Griffith cracks situated in an infinitely long elastic strip *International Journal of Fracture* **21** (1) 39–48 doi: 10.1007/BF00020153
[3] Nazarenko A M and Fil'shinskiy A G 1983 Interaction of stress waves with a periodic system of longitudinal shear cracks in a half-space *Akusticheskiy Zhurnal* **29** (3) 183–6
[4] Borodachev N M and Kol'tsov V A 1985 Determination of the stress intensity factors at the ends of two collinear transverse shear cracks under steady vibrations *Soviet Appied Mechanics* **21** (1) 96–102 doi: 10.1007/BF00887890
[5] Borodachov N M and Kol'tsov V A 1984 Determination of stress intensity factors at the ends of two collinear cracks with normal separation under steady-state vibrations *Izvestiya AN SSSR. Mekhanika Tverdogo Tela* **3** 119–26
[6] Fil'shinskiy M L 1988 Interaction of mechanical displacements waves with tunnel cracks of longitudinal shear in a piezoceramic half-space *Izvestiya AN Arm. SSR* **35** 50–55
[7] Zhang C 1990 Dynamic stress intensity factor of collinear antiplane cracks *Journal of Tongji University* **118** (4) 445–51
[8] Popov V G 1998 Interaction between elastic longitudinal shear waves and radially distributed fissures *International Applied Mechanics* **34** (2) 157–63
[9] Popov V G 1999 Interaction of plane elastic waves with systems of radial defects *Mechanics of Solids* **34** (4) 98–107
[10] Scarpetta E 2002 In-plane problem for wave propagation through elastic solids with a periodic array of cracks *Acta Mechanica* **154** (4) 179–87 doi: 10.1007/BF01170706
[11] Iovane G and Sumbatyan M A 2009 Periodic system of collinear cracks in an elastic porous medium *Mechanic of solids* **44** (3) 400–8 doi: 10.3103/S0025654409030091
[12] Sládek J and Sládek V 1987 A boundary integral equation method for dynamic crack problems *Engineering Fracture Mechanics* **27** (3) 269–77 doi: 10.1016/0013-7944(87)90145-7
[13] Gray L J and Chang S J 1992 Hypersingular integral formulation of elastic wave scattering *Engineering Analysis with Boundary Elements* **10** (4) 337–43 doi: 10.1016/0955-7997(92)90148-Z
[14] Yin-Bang Wang and Yu-Zhon Sun 2005 A new boundary integral equation method for cracked 2-D anisotropic bodies *Engineering Fracture Mechanics* **72** (13) 2128–43 doi: 10.1016/j.engfracmech.2005.01.007
[15] Popov V G 2013 Determination of the diffraction field during the interaction of a longitudinal shear wave with a system of defects *Visnyk Dnipropetrovs'koho Universytetu* **21** (5) 123–34
[16] Popov V G and UlanovskiI A E 1995 Comparative study of diffraction fields generated by elastic waves passing through imperfections of various nature *Mechanics of Solids* **30** (4) 93–102
[17] Belotserkovskii S M and Lifanov I K 1985 *Numerical Methods in Singular Integral Equations and Their Application in Aerodynamics, Theory of Elasticity, and Electrodynamics* (Moscow: Nauka) [in Russian]
[18] Krylov V I 1967 *Approximate Calculation of Integrals* (Moscow: Nauka) [in Russian]

Journal of Physics: Conference Series 2231 (2022) 012005 doi:10.1088/1742-6596/2231/1/012005

Strength, thermal resistance and electrical conductivity of aluminum-based composite wire

M Yu Murashkin[1*], D I Sadykov[1,2], A M Mavlyutov[1], D K Magomedova[1], and V U Kazykhanov[3]

[1]Saint Petersburg State University, St. Petersburg, Russia
[2]Saint Petersburg National Research University of Information Technologies, Mechanics and Optics, St. Petersburg, Russia
[3]Ufa State Aviation Technical University, Ufa, Russia

E-mail: *m.murashkin.70@gmail.com

Abstract. In this work, we demonstrate the possibility of producing a composite high strength and thermal resistant Al-based wire with ultrafine-grained (UFG) structure. Such composite wire consists of two aluminum alloys. Its clad is made of Al-1.17Mg-0.35Zr (wt.%) alloy, and the core is of Al-0.4Zr (wt.%). The specimens of Al alloys separately and as parts of a composite wire were subjected to thermomechanical treatment (TMT) that included annealing at 400°C, equal-channel angular pressing via "Conform" scheme (ECAP-C) and cold drawing (CD). The ratio between the core and clad materials in the composite wire was determined experimentally as well as calculated from the rule of mixtures (additivity). It has been shown that, an ultimate tensile strength over 360 MPa and a long-term operation temperature of 150°C can be achieved in a composite wire at a rational ratio between the chosen aluminum alloys with the UFG structures. Such combination of strength and thermal resistance is accompanied with an electrical conductivity of no less than 50% IACS in a composite wire.

1. Introduction

Today, the strongest Al-based materials widely used in electrical engineering are the alloys of Al-Mg-Si system [1]. A wire made of these alloys is generally applied in electrical cables and can exhibit an ultimate tensile strength up to 342 MPa with an electrical conductivity of 51.5% IACS [2]. New approaches to thermomechanical treatment of these alloys have been recently proposed, including the use of severe plastic deformation (SPD) techniques [3–5]. The formation of special microstructures provided by new schemes and regimes of TMT allowed for achieving an ultimate tensile strength (σ_{UTS}) over 350 MPa with an electrical conductivity of 53.0–56.5% IACS in Al-Mg-Si alloys. However, the thermal resistance of these materials is not high. The operating temperature is restricted to 90°C [1, 2], which significantly limits the area of their application in overhead conductors. The possibility of creating Al-based composite wires has been actively studied recently, as variation of their design, composition and treatment regimes opens new opportunities to control strength, thermal resistance and electrical conductivity [6–8]. For example, in [6] a composite wire consisting of a core made of steel 316L and a clad of AA6201 alloy of the Al-Mg-Si system was proposed. This wire after TMT with equal channel angular pressing demonstrates rather high strength (conditional yield stress of 424 MPa), but moderate electrical conductivity (42.97% IACS). Besides, steel in the construction of a composite wire levels down its advantage in weight as compared to Cu-based wires. In [8] an Al-based composite wire is produced. It consists of a core made of Al AA1350 with high electrical conductivity and a clad made of AA6201 alloy with enhanced strength. As is claimed in the work, the experimental strength values of this composite wire correlate well with the values calculated by the rule of mixtures (additivity). However, the proposed composite did not exhibit high strength, possibly due to lacking an optimal ratio between the composing alloys, and had insufficient thermal resistance as well as Al-Mg-Si alloys.

Generally, the problem of thermal resistance enhancement in Al-based overhead conductors is solved by applying Al-Zr alloys with a Zr content up to 0.4 wt.% [1]. However, the ultimate tensile strength (σ_{UTS}) of these conducting materials does not normally exceed 200 MPa [9, 10]. As shown in [11], the properties of Al-Zr may be improved through additional alloying with 0.4 wt.%Mg and TMT with ECAP-C. As a result, ultrafine-grained structure was formed in the material, thus providing rather high strength ($\sigma_{UTS} = 267$ MPa), electrical conductivity 57.1% IACS and operating temperature up to 150°C.

In the present study the possibility of creating a strong and thermally resistant composite wire with a UFG structure on the basis of Al-Zr and Al-Mg-Zr alloys is considered. The wire core is made of Al-Zr alloy with high electrical conductivity, the wire clad is of high-strength alloy Al-Mg-Zr. To achieve higher strength in a clad, the content of Mg was increased more than 2.5 times in the Al-Mg-Zr alloy, as compared to the alloy in [11].

2. Material and processing

Aluminum alloys of the Al-Zr and Al-Mg-Zr systems were selected as materials for the study. Their chemical compositions are shown in table 1.

Table 1. Chemical content of the Al alloys (wt.%).

Alloys	Zr	Mg	Fe	Si	Zn	Al
Al-Zr	0.39	—	0.24	0.02	0.02	99.25
Al-Mg-Zr	0.34	1.17	≤ 0.10	0.04	0.01	98.40

The Al-Zr alloy in the form of rods 9.5 mm in diameter was processed via continuous casting and rolling [12]. The material for investigation was supplied by UC RUSAL (Moscow, Russia).

The Al-Mg-Zr alloy was cast into rods with a diameter of 22 mm at the National University of Science and Technology MISiS (Moscow, Russia). Part of the cast rods were subjected to rolling to a diameter of 16 mm. To produce the required geometry and implement the subsequent TMT, the cast billets were rolled in a special unit by HANKOOK M-TECH INDUSTRIES (South Korea), and then drawn at room temperature to a diameter of 10 mm.

After these technological operations the rods of both alloys were TMT treated, including heat treatment (HT) – annealing at 400°C for at least 72 hours, 4 ECAP-C cycles via the Bc route at RT and final cold drawing (CD) to a diameter of 3.2 mm. CD resulted in the overall reduction/relative compression of $\sim$ 88%. Such TMT scheme was previously implemented in [11] to produce wires with a UFG structure from Al-0.4Mg-0.2Zr (wt.%).

A composite rod-shaped specimen was produced in two stages. At the first stage a tube with a targeted wall thickness was made from an initial cast Al-1.17Mg-0.35Zr rod with a diameter of 22 mm by mechanical treatment. Then an Al-0.4Zr rod with a corresponding diameter was inserted into it (figure 1).

At the second stage a composite specimen was subjected to rolling and drawing to shape a required geometry. Then the TMT specimen was processed into wire samples with a diameter

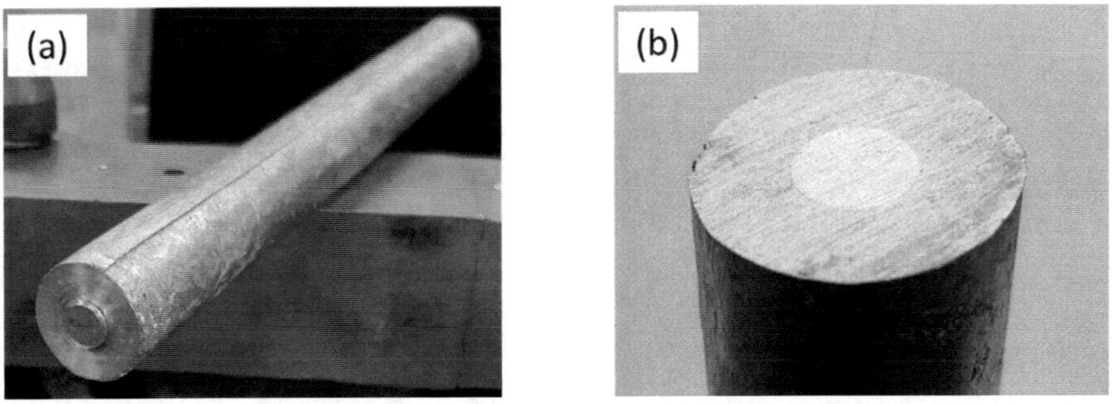

Figure 1. Composite specimen of Al alloys: (*a*) general view and (*b*) cross section.

of 3.2 mm. The schemes and regimes of TMT used to produce a wire are identical to those implemented for each of its component alloys separately.

3. Experimental procedures

Microstructure studies were performed using a transmission electron microscope (TEM) JEOL JEM 2100 at an accelerating voltage of 200 kV. Observations were made in both bright and dark field imaging modes and selected area electron diffraction (SAED) patterns.

Cylindrical tensile specimens with a gauge length of 15 mm and gauge diameter of 3 mm were machined from the rods before and after ECAP-C processing. Wire samples after TMT had 250 mm of a gauge length and were subjected to tensile testing [9]. Yield strength ($\sigma_{0.2}$), ultimate tensile strength (σ_{UTS}) and elongation to failure (δ) were estimated from the recorded engineering stress-engineering strain curves. Electrical resistivity of the material under study was measured in straightened samples of at least 1 m long in a measured part [9]. Thermal resistance of a wire was determined on the basis of σ_{UTS} variation after annealing at 230°C for 1 hour in accordance with [9, 10].

4. Results and Discussion

4.1. Microstructure and properties of Al-Zr and Al-Mg-Zr alloys after TMT

At the first stage of TMT the samples were heat-treated at 400°C. The heat treatment duration was selected experimentally. The process was terminated after stabilization of the electrical conductivity of both materials.

The microstructure studies by TEM showed the formation of particles with a size of 15–20 nm (figure 2) in both alloy as a result of heat treatment. The analysis of the SAED images (figure 2*a*) allowed referring the observed particles to a metastable Al_3Zr (L12) phase. As is known, the formation of this type of particles guarantees the increase of Al thermal resistance to 150–240°C [1]. The solid solution decomposition accompanied with the formation of nanosized particles and decrease in the concentration of Zr in Al [13] led to improvement of the electrical conductivity of the alloys. After terminating the heat treatment, the electrical conductivity of Al-0.4Zr and Al-1.17Mg-0.35Zr alloys was 59.8 and 49.8% IACS respectively (table 2). The noted difference in the level of their electrical conductivity is explained by the presence of Mg atoms in the solid solution of Al-Mg-Zr alloy after heat treatment. According to [14] the content of 1 wt.% Mg in Al decreases its electrical conductivity by 5.6 nΩm.

After the second stage of TMT, a UFG structure formed in the alloys as a result of ECAP-C

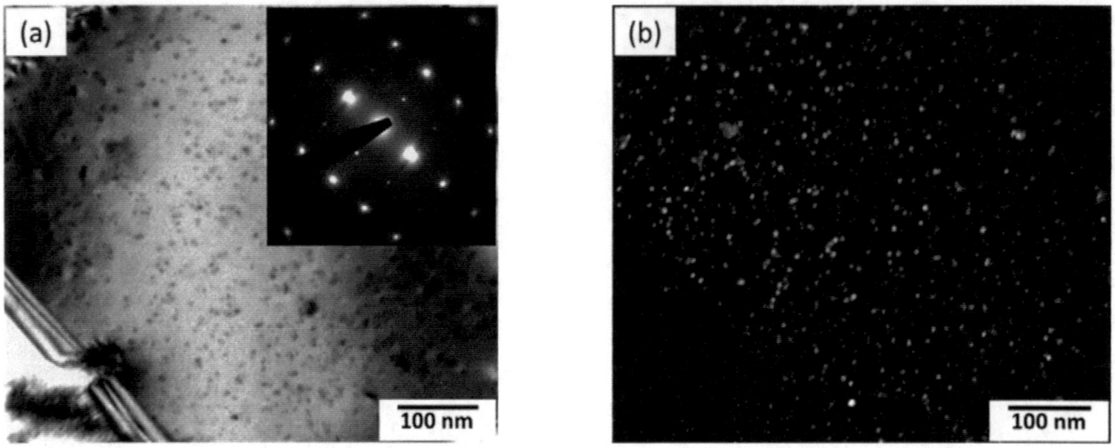

Figure 2. TEM images of AlMgZr microstructure after HT at 400oC (*a*) taken in bright field (BF) and (*b*) dark field (DF) modes. The insert to image (*a*): the corresponding SAED pattern (the crystal is close to the [112] zone axis orientation).

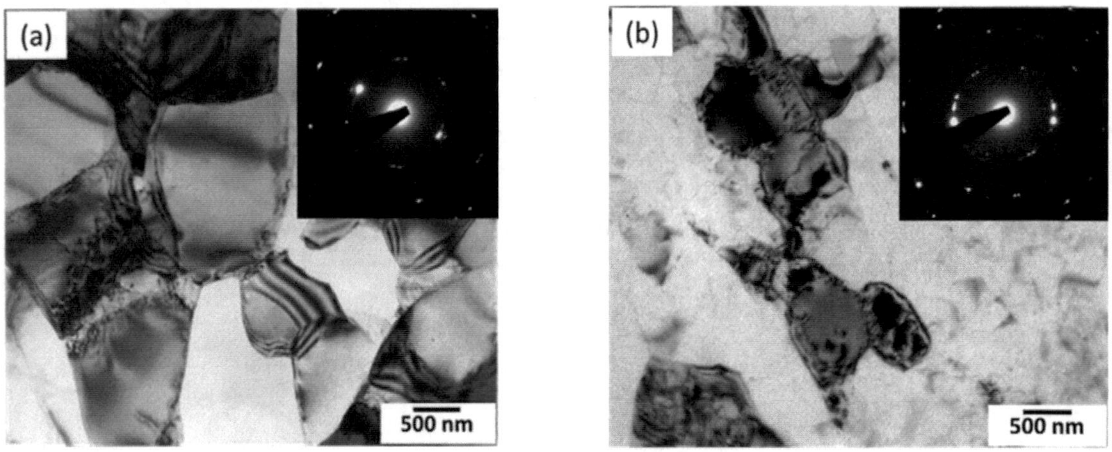

Figure 3. BF TEM images of microstructure of Al alloys after HT and ECAP-C processing: (*a*) Al-0.4Zr alloy and (*b*) Al-1.17Mg-0.35Zr alloy. The inserts to images (*a*, *b*): the corresponding SAED patterns.

(figure 3). In Al-0.4Zr and Al-1.17Mg-0.35Zr alloys the average grain size was 1570 ± 110 nm and 700 ± 40 nm, respectively. The typical SAED images (figure 3) testify to the formed states referring to microstructures of a grain type with mainly high-angle grain boundary misorientations [15]. The microstructure refinement caused a significant enhancement in the strength of alloys, and some reduction in the electrical conductivity as compared to the state after heat treatment (table 2). A slight change in the electrical conductivity is conditioned by insignificant influence of grain boundary and lattice dislocations on it even in the materials with UFG structure [3].

At the third stage of TMT cold drawing (CD) was applied to the UFG samples to achieve a desired for cable products shape and size. Besides, as previously stated, cold drawing contributed to additional hardening of UFG Al-based alloys due to further microstructure refinement [4, 11].

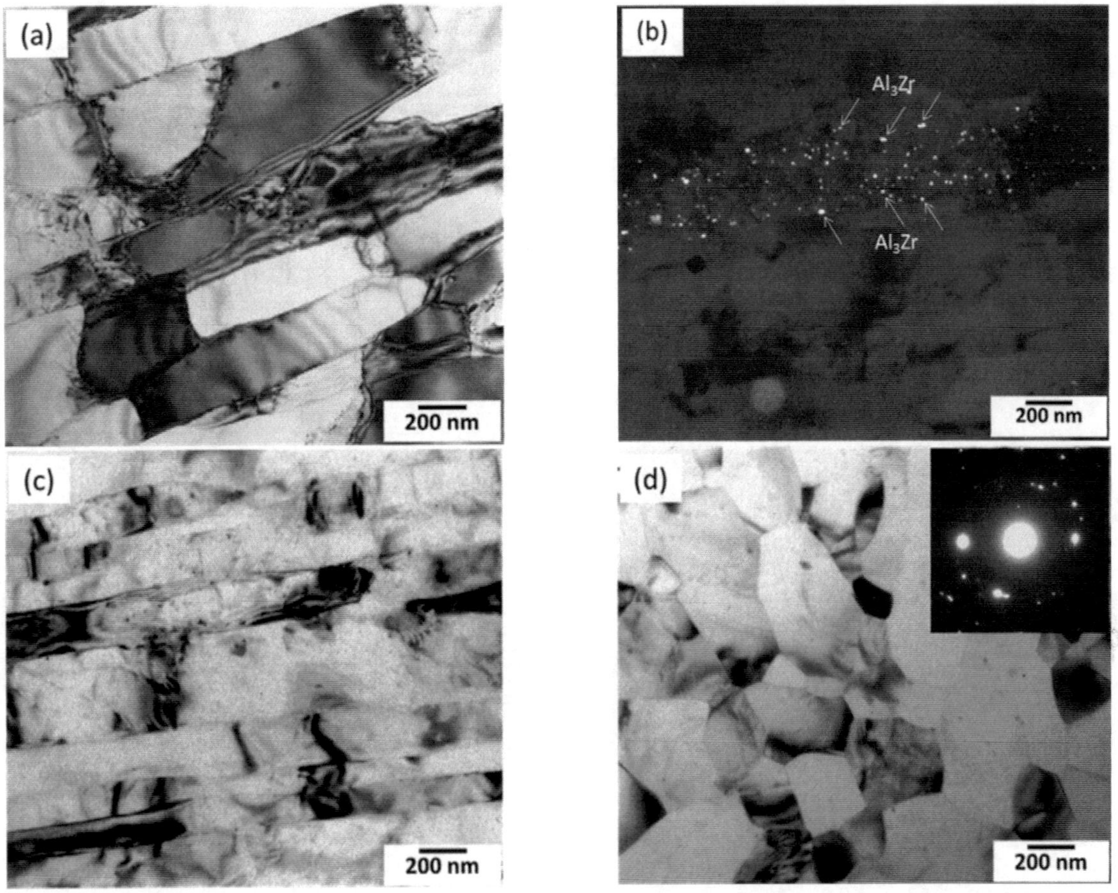

Figure 4. BF TEM images of microstructure of Al alloys after HT and ECAP-C processing: (*a*) Al-0.4Zr alloy and (*b*) Al-1.17Mg-0.35Zr alloy. The inserts to images (*a*, *b*): the corresponding SAED patterns.

After CD, ultrafine grains formed as a result of ECAP-C (figure 3) are elongated along the wire sample axis (figures 3*a,c*). The length of some grains achieves 1 μm. The grains are equiaxed in the cross section of a wire (figures 3*b,d*). Their average size in the Al-0.4Zr and Al-1.17Mg-0.35Zr alloys is 470 ± 85 nm and 250 ± 50 nm, respectively, which correlates well with the width of grains elongated in the longitudinal section (figures 3*a,c*). No significant changes neither in sizes nor in volume fraction of Al_3Zr phase particles were observed in both alloys as a result of ECAP-C and CD.

Use of TMT ensured an ultimate tensile strength of 225 MPa and 366 MPa and electrical conductivity of 59.3 and 48.3% IACS in UFG wires of Al-0.4Zr and Al-1.17Mg-0.35Zr alloys. A relative elongation of samples meets the requirements imposed on wires from heat-resistant Al alloys and exceeds 2% [9]. It is important to note that an Al-1.17Mg-0.35Zr wire in the UFG state has higher strength than those in the wires made of Al-Mg-Si alloys both after conventional treatment and after TMT including SPD [2–5].

4.2. Strength, electrical conductivity and thermal resistance of a composite wire

The experimental data on the properties of wire samples with a UFG structure produced of Al-0.4Zr and Al-1.17Mg-0.35Zr alloys given in table 2 were used to calculate the content of

Table 2. Mechanical properties and electrical conductivity of Al alloys and Al-based composite wire.

Treatment/State	$\sigma_{0.2}$, MPa	σ_{UTS}, MPa	δ, %	ρ, nΩm	% IACS
Al-Zr alloy wire					
HT	—	—	—	28.87	59.7
HT+ECAP-C	178 ± 5	193 ± 2	19.5 ± 1.6	28.93	59.6
HT+ECAP-C+CD	199 ± 7	226 ± 4	5.5 ± 0.3	29.09	59.3
Al-Mg-Zr alloy wire					
HT	173 ± 5	216 ± 4	19.6 ± 0.6	34.65	49.8
HT+ECAP-C	257 ± 5	267 ± 3	10.4 ± 1.0	35.08	49.1
HT+ECAP-C+CD	321 ± 11	366 ± 6	2.1 ± 0.1	35.70	48.3
Al-based composite wire					
HT+ECAP-C+CD^{calc}	—	345	—	—	—
HT+ECAP-C+CD^{exp}	307 ± 15	354 ± 4	3.2 ± 0.3	34.07	50.6

the alloys in a composite wire, which will ensure its strength at a level or higher than that in Al-Mg-Si alloys (ultimate tensile strength ±342 MPa). The calculation was made according to the rule of mixtures:

$$\sigma_{\text{c}} = \sigma_{\text{core}} V_{\text{core}} + \sigma_{\text{clad}}(1 - V_{\text{core}}),$$

where σ_{c} is the ultimate tensile strength of a composite wire, σ_{core} is the ultimate tensile strength of a core, V_{core} is the volume fraction of a core in a conductor, σ_{clad} is the ultimate tensile strength of a clad material.

As was stated, the chosen level of strength in a composite wire could be achieved by fulfilling the following condition: $V_{\text{clad}}/V_{\text{c}} \geq 0.82$, where Al-1.17Mg-0.35Zr alloy in the UFG state was used as a material of clad.

In accordance with the estimates, an experimental sample of a composite wire was produced. The chosen design, in which a clad is made of a stronger material, should ensure higher resistance to surface defects during exploitation capable of leading to early failure [16]. The same conclusion on the design of a wire was made by the authors in [8].

The properties of a composite wire produced by TMT are listed in table 2. The volume fraction of high-strength Al-1.17Mg-0.35Zr alloy is 0.83 in it. The analysis of the obtained data showed that the estimated and experimentally obtained values of the ultimate tensile strength correlate well. Their divergence is no more than 3%. The level of strength and electrical conductivity achieved in the composite wire is comparable and even exceeds similar properties in the conductors made of Al-Mg-Si [2–5].

Special annealing performed at 230°C for 1 hour, the regime of which is recommended to determine the level of thermal resistance of conducting aluminum alloys [14], revealed that the composite wire can be used at operating temperatures of 150°C. Its ultimate tensile strength reduces by 10% from 354 MPa to 322 MPa, and the electrical conductivity increases to 50.9%IACS as a result of annealing. Judging by the stated changes in the properties, the stability of a UFG structure formed during TMT and the strength of the alloys forming a composite wire are supposedly ensured by nanosized particles of a metastable Al_3Zr phase (figure 2).

5. Conclusion

As a result of the studies, the possibility of creating a composite high-strength heat-resistant conductor based on alloys of Al-Zr and Al-Mg-Zr systems was demonstrated.

Based on the obtained experimental data and calculations from the rule of mixtures (additivity), the optimal ratio of the content of the selected aluminum alloys in the composite conductor was determined.

A unique TMT was introduced, the implementation of which led to the achievement of an ultimate tensile strength of 366 MPa, an electrical conductivity of 50.4% IACS, and an operating temperature of up to 150°C in the composite material. This level of properties is ensured by the formation of a UFG structure in the conductor and the presence of nanosized particles of a metastable Al_3Zr phase in it.

In terms of strength and heat resistance, the composite wire with a UFG structure is noticeably superior to alloys of an Al-Mg-Si system widely used in electrical engineering.

Acknowledgments

The present work was supported by the Russian Science Foundation (grant No. 17-19-01311).

References

[1] Kiessling F, Nefzger P, Nolasco J F, and Kaintzyk U 2003 *Overhead Power Lines: Planning, Design, Construction* (Berlin: Springer)

[2] EN 50183 2002 *Overhead Power Line Conductors – Bare Conductors of Aluminium Alloy with Magnesium and Silicon Content* (European Standard)

[3] Murashkin M Yu, Sabirov I, Sauvage X, and Valiev R Z 2016 Nanostructured Al and Cu alloys with superior strength and electrical conductivity *Journal of Materials Science* **51** (1) 33–49 doi: 10.1007/s10853-015-9354-9

[4] Medvedev A, Arutyunyan A, Lomakin I, et al. 2018 Fatigue properties of ultra-fine grained Al-Mg-Si wires with enhanced mechanical strength and electrical conductivity *Metals* **8** (12) 1034 doi: 10.3390/met8121034

[5] Khangholi S N, Javidani M, Maltais A, and Chen X-G 2021 Effects of natural aging and pre-aging on the strength and electrical conductivity in Al-Mg-Si AA6201 conductor alloys *Materials Science and Engineering: A* **820** 141538 doi: 10.1016/j.msea.2021.141538

[6] Qi Y, Lapovok R, and Estrin Y 2016 Microstructure and electrical conductivity of aluminium/steel bimetallic rods processed by severe plastic deformation *Journal of Materials Science* **51** (14) 6860–75 doi: 10.1007/s10853-016-9973-9

[7] Lapovok R, Popov V V, Qi Y, et al. 2020 Architectured hybrid conductors: Aluminium with embedded copper helix *Materials & Design* **187** 108398 doi: 10.1016/j.matdes.2019.108398

[8] Yang C, Masquellier N, Gandiolle C, and Sauvage X 2020 Multifunctional properties of composition graded Al wires *Scripta Materialia* **189** 21–4 doi: 0.1016/j.scriptamat.2020.07.052

[9] IEC 62004:2007 *Thermal Resistant Aluminum Alloy Wire for Overhead Line Conductors*

[10] ASTM B941-05 2009 *Standard Specifcation for Heat Resistant Aluminum-Zirconium Alloy Wire for Electrical Purposes*

[11] Murashkin M Yu, Medvedev A E, Kazykhanov V U, et al. 2016 Microstructure, strength, electrical conductivity and heat resistance of an Al-Mg-Zr alloy after ECAP-conform and cold drawing *Reviews on Advanced Materials Science* **47** 16–25

[12] Schoerner R J 1972 *Method of Fabricating Aluminum Alloy Rod*, US 3670401

[13] Valiev R Z, Islamgaliev R K, and Alexandrov I V 2000 Bulk nanostructured materials from severe plastic deformation *Progress in Materials Science* **45** (2) 103–89 doi: 10.1016/S0079-6425(99)00007-9

[14] Olafsson P, Sandstrom R, and Karlsson Å 1997 Comparison of experimental, calculated and observed values for electrical and thermal conductivity of aluminium alloys *Journal of Materials Science* **32** (16) 4383–90 doi: 10.1023/A:1018680024876

[15] Knych T, Piwowarska M, and Uliasz P 2011 Studies on the process of heat treatment of conductive AlZr alloys obtained in various productive processes *Archives of Metallurgy and Materials* **56** (3) 685–92 doi: 10.2478/v10172-011-0075-z

[16] Kalombo R B, Araújo J A, Ferreira J L A, et al. 2015 Assessment of the fatigue failure of an All Aluminium Alloy Cable (AAAC) for a 230 kV transmission line in the Center-West of Brazil *Engineering Failure Analysis* **61** 77–87 doi: 10.1016/j.engfailanal.2015.08.043

A model of mechanics of growing solids for the stress-strain analysis of wound composites reinforced with thin unidirectional fibers

K E Kazakov[1*] **and D A Parshin**[2**]

[1]Bauman Moscow State Technical University, Moscow, Russia
[2]Ishlinsky Institute for Problems in Mechanics of the Russian Academy of Sciences, Moscow, Russia

E-mail: *kazakov-ke@bmstu.ru, **parshin@ipmnet.ru

Abstract. A mathematical model of the mechanics of growing solids has been developed that allows us to retrace the elastic stress-strain state progressing for cylindrical composites in the process of their slow layer-by-layer manufacturing by force winding of elementary material layers reinforced with sufficiently thin fibers in the circumferential direction, with an arbitrary density of their laying and an arbitrary pre-tension force. The obtained analytical relationships enable to set different control problems for a composite layer being formed including those for the characteristics of its contact interaction with the mandrel used.

1. Introduction

The use of composite materials can attain satisfactory mechanical properties with noticeable benefit in weight and size of the resulting products. This explains the widespread implementation of such materials in different branches of the national economy. Meanwhile, there is another distinguishing feature of composite materials. This is the existence of fundamental possibilities for the systematic formation of the desired stress-strain characteristics of the resulting products directly during the technological process of their production. It comes to all kinds of additive processes (in a generalized sense) for the manufacture of composite products, i.e. the layer-by-layer manufacturing processes. However, in order to take advantage of these opportunities, it is necessary to figure out the factors that cause the stress appearance and evolvement in composites in various additive manufacturing processes, as well as to elaborate adequate mathematical models for descripting these factors. Note that standard approaches of continuum mechanics are not valuably adequate for such purposes because these approaches do not take into account the very specifics of kinematics of solids that change their shape and size due to the replenishment with new elements of material and the deformation simultaneously. The indicated solids use to be termed as the growing solids. To describe the mechanical behavior of the growing solids, one should develop fundamentally novel approaches. This goal is dealt with in the works of the scientific school in mechanics of growing solids that has developed over the past twenty years under the initial leadership of Professor A.V. Manzhirov (see, e.g., [1–10]). It is on these approaches that the here presented research is based.

2. The problem concretization

One of the well-known additive technologies for the manufacture of composite products is the layer-by-layer winding of fiber-reinforced plies of material onto the surface of a mandrel of a certain shape. The wound plies pre-tension causes the appearance and evolvement of some significant technological stresses throughout the product being formed by this method. The nature of the current and resulting picture of the stress-strain state will depend on the specific setting parameters of the considered manufacture process. The present work is targeted at building and analysing a fairly simple mathematical model of growing solids mechanics for describing the technological process of manufacturing a cylindrical circular product by the method of power layer-by-layer slow winding of composite material plies reinforced with unidirectional fibers.

3. Formulation of the research goal

In the above mentioned method, the process of sequential laying on the manufactured cylindrical product surface the elementary plies of material having a sufficiently small but variable relative thickness is modeled. The single wound material ply contains, radially, the only turn of the reinforcing fiber whose accurate radial position within the thickness of the binder used is indefinite. In the longitudinal direction, the turns are arranged sequentially one after another at right angles to the longitudinal axis, but without rigidly fixed steps and concrete reference to the turns position in the previously laid elementary plies. In the situation, the axial longitudinal section of the resulting solid will be penetrated in the vicinity of each point in the normal (hoop) direction by a large number of quite randomly distributed fibers, and it is correct to assume that the composite cylindrical product being formed in the considered technological process will averaged acquire transversally, relative to the hoop direction, isotropic mechanical properties [11].

As mentioned above, it is allowed to vary the thickness of the laid material plies throughout winding but varying does not depend on the frequency of the plies winding onto the manufactured composite solid layer. So the entire thickness h of the latter remains, at an average, the same in any axial longitudinal section during the whole manufacturing process, and, accordingly, this process proceeds axisymmetrically.

In view of the stipulated low winding speed, and also requiring the absence of rapid changes in the winding speed during the process, we will consider the problem on deformation of the layer-by-layer manufactured composite solid in the quasi-static approximation. Since this solid is sufficiently extended in the axial direction we will study the case of a plane stain state. Note that a quasi-static statement of the problem on deformation of a growing solid does not at all imply the stationarity of this solid's stress-strain state. This state will change over time t due to continuous involving in the deformation process new elements of the used material being in its own state, not related to the state of the already formed solid part, and due to additional loading the already formed solid with forces carried by those elements. At a low winding speed and a relative small thickness of each separately laid elementary ply of additional material, the following model facilitation is also permissible. Firstly, it is possible to model the winding process under study as a process of continuous growth of a deformable solid, when an infinitely thin material layer is being applied onto the solid surface for every infinitely small period of time. Secondly, the effects associated with the strain delay in the entire material layer obtained by winding may not be taken into account because the lower the rate of replenishment of a growing solid with new material, the less the strain delay affects the development stresses in this solid [12], so it is possible to analyse only the elastic stress-strain state.

We will analyse the case of a small strain of the considered layer-by-layer formed composite solid. In this case, the function $h(t)$ that describes the change in the thickness of the entire composite layer growing on the mandrel over the time t can be considered as set by virtue of the implemented winding program. This function obviously increases monotonically. It is justified by the fact that the thickness of the growing layer in the process under consideration changes over time, generally speaking, due to two factors — attachment of new material to the surface of the growing solid and deformation of the latter. The smaller its strain, the more insignificant the second factor is compared to the first.

In the developed mathematical model, we consider the malleability of the used mandrel as negligible. This means that the inner radius a of the being manufactured layer, which coincides with the radius of the mandrel working surface, do not change over time. This simplification is not essential for the developed model and one may abandon it at the cost of complicating all the resulting dependencies being to construct by solving the respective boundary-value problem of mechanics.

4. Mathematical model

Let then be θ the hoop angle coordinate, z the axial linear coordinate, and r the radial linear coordinate in the reference frame associated with the rigid mandrel used for the winding process. In accordance with the above, the direction that locally defines the isotropy axis at a given point coincides with the direction of change of the angle coordinate θ. In this case, the to this direction perpendicular directions of change of the linear coordinates z and r set the local rectangular basis in the plane of isotropy containing this point.

Denote by V_θ, V_z, and V_r the components, in the considered curved coordinate system, of the velocity vector field obeying the growing composite layer deformation law. The condition of the axially symmetric plane strain impose the following restrictions on the velocity field:

$$V_\theta = V_\theta(r,t), \quad V_z \equiv 0, \quad V_r = V_r(r,t).$$

Neglecting the twist deformation of the considered growing cylindrical solid in the introduced system coordinates we will additionally have the condition of mirror symmetry for the field velocities:

$$V_\theta \equiv 0.$$

The velocity field satisfying the above specified conditions corresponds to the strain velocity tensor which has only two physical components in the considered curved orthogonal coordinate system that are distinct from zero, i.e. corresponding diagonal components

$$\xi_\theta = \frac{V_r}{r}, \quad \xi_r = \frac{\partial V_r}{\partial r}.$$

They are to be connected by equations of state with the diagonal physical components of the stress rate tensor that are the only non-zero components of this tensor in the coordinate system under consideration. For the current case of the transversal isotropy of an elastic composite material obtained by winding (see Section 3), we have

$$\frac{\partial}{\partial t}\begin{pmatrix} \sigma_\theta \\ \sigma_z \\ \sigma_r \end{pmatrix} = \begin{pmatrix} C_{\theta\theta} & C_{\theta r} \\ C_{\theta r} & C_{zr} \\ C_{\theta r} & C_{rr} \end{pmatrix}\begin{pmatrix} \xi_\theta \\ \xi_r \end{pmatrix},$$

where $C_{\theta\theta}$, $C_{\theta r}$, C_{zr}, and C_{rr} are the constants of the material.

The standard equation of equilibrium in stresses remains, obviously, actual in the developed model. As a consequence, we can write the following equation for the rates of stress change in the considered system of curved coordinates:

$$r\,\frac{\partial^2 \sigma_r}{\partial t\,\partial r} = \frac{\partial \sigma_\theta}{\partial t} - \frac{\partial \sigma_r}{\partial t}, \quad a < r < a + h(t).$$

Coupling of the being manufactured composite layer with the used rigid mandrel is provided by the following obvious boundary condition:

$$V_r\big|_{r=a} = 0.$$

The variable outer surface of the layer formed, on which additional elementary plies of material are superimposed throughout the winding, in the constructed model of a continuously growing solid, should move in space in the direction normal to the layer at a speed of

$$v(t) = \frac{dh(t)}{dt}$$

determined by the implemented winding program. It can be shown (see, for example, [13]) that on the outer surface of a growing solid of the considered shape, a boundary condition must be set that determines the radial stress initial rate of change in that elementary ply of new material which has just joined this surface and has become part of the growing solid:

$$\frac{\partial \sigma_r}{\partial t} = -\frac{\sigma_0(t)v(t)}{a+h(t)}, \quad r = a + h(t)$$

where the function $\sigma_0(t)$ sets the stress of pre-tension in the hoop direction in the elementary ply of material joined at the time t, and it is considered as known by virtue of the implemented winding program:

$$\sigma_\theta\big|_{r=a+h(t)} = \sigma_0(t), \quad \sigma_z\big|_{r=a+h(t)} = \sigma_r\big|_{r=a+h(t)} = 0.$$

5. Results discussion

Formulating the condition on the outer surface of the considered growing composite layer closes the boundary-value problem for determining the rate characteristics of the stress-strain state changing inside this layer in the simulated process of its slow power winding. We can construct an exact solution of this problem being non-classical for solid mechanics. The solution is not given here due to the limited scope of the article. The corresponding analytical dependencies give opportunities to state and solve various technological problems on the control of these characteristics, i.e. in order to minimize the damage hazard for the manufactured composite product including the damage already in manufacturing process. The developed model also makes it possible to predict the contact pressure evolvement between the composite layer being manufactured and the mandrel. This is of extremal importance for wear and stability problems analysis. In addition, the obtained dependencies allow us to solve the problem of finding residual stresses in the finished composite layer after its separation from the mandrel. Appearance of residual stresses is inevitable in any additive manufacturing process: they appear due to the very particularity of such processes as a result of strain incompatibility for various elements of material that sequentially make up any additively formed solid [14].

Acknowledgments

The study was carried out within the framework of the Russian State Assignment under Contract No. AAAA-A20-120011690132-4 with the Ministry of Science and Higher Education.

References

[1] Manzhirov A V 2017 Fundamentals of mechanical design and analysis for AM fabricated parts *Procedia Manufacturing* **7** 59–65 doi: 10.1016/j.promfg.2016.12.017

[2] Manzhirov A V 2017 Advances in the theory of surface growth with applications to additive manufacturing technologies *Procedia Engineering* **173** 11–6 doi: 10.1016/j.proeng.2016.12.008

[3] Manzhirov A V 2017 Mechanical design of AM fabricated prismatic rods under torsion *MATEC Web of Conferences* **95** 12002 doi: 10.1051/matecconf/20179512002
[4] Manzhirov A V 2017 Mechanical analysis of an AM fabricated viscoelastic shaft under torsion by rigid disks *Lecture Notes in Engineering and Computer Science* **2230** 856–60
[5] Manzhirov A V 2018 Theory of continuous additive growth for solids *Materials of IUTAM Symposium on Mechanical Design and Analysis for AM Technologies 2018 (Moscow). Ed. by A V Manzhirov and E V Murashkin* (Moscow: IPMech RAS) 27–9
[6] Manzhirov A V 2018 What is inside AM fabricated parts? The nature of residual stresses and shape distortion *International Journal of Advancements in Technology* **9** 38 doi: 10.4172/0976-4860-C1-001
[7] Manzhirov A V 2018 Some problems in mechanics of growing solids with applications to AM technologies *Journal of Physics: Conference Series* **991** 012056 doi: 10.1088/1742-6596/991/1/012056
[8] Manzhirov A V and Mikhin M N 2018 2D problems of surface growth theory with applications to additive manufacturing *Journal of Physics: Conference Series* **991** 012057 doi: 10.1088/1742-6596/991/1/012057
[9] Parshin D A 2019 Solving a problem of theory of viscoelasticity for continuously accreted solids with the satisfaction of static boundary conditions in the integral sense *Journal of Physics: Conference Series* **1203** 012006 doi: 10.1088/1742-6596/1203/1/012006
[10] Parshin D A 2020 Mathematical modelling of the process of internal technological stress fields development in cylindrical solids being layer-by-layer made of ageing viscoelastic materials *Journal of Physics: Conference Series* **1479** 012145 doi: 10.1088/1742-6596/1479/1/012145
[11] Ashkenazi E K and Ganov E V 1980 *Anisotropy of Structural Materials: Handbook. 2nd ed* (Leningrad: Mashinostroenie) [in Russian]
[12] Parshin D A 2018 Impact of the rate of the additive process of forming a heavy structure deforming in creep on the development of its technological stresses *AIP Conference Proceedings* **1959** 070029 doi: 10.1063/1.5034704
[13] Parshin D A 2021 Technological controlling the contact pressure onto a cylindrical substrate from a coating being formed additively on its outer surface *AIP Conference Proceedings* **2340** 040005 doi: 10.1063/5.0047838
[14] Bychkov P S, Kozintsev V M, Manzhirov A V, and Popov A L 2017 Determination of residual stresses in products in additive production by the layer-by-layer photopolymerization method *Mechanics of Solids* **52** (5) 524–29 doi: 10.3103/S0025654417050077

The evolution of creep curves of high impact polystyrene and polyurethane specimens after aging

A R Arutyunyan
St. Petersburg State University, St. Petersburg, Russia
E-mail: a.arutyunyan@spbu.ru

Abstract. Specimens of high impact polystyrene and polyurethane were studied for creep at room temperature and constant tensile force without aging and after aging according to various programs. For polyurethane specimens after aging, an increase in creep time for a given deformation by 3 times compared to specimens without aging is observed. For high impact polystyrene in the creep experiments after aging, a decrease of deformation by 6.5 times and the time to fracture by more than 24 times, compared with specimens without aging is shown. To describe the obtained experimental creep curves for polyurethane and polystyrene specimens without aging and after aging, modified versions of linear elastic–viscous models of Maxwell and Voigt written in the effective time scale are used. A good agreement between the theoretical and experimental creep curves is observed.

1. Introduction

Polymers and polymer–based materials are intensive implemented in almost all areas of engineering practice: industry, construction, aircraft, rocket science, shipbuilding and medicine. These applications frequently have a very high performance demand, which makes their long–term characteristics of paramount importance. At the same time, they were subjected to degradation due to environmental factors, including light, temperature, stress and others. In addition, the mechanical characteristics of polymer materials directly depend on their structure. During the aging, the degradation of polymer materials leads to significantly changes of their physical and mechanical properties [1–6]. As the result of degradation, the material becomes less deformable and brittle. The influence of aging on the mechanical properties of the polyurethane and polystyrene specimens at various environmental factors are investigated in works [7–13].

In the paper [7], the effects of weathering on the mechanical properties of the precipitated and compact polyurethane films are investigated. The natural weathering of polyurethane films carried out in earth, seawater and exposure to sunlight was compared with untreated samples. Stress–strain, elongation, modulus and ultimate tensile strength decreased with increasing of the weathering time. It was found, that the treatments in seawater improved the mechanical properties of the elastomers.

Physical and mechanical properties of thermoplastic polyurethane were investigated in [8] after immersion in water at a temperature of 70°C up to 6 months. The mechanical properties in bulk material, obtained from tensile test, were affected by aging. Elastic modulus and stress at 200% of strain of the studied polyurethane were decreased after sufficient exposure to moisture. The reversibility of mechanical and physical properties after moisture exposure was assessed. The polymer degradation was found as irreversible phenomenon.

The effects of environmental aging on the mechanical performance of elastomeric polyurethane were investigated in [9] using two accelerated aging techniques: ultraviolet and hygrothermal. Samples were subjected to ultraviolet and hygrothermal exposure for a period of 5 months. The stress and strain to failure, tearing energy, and storage modulus were evaluated at different intervals for both aging techniques. It was found that the ultraviolet exposure caused severe degradation of the polyurethane in comparison with the hygrothermal exposure. A reduction of more than 98% in the tearing energy was observed for the ultraviolet exposed samples after 5 months compared with only a 35% reduction in the tearing energy for the hygrothermal–exposed samples. A similar trend was observed for tear strength and storage modulus.

In the work [10], the degradation behavior of some polyurethane membranes exposed to artificial weathering environment for different exposure time was studied. It is observed that the mechanical properties of the polyurethane membranes change as function of ageing time. A slight decrease of the mechanical properties namely, tensile strength and elongation at break was remarked. Experimental results show that the decomposition of polyurethane membranes is a complex process and is the results of a multitude of physical and chemical phenomena, which involves the combination of several steps such as chain scission, re-arrangement of them and crosslinking.

Experiments on the alternation of deep compression of rectangular polyurethane specimens and long–term climatic and deformation aging were carried out in [11]. According to the obtained experimental results in the process of long aging of about nineteen years, the hardening of material is no monotonic. The maximum value of hardening (according to stress value) is more than 30 times compared to the specimens without aging. With increasing of aging time, the material become softening.

In paper [12] the experiments on alternation of fatigue, creep, long–term climatic and deformation aging of polyurethane specimens were performed. The specimens were exposed in ice during different periods, and then the cyclic loadings were conducted. According to the obtained results, at first a significant hardening during cyclic loadings is observed. The maximum number of cycles to fracture is observed at 11 days of exposure in ice and is 8 times more compared for specimens without aging. Further exposure in ice leads to decrease of number of cycles to fracture. For 23 days, it is approximately equal to number of cycles to fracture for specimens without aging. Consequently, is found that the aging process is no monotonic.

The effect of long–term natural aging at room temperature during 40 years on the mechanical properties of six marks of high impact polystyrene was investigated in [13]. Stress–strain curves in tensile experiments at two different loading speeds are obtained. For all marks of high impact polystyrene after aging, the effect of hardening and increasing the yield strength is observed. Also, for many marks of polystyrene, a decrease of the deformation at fracture is noted.

In this paper, experimental studies of deformation and climatic aging of polyurethane in creep experiments and experiments on long–term natural aging of high impact polystyrene during creep are carried out. To describe the obtained experimental results, the modified Maxwell and Voigt equations written on the effective time scale are considered.

2. Modification of linear elastic–viscous models using the concept of effective time

To specify the experimental creep curves, let's consider the problem of stretching of specimen of an elastic–viscous aged material under the influence of a constant load P. To describe the process of deformation aging parameter α is introduced [6]

$$d\alpha = f_1(\alpha, \varepsilon, T, t)dt + f_2(\alpha, \varepsilon, T, t)d\varepsilon, \tag{1}$$

where ε is the value of deformation, t is the real time, T is temperature.

The parameter α is considered as an effective time by which it is possible to describe the processes of climatic and deformation aging. According to equation (1) for instant, active loads, the parameter α corresponds to the 'deformation time' ε. In the unloading state, this parameter describes the kinetics of chemical aging processes and is reduced to usual time t. With this interpretation, the concept of 'chemical time' can be introduced. Thus, the effective time parameter, in generally, is able to describe the interrelated deformation and physic–chemical processes and define their development in the scale of deformation and chemical time. This is the difference between this parameter and the well-known temperature–time and polarization parameters used in mechanics of polymer materials [14–17].

When calculating using formula (1), let's take the effective time parameter in the following form

$$d\alpha = k(\alpha_\infty - \alpha)t^m dt, \tag{2}$$

where k, α_∞, m are constants, α is material degradation parameter ($\alpha = N/N_0$, N_0 is initial number of chemical bonds, N is current number of destroyed chemical bonds).

Thus, equation (2) can be considered as the equation of a chemical reaction, and the parameter α has the meaning of chemical time.

Let's use the modified Maxwell equation, written in the effective time scale

$$\frac{d\varepsilon}{d\alpha} = \frac{1}{E}\frac{d\sigma}{d\alpha} + \frac{\sigma}{\eta}, \tag{3}$$

where σ is true stress, ε is deformation, T is temperature, E is elasticity modulus, η is viscosity coefficient.

The solution of the system of equations (2) and (3) under $\sigma = \text{const}$ and initial conditions $t = 0$, $\alpha = \alpha_0$, $\varepsilon = \sigma_0/E_0$, can be written as

$$\frac{\varepsilon}{\sigma_0} = \frac{1}{E_0}\left\{1 + \frac{\alpha_\infty - \alpha_0}{\tau}\left[1 - \exp\left(-\frac{k}{m+1}t^{m+1}\right)\right]\right\}, \tag{4}$$

where τ is relaxation time.

The modified Voigt equation, written in the effective time scale has the following form

$$\frac{d\varepsilon}{d\alpha} + \frac{E}{\eta}\varepsilon = \frac{\sigma}{\eta}. \tag{5}$$

Solving of the system of equations (2) and (5) under $\sigma = \text{const}$ and following initial conditions $t = 0$, $\alpha = \alpha_0$, $\varepsilon = \sigma_0/E_0$, we can receive

$$\frac{\varepsilon}{\sigma_0} = \frac{1}{E_0}\left(1 - \exp\left\{-\frac{1}{\tau}\left[\alpha_\infty - (\alpha_\infty - \alpha_0)\exp\left(-\frac{k}{m+1}t^{m+1}\right)\right]\right\}\right). \tag{6}$$

3. Experimental investigations of aging of polyurethane specimens in creep experiments

To investigate the aging of polyurethane specimens a round reinforced polyurethane drive belt Continental Contitech with the diameter of 4 mm was used. Fatigue experiments were conducted on a desktop servo–hydraulic fatigue test machine Si-Plan SH-B and creep experiments on Shimadzu AGX-50 plus tearing machine at room temperature. The aging of polyurethane specimens was performed according to the following program: cyclic loadings with a displacement amplitude of 4 mm and a loading frequency of 10 Hz up to the 105000 cycles; natural aging for one year in laboratory conditions; cyclic experiments with the same loading parameters; natural aging for two years in laboratory conditions; creep experiments under constant load. Some of the specimens were additionally aged during one year, and then were also tested for creep under the same conditions.

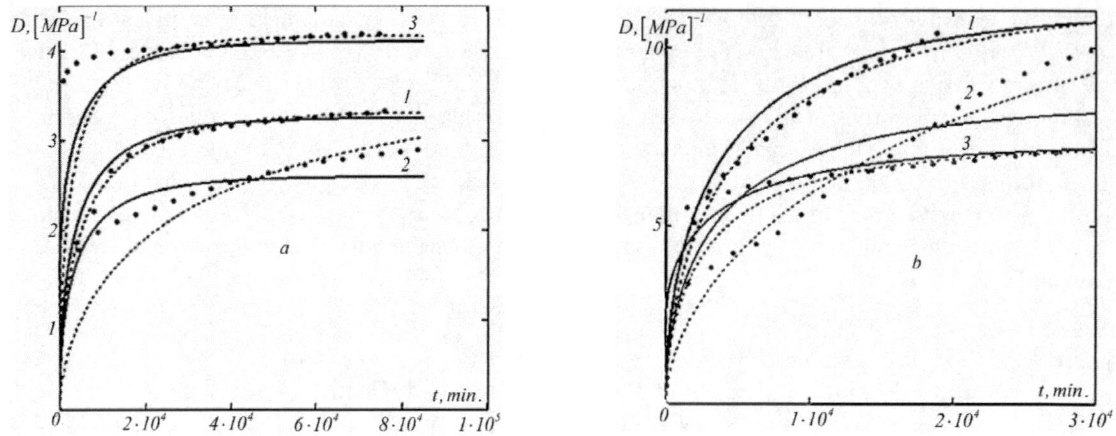

Figure 1. Theoretical compliance curves according to equations (4), (6) and experimental points for polyurethane specimens at a load of 70 N (a) and 80 N (b) without aging (curves *1*), with aging (curves *2*) and additional aging (curves *3*). Solid curves corresponds to the modified Maxwell model; dotted curves are the solution according to the modified Voigt model.

The compliance curves ($D = \varepsilon/\sigma_0$) according to the modified Maxwell model (4), Voigt model (6) and experimental creep data for polyurethane specimens without aging (curves *1*), with aging (curves *2*) and additional aging (curves *3*) at loads of 70 and 80 N are shown on figure 1. In the calculations, the following coefficient values are used:

- at a load of 70 N:
  - Maxwell model:
    $\alpha_0 = 0$, $\alpha_\infty = 1$, $m = -0.4$, $k = 4 \times 10^{-3}\,\mathrm{s}^{-1}$,
    for specimens without aging $\tau = 0.0107\,\mathrm{s}$, $E_0 = 29\,\mathrm{MPa}$;
    for specimens after aging $\tau = 0.015\,\mathrm{s}$, $E_0 = 26\,\mathrm{MPa}$;
    for specimens after additional aging $\tau = 0.5\,\mathrm{s}$, $E_0 = 0.73\,\mathrm{MPa}$;
  - Voigt model:
    $\alpha_0 = 0$, $\alpha_\infty = 1$, $m = -0.4$, $k = 2.3 \times 10^{-5}\,\mathrm{s}^{-1}$,
    for specimens without aging $\tau = 0.007\,\mathrm{s}$, $E_0 = 0.3\,\mathrm{MPa}$;
    for specimens after aging $\tau = 0.002\,\mathrm{s}$, $E_0 = 0.27\,\mathrm{MPa}$;
    for specimens after additional aging $\tau = 0.005\,\mathrm{s}$, $E_0 = 0.24\,\mathrm{MPa}$;
- at a load of 80 N:
  - Maxwell model:
    $\alpha_0 = 0$, $\alpha_\infty = 1$, $m = -0.4$, $k = 4 \times 10^{-3}\,\mathrm{s}^{-1}$,
    for specimens without aging $\tau = 0.0107\,\mathrm{s}$, $E_0 = 8.5\,\mathrm{MPa}$;
    for specimens after aging $\tau = 0.015\,\mathrm{s}$, $E_0 = 8\,\mathrm{MPa}$;
    for specimens after additional aging $\tau = 0.5\,\mathrm{s}$, $E_0 = 0.41\,\mathrm{MPa}$;
  - Voigt model:
    $\alpha_0 = 0$, $\alpha_\infty = 1$, $m = -0.4$, $k = 2.3 \times 10^{-5}\,\mathrm{s}^{-1}$,
    for specimens without aging $\tau = 0.007\,\mathrm{s}$, $E_0 = 0.087\,\mathrm{MPa}$;
    for specimens after aging $\tau = 0.002\,\mathrm{s}$, $E_0 = 0.065\,\mathrm{MPa}$;
    for specimens after additional aging $\tau = 0.005\,\mathrm{s}$, $E_0 = 0.138\,\mathrm{MPa}$.

From figure 1 it can be seen that the obtained experimental curves are in good agreement with the corresponding theoretical creep curves for polyurethane specimens. Experimental data shown that after one year of additional aging, polyurethane specimens are continued hardening. For specimens after aging, an increase in creep time for a given deformation by 3 times compared to specimens without aging is observed.

4. Experimental investigations of aging of high impact polystyrene specimens in creep experiments

High impact polystyrene (HIPS) is a polymer–polymer matrix composites with a polystyrene matrix, in which there are rubber–like inclusions with size of 0.5–5 microns and a volume concentration in the range of 20–40%.

Creep experiments were carried out for the HIPS of UPS 03L mark at constant tensile stresses equal to 0.8 and 0.9 of the ultimate tensile stress (σ_f). Creep experiments were carried out in 1979 during 30 days. At such time period at stress of $0.8\sigma_f$, no one of the specimens does not reached the limit state. During creep at stress of $0.9\sigma_f$, HIPS specimens were fractured by the end of the 10th hour.

After 40 years of natural aging in the laboratory conditions, the specimens were tested again for creep under the same loading parameters. After forty years of natural aging, a significant decrease in the time to fracture is observed: at stress of $0.8\sigma_f$ time to fracture is equal only to 402 min, at stress of $0.9\sigma_f$ it is only 9.6 min. Also, in almost all cases, the resulting deformation is greatly reduced by up to 6.5 times compared with specimens without aging.

On figures 2 and 3 the compliance curves according to the modified Maxwell model (4), Voigt model (6) and experimental creep data for specimens of the HIPS of UPS 03L mark without aging and after natural aging at stresses of 0.8 and 0.9 of the ultimate tensile stress are shown.

In the calculations, the following coefficient values are used:

- at stress $\sigma = 0.8\sigma_f = 26.64\,\text{MPa}$:
  - Maxwell model:
    $\alpha_0 = 0$, $\alpha_\infty = 1$, $m = -0.2$, $k = 6.5 \times 10^{-4}\,\text{s}^{-1}$,
    for specimens without aging $\tau = 0.00625\,\text{s}$, $E_0 = 150\,\text{MPa}$;
    for specimens after aging $\tau = 0.00085\,\text{s}$, $E_0 = 50\,\text{MPa}$;
  - Voigt model:
    $\alpha_0 = 0$, $\alpha_\infty = 1$, $m = -0.3$, $k = 1.2 \times 10^{-5}\,\text{s}^{-1}$,
    for specimens without aging $\tau = 0.002\,\text{s}$, $E_0 = 1.1\,\text{MPa}$;
    for specimens after aging $\tau = 0.0042\,\text{s}$, $E_0 = 0.1\,\text{MPa}$;
- at stress $\sigma = 0.9\sigma_f = 29.97\,\text{MPa}$:
  - Maxwell model:
    $\alpha_0 = 0$, $\alpha_\infty = 1$, $m = -0.5$, $k = 5.7 \times 10^{-3}\,\text{s}^{-1}$,
    for specimens without aging $\tau = 0.00625\,\text{s}$, $E_0 = 150\,\text{MPa}$;
    for specimens after aging $\tau = 0.00085\,\text{s}$, $E_0 = 50\,\text{MPa}$;
  - Voigt model:
    $\alpha_0 = 0$, $\alpha_\infty = 1$, $m = -0.5$, $k = 4.5 \times 10^{-5}\,\text{s}^{-1}$,
    for specimens without aging $\tau = 0.002\,\text{s}$, $E_0 = 1.1\,\text{MPa}$;
    for specimens after aging $\tau = 0.0042\,\text{s}$, $E_0 = 0.1\,\text{MPa}$.

Based on the obtained results, it can be concluded that the Voigt model described the experimental results on the creep of high impact polystyrene more accurately compared to the Maxwell model.

5. Conclusions

Experimental data on the aging of polyurethane specimens during creep experiments with alternating of cyclic loadings and prolonged deformation aging (during 1–2 years) were obtained.

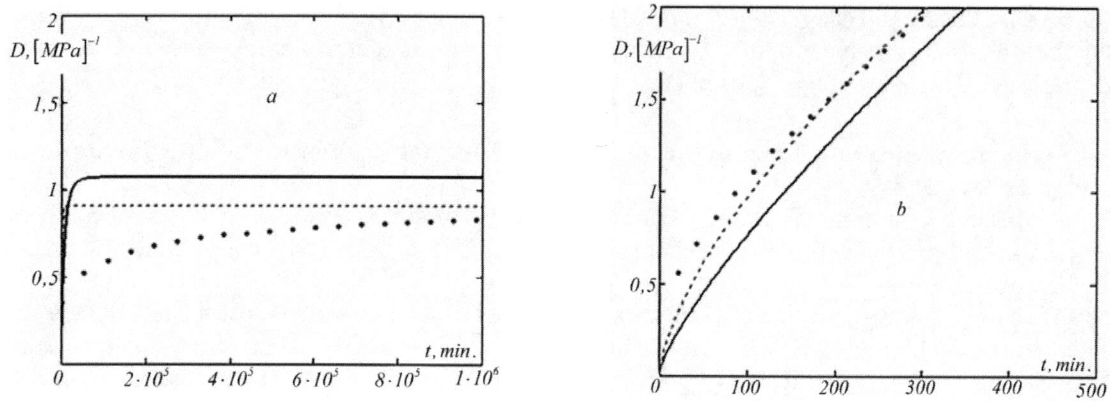

Figure 2. Theoretical compliance curves according to equations (4), (6) and experimental points for specimens of the HIPS of UPS 03L mark with a constant tensile stress of $0.8\sigma_\mathrm{f} = 26.64$ MPa: (a) without aging, (b) after natural aging during 40 years. Solid curves are the solution according to the Maxwell model; dotted curves are the solution according to the Voigt model.

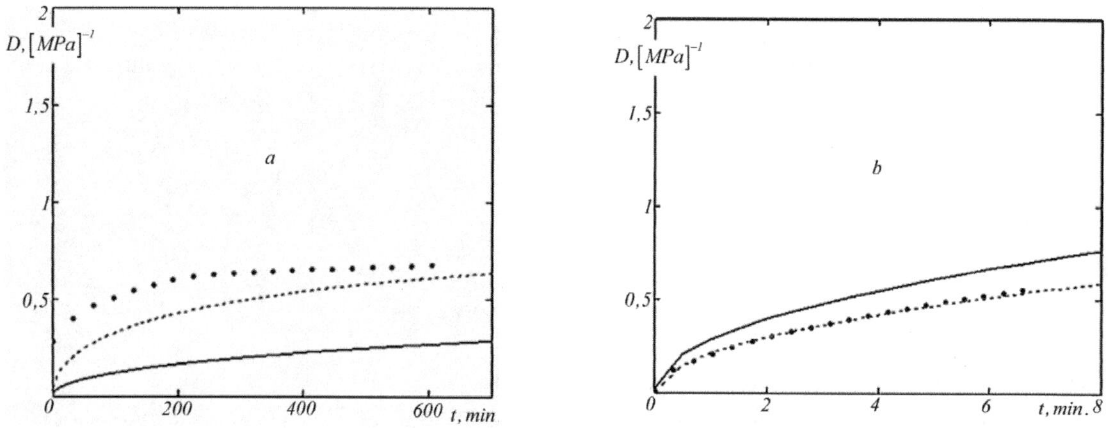

Figure 3. Theoretical compliance curves according to equations (4), (6) and experimental points for specimens of the HIPS of UPS 03L mark with a constant tensile stress of $0.9\sigma_\mathrm{f} = 29.97$ MPa: (a) without aging, (b) after natural aging during 40 years. Solid curves are the solution according to the Maxwell model; dotted curves are the solution according to the Voigt model.

For specimens after aging, an increase in creep time for a given deformation by 3 times compared to specimens without aging is observed. The influence of long-term natural aging over 40 years of specimens of high impact polystyrene of UPS 03L mark on creep also has been investigated. In the creep experiments after aging, a decrease of deformation by 6.5 times and the time to fracture by more than 24 times, compared with specimens without aging is observed. To describe the experimental creep curves of polyurethane and polystyrene specimens after aging and without aging, modified Maxwell and Voigt equations written in the effective time scale are used. The obtained theoretical creep curves are compared with the corresponding experimental results. There is a good agreement between theoretical and experimental compliance curves. It was received that the Voigt model described the experimental results on the creep of high impact polystyrene and polyurethane specimens more accurately compared to the Maxwell model.

References

[1] Struik L C E 1978 *Physical Aging in Amorphous Polymers and Other Materials* (Amsterdam, Oxford, New York: Elsevier Sci. Publ. Comp.)
[2] Nadai A 1963 *Theory of Flow and Fracture of Solids* (New York: McGraw-Hill)
[3] Filatov I S 1983 *Climatic Stability of Polymeric Materials* (Moscow: Nauka) [in Russian]
[4] Bruijn de J C M 1992 *The Failure Behavior of High Density Polyethylene with an Embrittled Surface Layer due to Weathering* (Delft: Delft University press)
[5] Bochkarev R N and Filatov I S 1990 *Aging of Materials Based on Polyvinyl Chloride in a Cold Climate* (Novosibirsk: Nauka Sib. Otd.) [in Russian]
[6] Arutyunyan R A 2004 *The Problem of Deformation Aging and Prolonged Fracture in Mechanics of Materials* (St. Petersburg: Izdat. SPbGU) [in Russian]
[7] Oprea S and Oprea V 2002 Mechanical behavior during different weathering tests of the polyurethane elastomers films *European Polymer Journal* **38** (6) 1205–10 doi: 10.1016/S0014-3057(01)00280-4
[8] Boubakri A, Elleuch K, Guermazi N, and Ayedi H F 2009 Investigations on hygrothermal aging of thermoplastic polyurethane material *Materials & Design* **30** (10) 3958–65 doi: 10.1016/j.matdes.2009.05.038
[9] Aglan H, Calhoun M, and Allie L 2008 Effect of UV and hygrothermal aging on the mechanical performance of polyurethane elastomers *Journal of Applied Polymer Science* **108** (1) 558–64 doi: 10.1002/app.27049
[10] Gradinaru L M, Vlad S, Spiridon I, and Petrescu M 2019 Durability of polyurethane membranes in artificial weathering environment *Polymer Testing* **80** 106144 doi: 10.1016/j.polymertesting.2019.106144
[11] Arutyunyan A R and Arutyunyan R A 2017 Damage and long-term strength of compressible elastic viscous ageg media *Marine Intellectual Technologies* No. 2(36) v2 76–83
[12] Arutyunyan A R 2020 Influence of aging on fatigue and creep properties of polyurethane *Material Physics and Mechanics* **44** (2) 256–63 doi: 10.18720/MPM.4422020_10
[13] Arutyunyan A R 2020 Experimental and theoretical studies of the effect of long-term natural aging on the mechanical properties of high impact polystyrene *Journal of Physics: Conference Series* **1474** 012004 doi: 10.1088/1742-6596/1474/1/012004
[14] Gul V E 1971 *Structure and Strength of Polymers* (Moscow: Khimiya) [in Russian]
[15] Bartenev G M and Zuev Yu S 1964 *Strength and Destruction of Highly Elastic Materials* (Moscow-Leningrad: Khimiya) [in Russian]
[16] Moskvitin V V 1972 *Resistance of Visco-Elastic Materials* (Moscow: Nauka) [in Russian]
[17] Valanis K C 1975 On the foundation of the endochronic theory of viscoplasticity *Archiwum Mechaniki Stosowanej* **27** (5-6) 857–68

On the wave motion on the surface of a viscous stratified liquid

Yu Chashechkin[1*] **and A Ochirov**[1,2**]

[1]Ishlinsky Institute for Problems in Mechanics RAS, Moscow, Russia
[2]P.G. Demidov Yaroslavl State University, Yaroslavl, Russia

E-mail: *chakin@ipmnet.ru, **otchirov@mail.ru

Abstract. The problem of propagation of gravitational waves along the surface of a semi-infinite viscous stratified liquid is considered. Wave and related non-wave components of the flow are detected. The dispersion relations of the detected types of motion are investigated.

1. Introduction

The issues of constructing a solution to the problem of propagation of capillary waves and related components in a viscous liquid were discussed earlier [1]. However, in real conditions, both internal energy and density, which affect the development of the structure of droplet flows [2], are unevenly distributed. A large number of questions were raised in the review work [3]. The heterogeneity of the density distribution (stratification) affects the propagation of waves [4]. A rigorous technique for constructing an exact solution to the linearized problem of internal wave emission in a viscous continuously stratified liquid was developed in [5]. In the study [6] numerical simulation of the generation of internal waves in a two-layer stratified liquid by a submarine is carried out. Internal waves in a stratified fluid initiated by submerged body in the study [7] were also studied numerically and experimentally. Paper [8] is devoted to the interaction of a shear flow in stratified fluid and its solitary-wave solutions. However, the problem of propagation of gravitational waves and their accompanying thin components along the free surface of a viscous continuously stratified liquid has not been discussed before. This study is devoted to this issue.

2. Problem statement

Let us consider the plane problem of surface wave propagation in a viscous Newtonian stratified fluid. We assume that the fluid is semi-infinite and unlimited in horizontal directions. The kinematic viscosity of the liquid is denoted by the symbol ν, the pressure by the symbol p. In the equilibrium state, the liquid occupies the lower half-space $z < 0$ in the Cartesian coordinate system Oxz, in which the vertical axis is opposite to acceleration $\mathbf{g}$. We assume that the gravitational wave propagates along the surface, which determines the deviation of the free surface from the equilibrium position $\zeta(x, t)$. The task is to determine the wave and non-wave components of motion resulting from the propagation of gravitational surface waves. The stratification of a liquid will be considered linear, which is true for describing weakly stratified liquids. The mathematical formulation of the problem of determining the velocity field $\mathbf{u}$ in a liquid and the dispersion relations is as follows:

$$z<\zeta: \quad \rho\left[\frac{\partial \mathbf{u}}{\partial t}+(\mathbf{u}\cdot\nabla)\mathbf{u}\right]=-\nabla p+\rho\nu\Delta\mathbf{u}+\rho\mathbf{g}; \tag{1}$$

$$\frac{\partial \rho}{\partial t}+\mathbf{u}\cdot\nabla\rho=0, \quad \nabla\cdot\mathbf{u}=0, \tag{2}$$

$$z<\zeta: \quad (p-p_0)n_i-\sigma'_{ik}n_k=0, \quad v-u\frac{\partial\zeta}{\partial x}=\frac{\partial\zeta}{\partial t}, \tag{3}$$

$$\mathbf{n}=\left(-\frac{\partial\zeta}{\partial x}\mathbf{e}_x+\mathbf{e}_z\right)\left[1+\left(\frac{\partial\zeta}{\partial x}\right)^2\right]^{-1/2}, \quad \sigma'_{ik}=\rho\nu\left(\frac{\partial u_i}{\partial x_k}+\frac{\partial u_k}{\partial x_i}\right).$$

Here v and u are the vertical and horizontal components of the fluid velocity, $\mathbf{e}_x$ and $\mathbf{e}_z$ are orts of the axes Ox and Oz, respectively, p is hydrodynamic pressure in the fluid, p_0 is atmospheric pressure, $\mathbf{n}$ is ort of the external normal to the interface and σ'_{ik} is tensor of viscous stresses in a liquid.

Let us consider the hydrodynamic pressure in the fluid as the sum of the atmospheric, hydrostatic and wave parts $\tilde{p}$:

$$p=p_0+g\int_z^{\zeta}\rho(x,\xi,t)d\xi+\tilde{p}.$$

And the density of the liquid is represented as a function:

$$\rho(x,z,t)=1+q(x,z,t).$$

Due to the fact that the flow is two-dimensional and the fluid is incompressible it is possible to introduce a stream function ψ. The flow velocity components in relation to the stream function ψ have to be

$$u=\frac{\partial\psi}{\partial z}, \quad v=-\frac{\partial\psi}{\partial x}.$$

Thus the equations of motion and boundary conditions may be rewritten as follows:

$$z<\zeta: \quad \frac{\partial q}{\partial z}\left(\frac{\partial L[\psi]}{\partial z}+H_{\{x,z\}}\left[\psi,\frac{\partial\psi}{\partial z}\right]\right)+\frac{\partial q}{\partial x}\left(-g+\frac{\partial L[\psi]}{\partial x}+H_{\{x,z\}}\left[\psi,\frac{\partial\psi}{\partial x}\right]\right)$$
$$+(1+q)(\Delta L[\psi]+H_{\{x,z\}}[\psi,\Delta\psi])=0, \tag{4}$$

$$\frac{\partial q}{\partial t}+H_{\{x,z\}}[\psi,q]=0, \tag{5}$$

$$z=\zeta: \quad \frac{\partial\zeta}{\partial t}+\frac{\partial\zeta}{\partial x}\frac{\partial\psi}{\partial z}+\frac{\partial\psi}{\partial x}=0, \tag{6}$$

$$\frac{2\nu\partial_x q[\partial_{xz}\psi-\partial_x\zeta(\partial_x\zeta\partial_{xz}\psi+\partial_{xx}\psi-\partial_{zz}\psi)]}{1+(\partial_x\zeta)^2}$$
$$+(1+q)[-\partial_z L[\psi]+H_{\{x,z\}}[\partial_z\psi,\psi]+\partial_x\zeta(-g+\partial_x L[\psi]+H_{\{x,z\}}[\psi,\partial_x\psi])]$$
$$-\frac{2\nu}{[1+(\partial_x\zeta)^2]^2}(\partial_{xx}\zeta\{[-1+(\partial_x\zeta)^2]\partial_{zz}\psi+4\partial_x\zeta\partial_{xz}\psi-[-1+(\partial_x\zeta)^2]\partial_{xx}\psi\}$$
$$+[1+(\partial_x\zeta)^2][-\partial_{xxz}\psi+\partial_x\zeta(-\partial_{xzz}\psi+\partial_x\zeta\partial_{xxz}\psi+\partial_{xxx}\psi)])=0, \tag{7}$$

$$\frac{[-1+(\partial_x\zeta)^2](\partial_{xx}\psi-\partial_{zz}\psi)-4\partial_x\zeta\partial_{xz}\psi}{1+(\partial_x\zeta)^2}=0, \tag{8}$$

$$L[\alpha]=\frac{\partial\alpha}{\partial t}-\nu\Delta\alpha, \quad H_{\{x,z\}}[\alpha,\beta]=\frac{\partial\alpha}{\partial z}\frac{\partial\beta}{\partial x}-\frac{\partial\alpha}{\partial x}\frac{\partial\beta}{\partial z}.$$

Let us consider the problem in the Boussinesq approximation ($q(x,z,t) \ll 1$). We assume the wave motion infinitesimally small (this allows us to describe the function $q(x,z,t)$ as a sum of terms describing the initial stratification $r(z)$ and the disturbance introduced by the wave $s(x,z,t)$). Boundary conditions is written on the equilibrium surface according to the known procedure. The problem of the first order of smallness in the wave amplitude is as follows:

$$z<0: \quad \left(\frac{\partial}{\partial t} - \nu\Delta\right)\Delta\frac{\partial\psi}{\partial t} - g\frac{\partial r}{\partial z}\frac{\partial^2\psi}{\partial x^2} = 0, \tag{9}$$

$$z=0: \quad \frac{\partial\zeta}{\partial t} + \frac{\partial\psi}{\partial x} = 0, \quad \frac{\partial^2\psi}{\partial x^2} - \frac{\partial^2\psi}{\partial z^2} = 0, \tag{10}$$

$$\frac{\partial^2\psi}{\partial z\partial t} - \nu\Delta\frac{\partial\psi}{\partial z} - 2\nu\frac{\partial^3\psi}{\partial x^2\partial z} + g\frac{\partial\zeta}{\partial x} = 0. \tag{11}$$

3. Solving the problem of the first order of smallness

Let us consider a fluid with linear stratification $r = -z/\Lambda$ (Λ is a stratification scale). We will look for a solution in the form of harmonic oscillations $\psi = \Psi(x,z)e^{-i\omega t}$. In the case of surface gravitational waves, the frequency of the wave ω exceeds the Brunt-Väisälä frequency $N = \sqrt{g/\Lambda}$, and equation (9) is transformed into:

$$N^2\frac{\partial^2\Psi}{\partial x^2} - \omega^2\Delta\Psi + i\nu\omega\Delta\Delta\Psi = 0. \tag{12}$$

Consider the solution of equation (12) in the form:

$$\Psi = (e^{k_z z} + \beta e^{qz})Ae^{ik_x x}e^{-i\omega t}$$

and then we get the relation for k_z(for the q relation will be the same):

$$i\nu\omega k_z^4 - k_z^2(\omega^2 + 2i\nu\omega k_x^2) + k_x^2(\omega^2 - N^2 + i\nu\omega k_x^2) = 0. \tag{13}$$

Substituting the function ψ into the boundary conditions leads to the following dispersion relation:

$$-k_z\omega(\nu k_z^2 + i\omega) + k_x^2(ig + 3k_z\nu\omega) = 0. \tag{14}$$

Let us perform the dimensionalization of expressions (13) and (14). We will choose the Brunt-Väisälä frequency N as the frequency scale, and $L = (\nu g)^{1/3}/N$ as the length scale. Dimensionless variables are denoted by symbol $*$. In a dimensionless form, a small parameter $\delta = (\nu N^3/g^2)^{1/3}$ is naturally allocated:

$$-k_x^{*2} + \delta i\omega^* k_x^{*4} - 2\delta i\omega^* k_x^{*2}k_z^{*2} + \delta i\omega^* k_z^{*4} + k_x^{*2}\omega^{*2} - k_z^{*2}\omega^{*2} = 0, \tag{15}$$

$$ik_x^{*2} - \delta i\omega^{*2}k_z^* + \delta^2(3k_x^{*2}k_z^*\omega^* - k_z^{*3}\omega^*) = 0. \tag{16}$$

Up to the principal terms of the decomposition, we obtain a solution of the dispersion relations for the regular part:

$$k_z^* = \delta(\omega^{*2} - 1), \tag{17}$$

$$k_{x1}^* = \pm\delta\omega^*\sqrt{\omega^{*2} - 1}, \tag{18}$$

and for the singular part:

$$q_\pm^* = \frac{1}{2}\left[\delta(1+\omega^{*2}) \pm \sqrt{2}(1-i)\sqrt{\frac{\omega^*}{\delta}}\right], \tag{19}$$

$$k_{x2\pm}^* = \pm\frac{\omega^*\sqrt{2}}{2}\sqrt{\delta\left[\delta(1+\omega^{*2}) \pm \sqrt{2}(1-i)\sqrt{\frac{\omega^*}{\delta}}\right]}. \tag{20}$$

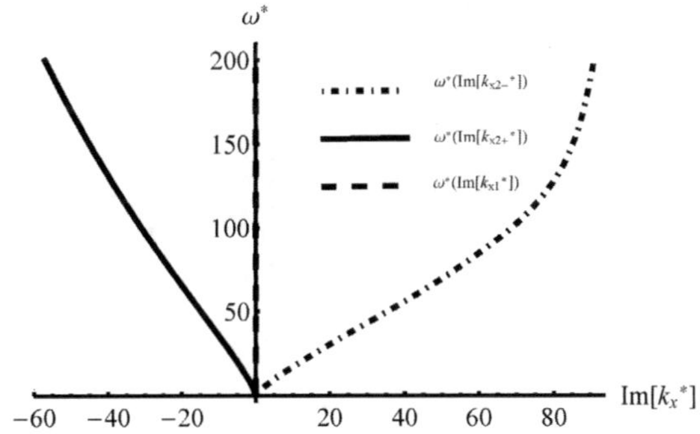

Figure 1. The decay decrements various wave modes.

The regular part is responsible for the wave component of motion, and the irregular part is responsible for the non-wave component of motion.

4. Analysis of the solution

Due to the absence of bounding surfaces in the formulation of the problem, not all solutions of dispersion relations obtained can be implemented in nature. Taking into account the type of the current function and the damping condition of the fluid motion at a point removed to infinity from the free surface $z \to -\infty$, the roots satisfying the condition are physically realizable:

$$\mathrm{Re}[k_z^*] > 0, \quad \mathrm{Re}[q^*] > 0.$$

It is obvious that for the regular part, the root realization condition holds for gravitational waves of any frequency ($\mathrm{Re}[k_z^*] > 0 \ \forall \omega^* > 1$). The realizability of singular motion components at $q = q_+$ holds for all frequencies and at $q = q_-$ satisfying the condition:

$$\delta(1 + \omega^{*2}) - \sqrt{2}(1 - i)\sqrt{\frac{\omega^*}{\delta}} > 0.$$

In the expression for k_x^* (18), (20) the '+' sign corresponds to a wave propagating in the positive direction of the Ox axis. The '−' sign in front of the expression refers to a wave propagating in the opposite direction. Here and further, we are talking only about a wave with a positive number k_x^*. Decay decrements of the spatial variable k_x^* for various wave modes are shown in figure 1.

The dispersion relations are shown in figure 2. Figure $2a$ shows the dimensionless dependencies $\omega^*(\mathrm{Re}[k_x^*])$ for all three possible roots (18), (20). Figure $2b$ shows the dimensionless dependencies $\omega^*(\mathrm{Re}[k_z^*])$, $\omega^*(\mathrm{Re}[q_\pm^*])$ for the possible roots (17), (19). Here and further all plots are constructed at dimensionless parameter $\delta = 0.01$.

There are only two physically realizable types of roots: k_{x1}^* (18) and k_{x2-}^* (20). Figure 2 shows that in this formulation of the problem, there are surface gravitational waves that do not generate internal waves. In this case starting from a certain frequency, surface singular solutions of q_- is realized.

Figure 3 demonstrates the phase velocity relations on the wavelength in the horizontal dirrection (Ox) and in the vertical direction (Oz). It can be seen that in Oz direction phase velocity of the regular solution exceeds those one of the singular solution. In the Ox direction at small wavelength phase velocity of the singular solution exceeds those one of the regular solution. Starting from a certain value (less then dimensionless parameter of the length) phase velocity of the regular solution dominates at the those one of the singular part.

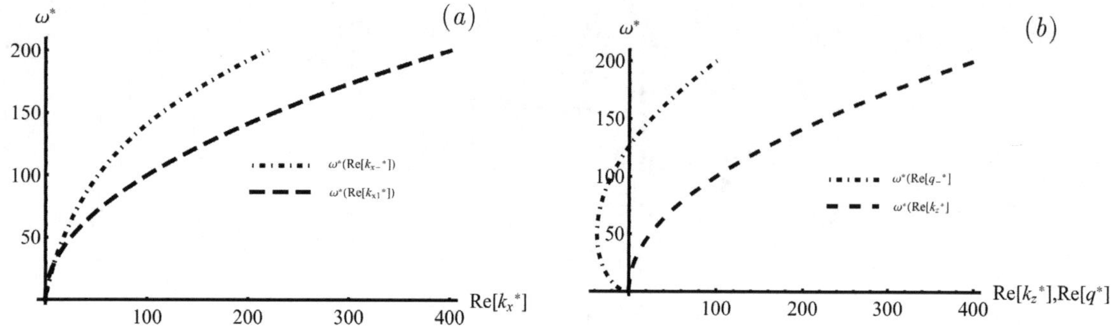

Figure 2. Dispersion relations for various wave modes (a) $\omega^*(\mathrm{Re}[k_x^*])$, (b) $\omega^*(\mathrm{Re}[k_z^*])$, $\omega^*(\mathrm{Re}[q_\pm^*])$.

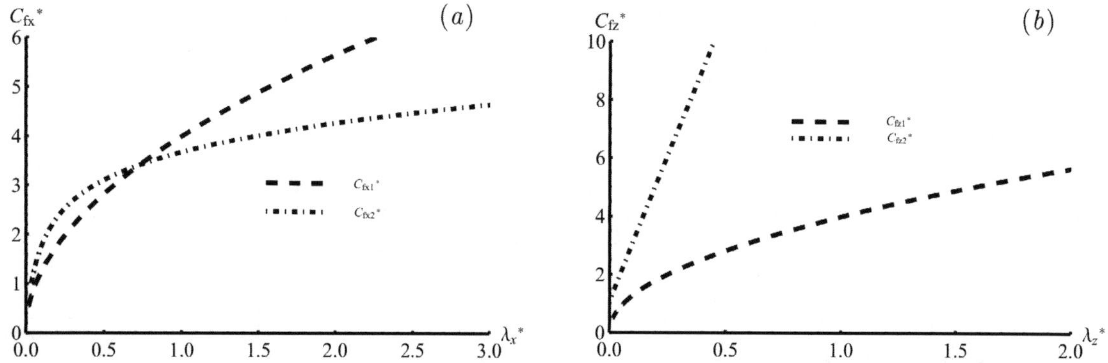

Figure 3. Phase velocity for various wave modes (a) $\omega^*/(\mathrm{Re}[k_x^*])$; (b) $\omega^*/(\mathrm{Re}[k_z^*])$, $\omega^*/(\mathrm{Re}[q_\pm^*])$.

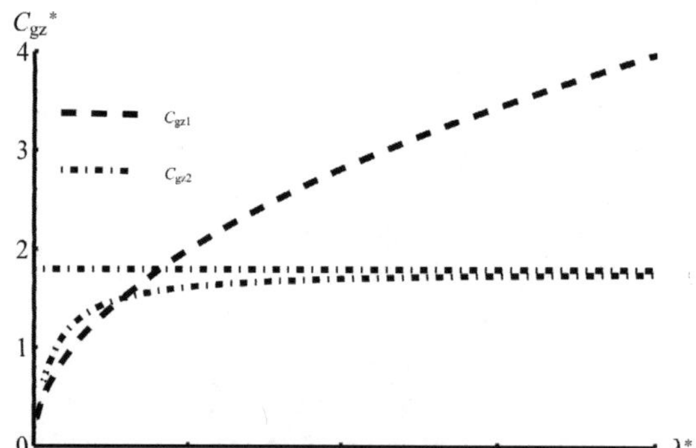

Figure 4. Group velocity for various wave modes at the vertical direction.

At figure 4 one can see the group velocity relation on the wavelength in the vertical direction. At small wavelength group velocity of the singular solution exceeds those one of the regular part. The group velocity of the singular solution The group velocity of the singular part of the solution rapidly tends to its asymptotic value. At the same time the group velocity of the regular part of the solution increases monotonically.

5. Conclusion

It was shown that surface gravity waves do not product internal waves. Dispersion relations of wave motion were obtained. In this problem there are both regular and singular part of the solution is realized. Phase velocity of both solutions also were obtained and demonstrated. At vertical direction phase velocity of the singular part exceed those one of the regular part for all wavelength. It was shown that into the depth direction (Oz) group velocity of the singular solution dominates at small wavelength. Starting from a certain value, the group velocity of the regular part exceeds the velocity of the singular part. The group velocity of the singular part asymptotically tends to a certain value.

Acknowledgments

The study was carried out with the support of the Russian Science Foundation (project No. 19-19-00598).

References

[1] Kistovich A V and Chashechkin Yu D 2007 Dynamics of gravity-capillary waves on the surface of a nonuniformly heated fluid *Izvestiya, Atmospheric and Oceanic Physics* **43** (1) 95–102 doi: 10.1134/S0001433807010112

[2] Chashechkin Yu 2021 Hydrodynamics of detachment, free falling and impact of drops *Journal of Physics: Conference Series* **1891** 012022 doi: 10.1088/1742-6596/1891/1/012022

[3] Sullivan P P and McWilliams J C 2010 Dynamics of winds and currents coupled to surface waves *Annual Review of Fluid Mechanics* **42** 19–42 doi: 10.1146/annurev-fluid-121108-145541

[4] Lighthill J 1978 *Waves in Fluids* (Cambridge: Cambridge University Press)

[5] Vasil'ev A Yu and Chashechkin Yu D 2006 Generation of beams of three-dimensional periodic internal waves by sources of various types *Journal of Applied Mechanics and Technical Physics* **47** (3) 314–23 doi: 10.1007/s10808-006-0058-4

[6] Chang Y, Zhao F, Zhang J, et al. 2006 Numerical simulation of internal waves excited by a submarine moving in the two-layer stratified fluid *Journal of Hydrodynamics, Ser. B* **18** (3) 330–36 doi: 10.1016/S1001-6058(06)60074-7

[7] Li Y, Chen K, Wang H, et al. 2021 Numerical and experimental timefrequency analysis of internal waves induced by a submerged body *Ocean Engineering* **239** 109779 doi: 10.1016/j.oceaneng.2021.109779

[8] Seadawy A R 2017 Two-dimensional interaction of a shear flow with a free surface in a stratified fluid and its solitary-wave solutions via mathematical methods *The European Physical Journal Plus* **132** (12) 1–11 doi: 10.1140/epjp/i2017-11755-6

Influence of metal texture and loading rate on heat release under quasi-static tension

Yu V Sudienkov[1,2*], I V Smirnov[2], and B A Zimin[3]
[1]Skolkovo Institute of Science and Technology, Moscow, Russia
[2]St. Petersburg State University, St. Petersburg, Russia
[3]Baltic State Technical University, St. Petersburg, Russia
E-mail: *y.sudenkov@yandex.ru

Abstract. This paper presents a study of the influence of the initial structure of metal materials obtained from plastic deformation methods on the processes of heat generation under quasi-static tension and low cyclic tests. Experimental data on the dependence of changes in the fraction of latent energy during deformation on the anisotropy of the initial structure of polycrystalline specimens of copper M1, duralumin D16, titanium VT6, as well as the grain size of copper M1 and magnesium alloy AZ31B are presented. It is shown that the thermodynamics of the plastic deformation process significantly depends on the anisotropy of the initial structure of the materials, brought about by the material processing used, as well as on the features of the lattice and the development of a meso-defect structure.

1. Introduction

Analysis of the mechanisms of structural rearrangements at different stages of the deformation process and changes in the energy balance as a result of these rearrangements is of significant interest in studying the behavior of metal materials under the action of external loads. The plastic flow of a solid occurs due to the nucleation, motion, and interaction of a large number of independent local deformation defects and the subsequent movement of these structural elements of deformation as a whole at various scale levels [1–6]. For example, as plastic deformation accumulates and the dislocation density increases, the number of obstacles to the movement of dislocations through the grains also increases. The growth of dislocations gradually decreases, while the degree of internal stresses increases. Therefore, further developed plastic deformation is associated with the collective evolutionary processes of meso-level defects. Because polycrystalline materials combine different orientations of grains with respect to the axes of the principal stresses, plastic incompatibilities arise at boundaries and junctions of the grains. In this regard, the state of the material before deformation has significance, because it can determine not only the degree of development of the dislocation processes, but also the mechanisms of deformation at the meso- and macroscales. Naturally, the characteristic parameters of the initial structure of the material stipulate largely the processes of structural rearrangement and the corresponding changes in energy balance during deformation. Thus, the study of the energy balance of plastic deformation can provide additional significant information for the analysis of micromechanisms that determine the elastoplastic deformation of metals and alloys, and in particular, the stage of plastic flow [7–9].

Analyzing heat release during deformation is one way to study the influence of the external conditions and structural parameters of a metallic material on changes in its energy balance, which is determined by the equality of the work of deformation (W) to the sum of the released heat (Q) and latent energy (L). The contribution of heat release to the energy balance is traditionally characterized by the Taylor–Quinney coefficient $\beta = Q/W$ [10–12]. Thus, the coefficient β is a measure of the efficiency of converting plastic work into heat [7–12]. Many experimental and theoretical works have shown that this approach can be an effective tool for estimating the energy costs of internal structural transformations and, consequently, the efficiency of the structure of a material. For example, the results of experimental studies of heat release processes during the testing of copper, magnesium alloy, titanium, steel, and aluminum alloys [7, 13–15] have demonstrated a significant dependence of heat release intensity on strain rates under conditions of quasi-static tension in metals and alloys, with no significant effect on the stress-strain diagram. Moreover, the elements of the energy balance strongly depend on the deformation history, which leads to various mechanisms of plastic deformation in the tested metallic material [15–23]. It has been shown in [15–17] that an analysis of the effect of heat generation during plastic deformation allows an estimation of the fraction of latent heat due to phase transformation. The dependence of the Taylor–Quinney coefficient on dynamic loading and its correlation with deformation micromechanisms has also been reported for different materials [18–20]. Thermal measurements of the fraction of plastic work converted to heat have also been used successfully in the study of energy storage and dissipation in low-cycle fatigue tests of metals and alloys [21–23].

The energy of cold work is stored internally through the creation and rearrangement of crystal imperfections, such as dislocations, twins, point defects, and stacking faults. Measurements of the energy storage rate as a function of plastic deformation for pre-strained steel specimens in a uniaxial tensile test [24–26] support the claim that the energy storage rate depends on the generation of long-range internal microstresses caused by the polycrystalline nature of the material. When the subsequent deformation of the pre-strained specimen follows the direction of pre-strain, the contribution of internal stresses due to the different orientations of neighboring grains and the inhomogeneous distribution of dislocations is not significant. Loading in the pre-strain direction causes the evolution of dislocation structures from high-energy configurations to low-energy ones. When the subsequent deformation of the pre-strained specimen follows in the direction perpendicular to the pre-strain, a redistribution of long-range internal microstresses occurs, as the grain structure must be rearranged in accordance with the new direction of tension. These factors can be of great importance for the practical application of specific deformation structures in the processing of metals and alloys to create a given set of mechanical properties for the materials, as well as for modeling the behavior of such materials under various conditions of exploitation. However, there are limited experimental studies on the effect of pre-strain on released heat and latent energy during the plastic deformation of metals and alloys. This paper presents the results of studies on the influence of the some known initial states of the structure of metals and alloys, where the states were the result of thermomechanical processing, on the processes of heat release, that is, on the nature of the changes in the energy balance during the elastoplastic deformation of the materials. The purpose of this paper is to provide further experimental data to refine theoretical models of the relationship between stored energy and plastic deformation mechanisms. Three groups of tests were carried out: quasi-static tension at different strain rates of pure copper and aluminum alloy specimens cut from along and across the rolling direction, low-cycle fatigue tensile tests of titanium alloy specimens cut from along and across the rolling direction, and quasi-static tension tests at different strain rates for pure copper and magnesium alloys with different grain sizes (specimens before and after equal channel angular pressing (ECAP) processing).

2. Materials and methods

Specimens were cut from unidirectionally cold rolled plates (copper M1, titanium alloy VT6, duralumin D16) along and across the rolling direction. The working part of the flat specimens had the dimensions of $50 \times 5 \times 2$ mm. Copper and duralumin were tested for tension at two different strain rates (0.0012 and 0.012 1/s). The titanium alloy was tested for low-cycle fatigue using a tensile mode at a frequency of 10 Hz and a fixed load oscillation from 0 to 2 kN.

Moreover, specimens were made of copper M1 and magnesium alloy AZ31B with different grain sizes. Copper M1 was annealed for 2 hours at 550°C to obtain a coarse-grained state. To obtain ultrafine-grained (UFG) copper, the annealed rods were processed using ECAP technology [27, 28]. The average grain size in the obtained states of copper specimen was $d = 45\,\mu$m and $d = 0.8\,\mu$m. In the case of the magnesium alloy, the as-received state (hot-rolled plate) and the state after ECAP processing in combination with different temperature conditions [29, 30] were considered. The average grain size of the alloy was $d = 14\,\mu$m before ECAP processing and $d = 1.7\,\mu$m after ECAP processing. Flat specimens for tensile tests were cut using electrical discharge machining, and they had a working section size of $15 \times 6 \times 1.2$ mm. The specimens were tested at two different strain rates of 0.001 and 0.01 1/s.

The specimens were tested on a universal testing machine, SHIMADZU AG-50kNX, at room temperature. The fatigue tests of the titanium alloy were carried out on an Instron ElectroPuls E3000 machine. Before testing, the specimens were carefully ground and covered with a thin layer of black acrylic paint.

The change in the temperature of the specimens was determined by infrared radiation using the thermal imaging camera ThermaCAM SC 3000 with a shooting frequency of 50 Hz. A more detailed description of the technique is presented in [13].

3. Results and discussion

The results of the uniaxial tensile tests for copper M1 and duralumin D16 specimens cut from along and across the rolling direction are shown in figures 1 and 2. The figures show the dependences $\sigma(\varepsilon)$ and $\Delta T(\varepsilon)$ at strain rates of 0.0012 and 0.012 1/s, as well as the dependences of the Taylor-Quinney coefficients $\beta(\varepsilon)$, obtained from the diagrams of $\sigma(\varepsilon)$ and $\Delta T(\varepsilon)$.

As can be seen from the figures, the intensity of heat release and the value of its contribution to the energy balance are influenced not only by the strain rate, but also by the anisotropy of properties due to texture as a result of rolling [1, 4]. The efficiency of converting plastic work into heat for the two types of specimens differs significantly in the deformation region up to approximately 5–6%, after which the changes occur with almost a similar trend. Thus, the defining contribution to the Taylor–Quinney coefficient is due to the initial stage of strain hardening. According to well-known concepts presented in [1–6], the development of plastic deformation accompanies a change in various types of defects, mainly dislocation substructures. For example, at the early stage of plastic deformation, applying a sufficient level of stresses activates dislocation generators [31], which then pile up at obstacles and grain boundaries. Non-interacting dislocation regions, which are dense formations of dislocations, occupy the largest volume fraction. The generators are predominantly independent of each other, and the piled-up dislocations can break through without significant complications. At subsequent stages of deformation, the number of dislocations increases, and the dislocation regions join to form a cellular dislocation structure. Then, the stage of the maximum value of strain hardening and the accumulation rate of the dislocation density moves to a stage of intense formation, motion, and annihilation of new structural meso-defects [4, 5]. Thus, with an increase in plastic deformation, the inter-dislocation distances decrease, and the forces of inter-dislocation interaction increase. For this reason, it can be concluded that the difference between the Taylor–Quinney coefficients during deformation relative to the different directions of rolling for the aluminum alloy and copper is due to the development of many non-interacting individual dislocations at the initial

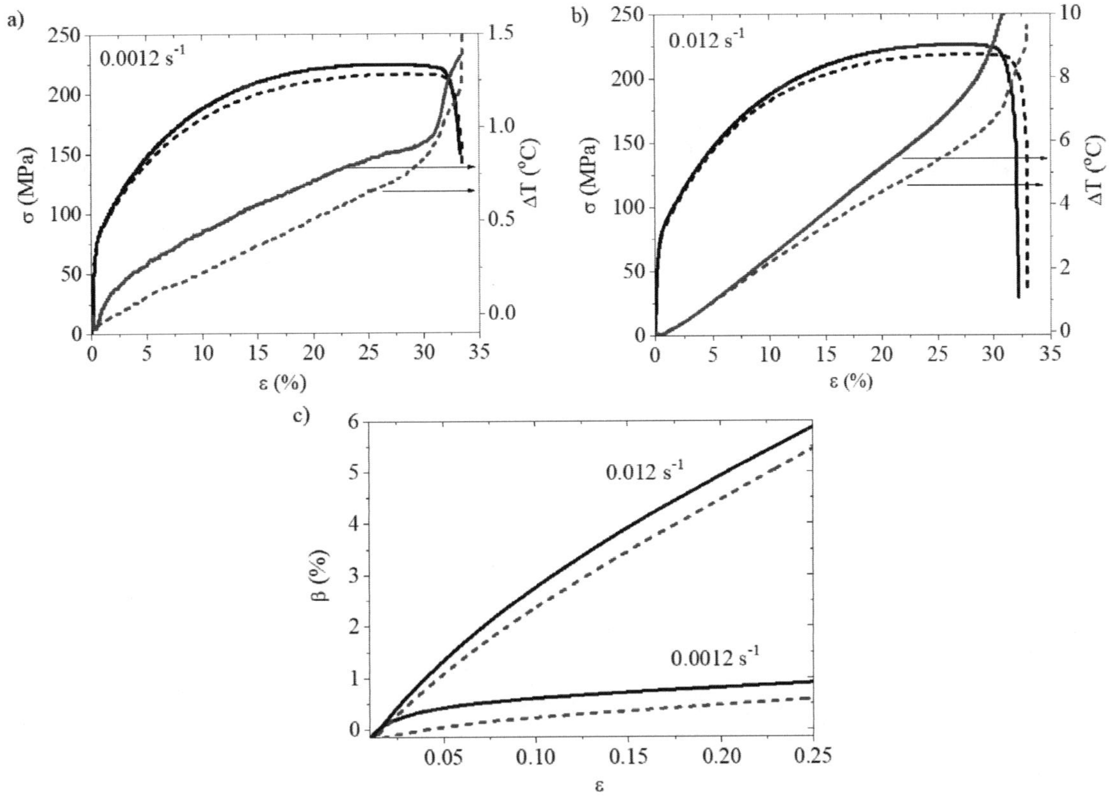

Figure 1. Dependencies $\sigma(\varepsilon)$ and $\Delta T(\varepsilon)$ for specimens of M1 copper at strain rates 0.0012 (a) and 0.012 1/s (b). Dependencies of Taylor-Quinney coefficients $\beta(\varepsilon)$ (c).

stage of plastic deformation.

Pre-deformation by rolling can also affect the anisotropy of mechanical properties under cyclic loading. For example, it was shown in [32, 33] that the cyclic performance of titanium alloy is sensitive to the relationship between the principal stress axis and the dominant basal plane texture. Therefore, in this work, the manifestation of energy dissipation for titanium alloy VT6 with different textures was considered. Figure 3 shows the results of the cyclic tests on the VT6 titanium specimens, which were cut from a sheet along and across its rolling direction. The low cyclic tests were carried out under stress control with loading frequency $f = 10\,\mathrm{Hz}$, and a variable load from 0 to 2000 N.

The results of the cyclic tests demonstrate a significant difference in heat release during the loading of specimens, with different directions of force relative to the texture direction. Loading parallel to the transverse direction results in less deformation and a lower Taylor–Quinney coefficient than for the same loading parallel to the longitudinal direction. Thus, the same cyclic tensile load requires different amounts of energy to bring about microstructural changes in the same material, but with different orientations of the specimen with respect to rolling. As noted in [32], this may be related to the hexagonal crystallographic form of the α phase and the availability of preferential slip systems. In the case of loading parallel to the rolling direction, prism planes have a favorable orientation for slip deformation. In the case of transverse loading, deformation occurs on the basal planes of the aligned hexagonal crystallographic prisms. The first case leads to shear failure, and the second leads to cup and cone fracture. The prismatic planes of the HCP crystallographic elements are oriented in a suitable manner for easy slip, thus

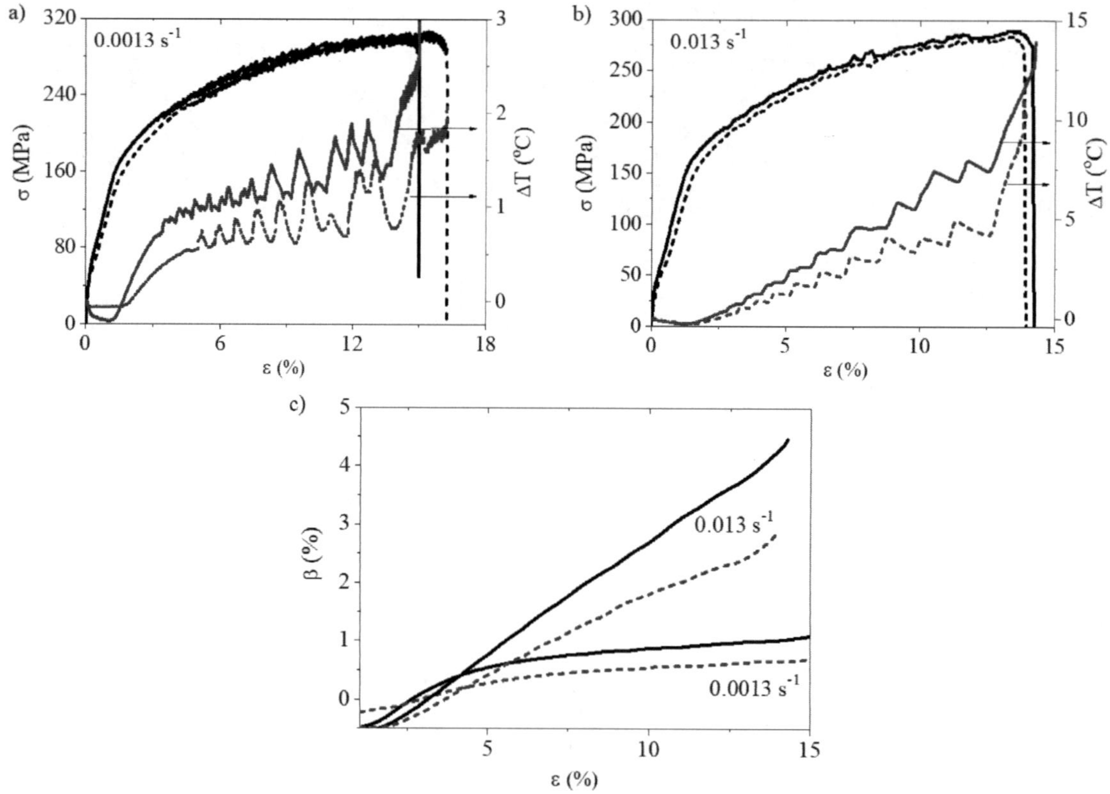

Figure 2. Dependencies $\sigma(\varepsilon)$ and $\Delta T(\varepsilon)$ for specimens of duralumin D16 at strain rates 0.0012 (a) and 0.012 1/s (b). Dependencies of Taylor-Quinney coefficients $\beta(\varepsilon)$ (c). (Solid line — along rolling, dashed line — across rolling).

prismatic slip is preferred [34]. Therefore, the material exhibits greater stiffness when deformed parallel to the transverse direction with respect to rolling.

Figure 4 shows the results obtained under uniaxial tension of copper with different grain sizes. The dependencies $\sigma(\varepsilon)$, $\Delta T(\varepsilon)$, and $\beta(\varepsilon)$ were obtained for two strain rates of 0.001 and 0.01 1/s. The results demonstrate the well-known occurrence of hardening after ECAP [27, 28], as well as a significant temperature dependence on both the strain rate and the characteristic scale of the microstructure.

When the grain is refined to 1 μm or less using methods of severe plastic deformation, a high initial strain hardening reduces the possibility of further strain hardening, the deformation mechanisms change, and the value of the relative elongation accumulates mainly in the stage after the maximum stress. The dependencies of the Taylor coefficient $\beta(\varepsilon)$, shown in figure 4c, illustrate an increase in heat release with an increase in plastic deformation, an increase in the strain rate, as well as with a decrease in the grain size and, consequently, with an increase in the density of grain boundaries. The growth of $\beta(\varepsilon)$ in the UFG specimens characterizes a decrease in the proportion of latent energy $L/W = 1 - \beta(\varepsilon)$ during deformation, that is, in the proportion of free energy due to structural defects. Moreover, the higher the strain rate, the more significant the decrease in the proportion of accumulated energy L/W. This allows for assuming that after grain refinement and accumulation of a certain pre-strain, the grain sizes and boundaries have little effect on the energy storage processes.

Figure 5 shows the results obtained under uniaxial tension of magnesium alloy specimens at

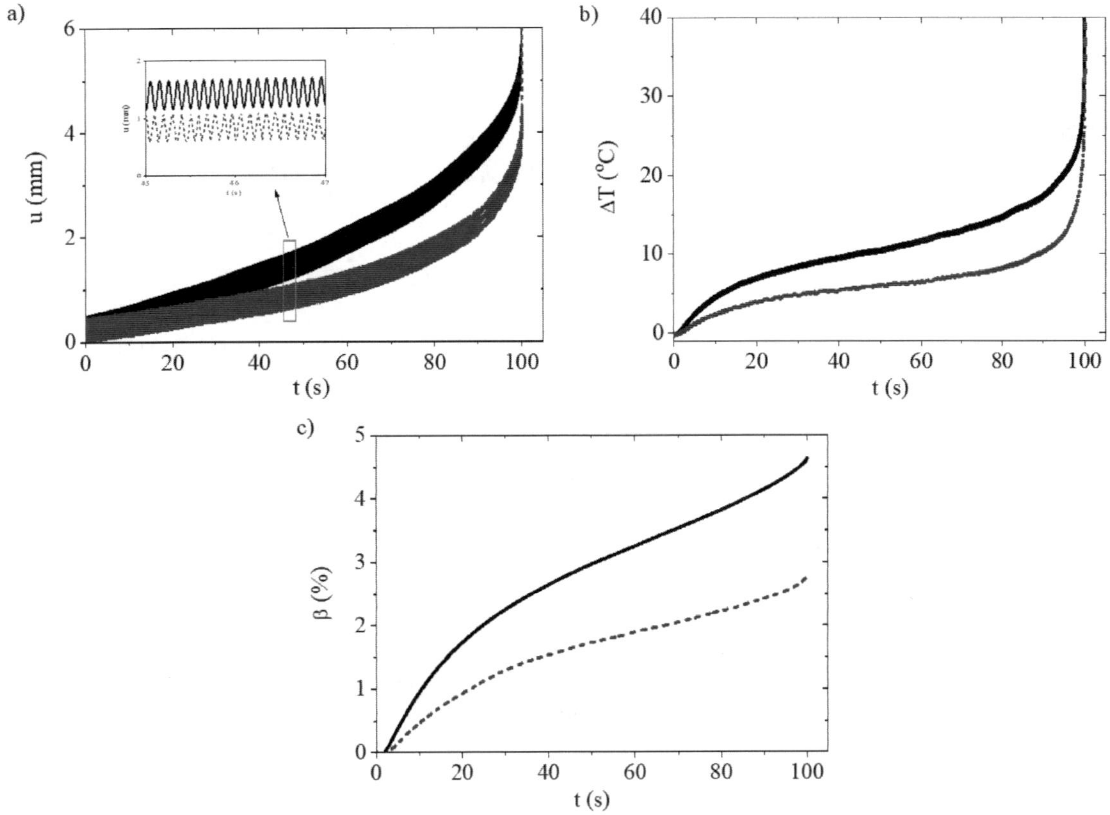

Figure 3. Diagrams of specimen elongation $u(t)$ (a), temperature change $\Delta T(t)$ (b), and Taylor coefficients $\beta(t)$ — (c) during cyclic tests of titanium. (Solid line — along rolling, dashed line — across rolling).

two strain rates of 0.001 and 0.01 1/s. In contrast to the tests of the FCC metal, the dependencies $\sigma(\varepsilon)$ of the UFG alloy AZ31B under tension demonstrate the presence of a significant linear hardening section after the stage of elastic response. Note that at the same time, there is a decrease in heat release for the UFG specimens of AZ31B during deformation (figure 5c) compared with the results of testing the alloy in the initial state. The fraction of latent energy during deformation of the UFG magnesium alloy increases, in contrast with the process of the accumulation of latent energy in the FCC metal (figure 4). Grain refinement in the magnesium alloy led to a change in the efficiency of converting plastic work into heat. Changes in the distribution of grain sizes, as well as the disorientation angle of the grain boundaries of the AZ31B magnesium alloy were considered in [29, 30]. It was shown that the peculiarities of the changes in the properties of the magnesium alloy after ECAP are due to the difference in dynamic recrystallization processes, which differs from the processes in FCC metals, and also to a significant contribution from the twinning mechanism to the deformation of metals with an hcp lattice. However, during the plastic deformation of magnesium alloys, both dislocations and twinning mechanisms are activated; the latter are a more dissipative mechanism, and they store less cold working energy than dislocations [35, 36]. Thus, an increase in the fraction of latent energy in UFG materials can be associated with an increase in the fraction of dislocation mechanisms during plastic deformation.

Thus, monitoring the processes of heat release during the plastic deformation of

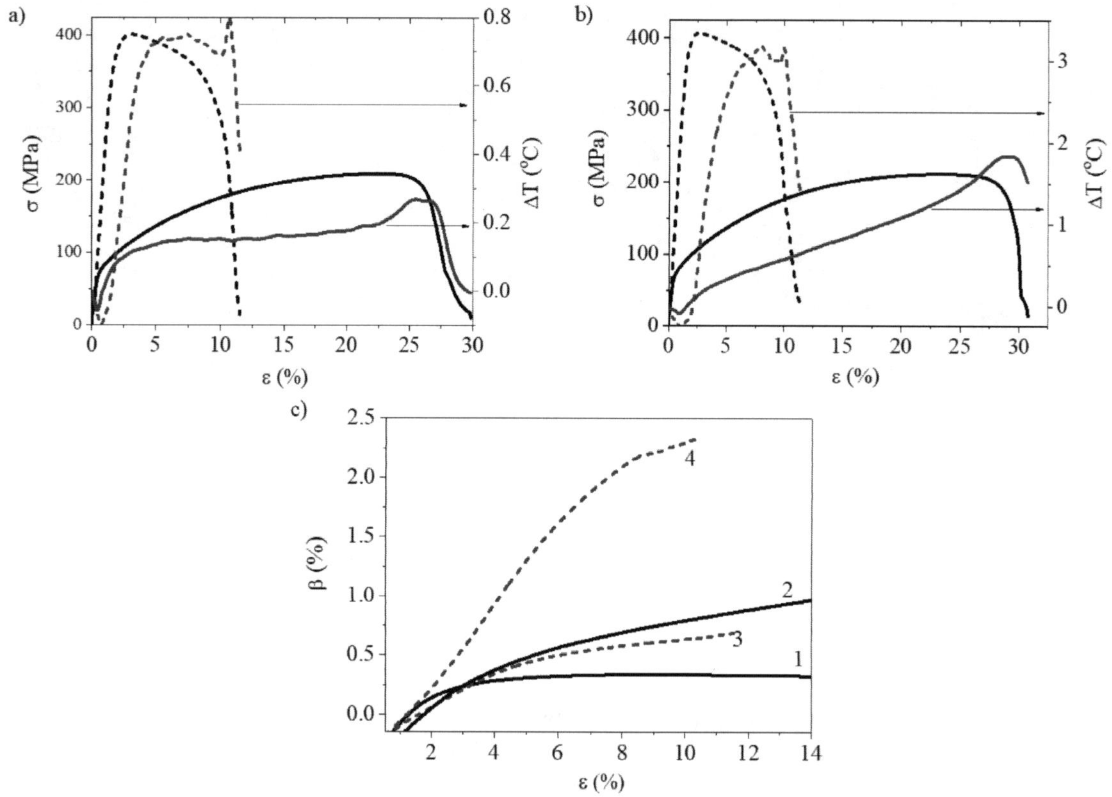

Figure 4. Dependences $\sigma(\varepsilon)$ and $\Delta T(\varepsilon)$ at uniaxial tension of copper specimens with strain rates of $\dot{\varepsilon} = 0.001\,1/\mathrm{s}$ (a) and $\dot{\varepsilon} = 0.01\,1/\mathrm{s}$ (b), and the corresponding dependences $\beta(\varepsilon)$ (c). Solid line — $d = 45\,\mu\mathrm{m}$, dotted line — $d = 0.8\,\mu\mathrm{m}$. *1* and *3* — $\dot{\varepsilon} = 0.001\,1/\mathrm{s}$, *2* and *4* — $\dot{\varepsilon} = 0.01\,1/\mathrm{s}$.

polycrystalline metals makes it possible to obtain an additional integral characteristic of the structural evolution processes. The difference between the process of accumulation of latent energy in metals with HCP and FCC lattices shows that processes at the microlevel largely control the nature of thermodynamic processes during the deformation of metallic materials.

Conclusion

The presented results show that the thermodynamics of the deformation process of polycrystalline metals and alloys significantly depends on both the structure of the lattice and the anisotropy of the material properties due to the texture, grain size, and boundaries between them. The obtained results supplement the data on the influence of the structure of metallic materials on the energy balance during plastic deformation and confirm the current understanding that the accumulation of latent energy in the process of plastic deformation is due to the evolution of defects at various structural levels.

Note that existing attempts at theoretical description using the energy balance equations for plastic deformation have not yet presented a model that allows one to take into account the diversity, relationship, and hierarchy of plastic flow mechanisms in polycrystalline metals and alloys [7, 8, 23, 24, 37, 38]. In our opinion, a physically reasonable derivation of the constitutive equations for elastoplastic deformation of materials that takes into account the accompanying

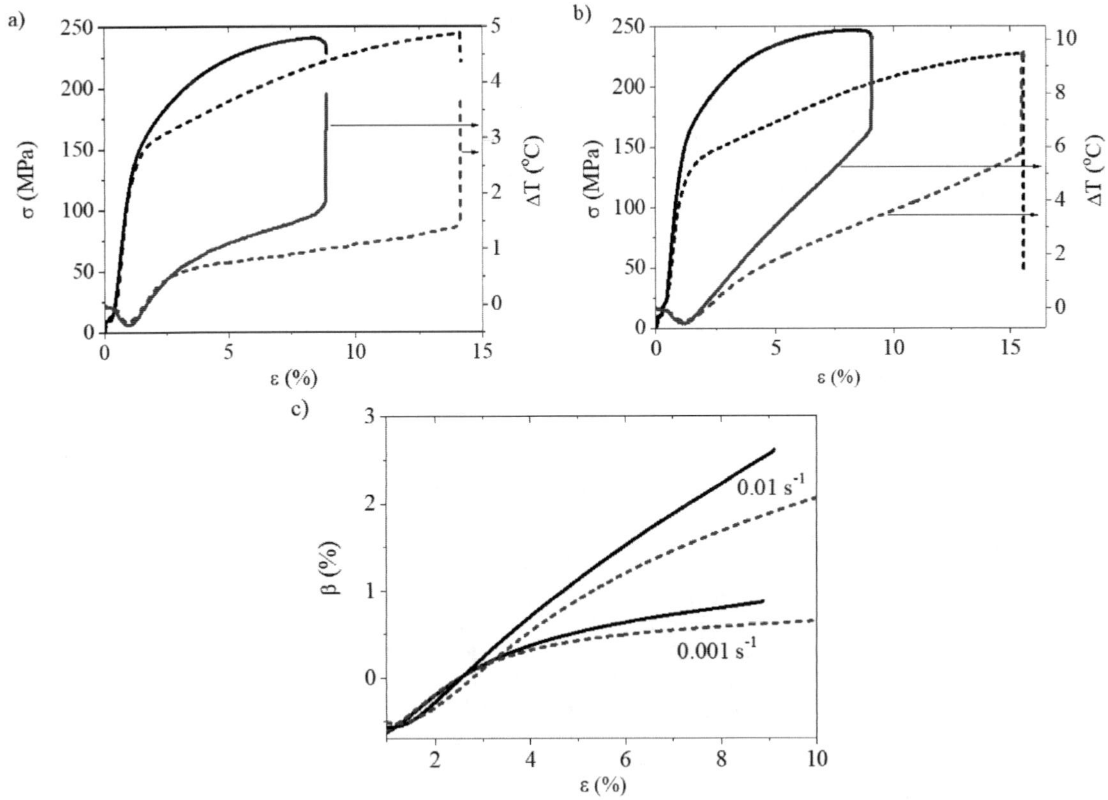

Figure 5. Dependences $\sigma(\varepsilon)$ and $\Delta T(\varepsilon)$ at uniaxial tension of alloy AZ31B with strain rates of $\dot{\varepsilon} = 0.001\,1/\mathrm{s}$ (a) and $\dot{\varepsilon} = 0.01\,1/\mathrm{s}$ (b), and the corresponding dependences $\beta(\varepsilon)$ (c). Solid line — $d = 14\,\mu\mathrm{m}$, dotted line — $d = 1.7\,\mu\mathrm{m}$.

phenomena (structural rearrangement, dissipation, thermoelastic effects) should be based on methods of the thermodynamics of irreversible processes [1–6], with further detailing of the contributions of micromechanisms, such as dislocations, disclinations, twins, point defects, and stacking faults, and the interaction of defects at the meso-scale of the structure.

Acknowledgments

The work was supported by the Russian Science Foundation (grant No. 21-19-00563).

References

[1] Kozlov E V, Trishkina L I, Popova N A, and Koneva N A 2011 Dislocation physics in the multilevel approach to plastic deformation *Physical Mesomechanics* **14** (5-6) 283–96

[2] Koneva N A, Trishkina L I, Popova N A, and Kozlov É V 2014 Accumulation of defects during plastic deformation of polycrystals of the meso- and microscale grain size *Russian Physics Journal* **57** (2) 187–96 doi: 10.1007/s11182-014-0224-8

[3] Trefilov V I, Moiseev V F, and Pechkovskiy E P 1987 *Strain Hardening and Fracture of Polycrystalline Materials* (Kiev: Naukova Dumka) [in Russian]

[4] Panin V E and Egorushkin V E 2015 Basic physical mesomechanics of plastic deformation and fracture of solids as hierarchically organized nonlinear systems *Physical Mesomechanics* **18** (4) 377–90 doi: 10.1134/S1029959915040104

[5] Panin V E, Korotaev A D, Makarov P V, and Kuznetsov V M 1998 Physical mesomechanics of materials *Russian Physics Journal* **41** (9) 856–84 doi: 10.1007/BF02508720

[6] Makarov P V, Schmauder S, Cherepanov O I, et al. 2001 Simulation of elastic–plastic deformation and fracture of materials at micro-, meso- and macrolevels *Theoretical and Applied Fracture Mechanics* **37** (1-3) 183–244 doi: 10.1016/S0167-8442(01)00078-7
[7] Bever M B, Holt D L, and Titchener A L 1973 The stored energy of cold work *Progress in Materials Science* **17** 5–177 doi: 10.1016/0079-6425(73)90001-7
[8] Oliferuk W and Raniecki B 2018 Thermodynamic description of the plastic work partition into stored energy and heat during deformation of polycrystalline materials *European Journal of Mechanics - A/Solids* **71** 326–34 doi: 10.1016/j.euromechsol.2018.04.002
[9] Hodowany J, Ravichandran G, Rosakis A J, and Rosakis P 2000 Partition of plastic work into heat and stored energy in metals *Experimental Mechanics* **40** (2) 113–23 doi: 10.1007/BF02325036
[10] Taylor G I and Quinney H 1934 The latent energy remaining in a metal after cold working *Proceedings of the Royal Society of London. Series A, Containing Papers of a Mathematical and Physical Character* **143** (849) 307–26 doi: 10.1098/rspa.1934.0004
[11] Bever M and Ticknor L 1953 The energy stored during the cold working of a gold-silver alloy *Acta Metallurgica* **1** (2) 116–22 doi: 10.1016/0001-6160(53)90048-0
[12] Zubelewicz A 2019 Century-long Taylor-Quinney interpretation of plasticity-induced heating reexamined *Scientific Reports* **9** 9088 doi: 10.1038/s41598-019-45533-0
[13] Zimin B A, Sventitskaya V E, Smirnov I V, and Sud'enkov Y V 2018 Influence of strain rate on heat release under quasi-static stretching of metals experiment *Physics of the Solid State* **60** (4) 758–63 doi: 10.1134/S1063783418040352
[14] Sud'enkov Y V 2020 Anharmonicity effect on heat release and metals hardening under quasistatic tensile stress *Physics of the Solid State* **62** (1) 168–73 doi: 10.1134/S1063783420010333
[15] Rusinek A and Klepaczko J R 2009 Experiments on heat generated during plastic deformation and stored energy for TRIP steels *Materials & Design* **30** (1) 35–48 doi: 10.1016/j.matdes.2008.04.048
[16] Zaera R, Rodríguez-Martínez J A, and Rittel D 2013 On the Taylor–Quinney coefficient in dynamically phase transforming materials. Application to 304 stainless steel *International Journal of Plasticity* **40** 185–201 doi: 10.1016/j.ijplas.2012.08.003
[17] de Oliveira H M R, Louche H, Grassi E N D, and Favier D 2020 Specific forward/reverse latent heat and martensite fraction measurement during superelastic deformation of nanostructured NiTi wires *Materials Science and Engineering: A* **774** 138928 doi: 10.1016/j.msea.2020.138928
[18] Rittel D, Zhang L H, and Osovski S 2017 The dependence of the Taylor–Quinney coefficient on the dynamic loading mode *Journal of the Mechanics and Physics of Solids* **107** 96–114 doi: 10.1016/j.jmps.2017.06.016
[19] Nieto-Fuentes J C, Rittel D, and Osovski S 2018 On a dislocation-based constitutive model and dynamic thermomechanical considerations *International Journal of Plasticity* **108** 55–69 doi: 10.1016/j.ijplas.2018.04.012
[20] Nieto-Fuentes J C, Osovski S, Venkert A, and Rittel D 2019 Reassessment of the dynamic thermomechanical conversion in metals *Physical Review Letters* **123** 255502 doi: 10.1103/PhysRevLett.123.255502
[21] Pandey K N and Chand S 2003 Deformation based temperature rise: A review *International Journal of Pressure Vessels and Piping* **80** (10) 673–87 doi: 10.1016/j.ijpvp.2003.07.001
[22] Battini D, Solazzi L, Lezzi A M, et al. 2020 Prediction of steel wire rope fatigue life based on thermal measurements *International Journal of Mechanical Sciences* **182** 105761 doi: 10.1016/j.ijmecsci.2020.105761
[23] Iziumova A, Vshivkov A, Prokhorov A, et al. 2016 The study of energy balance in metals under deformation and failure process *Quantitative InfraRed Thermography Journal* **13** (2) 242–56 doi: 10.1080/17686733.2016.1212527
[24] Benzerga A A, Bréchet Y, Needleman A, and Van der Giessen E 2005 The stored energy of cold work: Predictions from discrete dislocation plasticity *Acta Materialia* **53** (18) 4765–79 doi: 10.1016/j.actamat.2005.07.011
[25] Borbély A, Driver J H, and Ungár T 2000 An X-ray method for the determination of stored energies in texture components of deformed metals; application to cold worked ultra high purity iron *Acta Materialia* **48** (8) 2005–16 doi: 10.1016/S1359-6454(99)00457-7
[26] Oliferuk W, Maj M, and Raniecki B 2004 Experimental analysis of energy storage rate components during tensile deformation of polycrystals *Materials Science and Engineering: A* **374** (1-2) 77–81 doi: 10.1016/j.msea.2003.12.056
[27] Segal V M 1995 Materials processing by simple shear *Materials Science and Engineering: A* **197** (2) 157–64 doi: 10.1016/0921-5093(95)09705-8
[28] Nejnoto M, Horita Z, Furukawa M, and Langdon T G 1998 Equal-channel angular pressing: A novel tool for microstructural control *Metals and Materials* **4** (6) 1181–90 doi: 10.1007/BF03025992
[29] Figueiredo R B and Langdon T G 2010 Grain refinement and mechanical behavior of a magnesium alloy processed by ECAP *Journal of Materials Science* **45** (17) 4827–36 doi: 10.1007/s10853-010-4589-y

[30] Molnár P and Jäger A 2013 Multi-temperature equal channel angular pressing of Mg-3 wt%Al-1 wt%Zn alloy *Philosophical Magazine* **93** (27) 3612–26 doi: 10.1080/14786435.2013.817693
[31] Messerschmidt U and Bartsch M 2003 Generation of dislocations during plastic deformation *Materials Chemistry and Physics* **81** (2-3) 518–23 doi: 10.1016/S0254-0584(03)00064-6
[32] Bache M R, Evans W J, Suddell B, and Herrouin F R M 2001 The effects of texture in titanium alloys for engineering components under fatigue *International Journal of Fatigue* **23** (1) 153–9 doi: 10.1016/S0142-1123(01)00124-4
[33] Sinha S and Gurao N P 2017 The role of crystallographic texture on load reversal and low cycle fatigue performance of commercially pure titanium *Materials Science and Engineering: A* **691** 100–9 doi: 10.1016/j.msea.2017.03.043
[34] Shechtman D and Brandon D G 1973 Orientation dependent slip in polycrystalline titanium *Journal of Materials Science* **8** (9) 1233–7 doi: 10.1007/BF00549337
[35] Eswar Prasad K, Li B, Dixit N, et al. 2014 The dynamic flow and failure behavior of magnesium and magnesium alloys *JOM* **66** (2) 291–304 doi: 10.1007/s11837-013-0850-6
[36] Kingstedt O T and Lloyd J T 2019 On the conversion of plastic work to heat in Mg alloy AZ31B for dislocation slip and twinning deformation *Mechanics of Materials* **134** 176–84 doi: 10.1016/j.mechmat.2019.04.009
[37] Rosakis P, Rosakis A J, Ravichandran G, and Hodowany J 2000 A thermodynamic internal variable model for the partition of plastic work into heat and stored energy in metals *Journal of the Mechanics and Physics of Solids* **48** (3) 581–607 doi: 10.1016/S0022-5096(99)00048-4
[38] Plekhov O A, Panteleev I A, and Naimark O B 2007 Energy accumulation and dissipation in metals as a result of structural-scaling transitions in a mesodefect ensemble *Physical Mesomechanics* **10** (5-6) 294–301

Mathematical modeling of temperature stresses in a nonlocal thermoviscoelastic continuous medium

G N Kuvyrkin, I Yu Savelyeva, and D A Kuvshinnikova*
Bauman Moscow State Technical University, Moscow, Russia
E-mail: *kuvshinnikovadasha@gmail.com

Abstract. The improvement of the properties of structural and functional materials is associated with the synthesis of materials from structures with limiting values of properties (for example, extremely strong, refractory, thermostable, etc.). Such materials have an inhomogeneous structure, justified by the technological features of their preparation. In recent years, such materials have come to be called "structure–sensitive," and the study of their thermomechanical characteristics has become possible within the framework of generalized mechanics of a continuous medium. Mathematical models describing such materials belong to the class of non-local models proposed by A. K. Ehringen. Now, there are few works on the use of nonlocal viscoelasticity, while many materials exhibit viscoelastic properties, even nonlocal viscoelastic properties, for example, solid propellants, composite materials, etc. The paper considers a mathematical model of a nonlocal thermoviscoelastic continuous medium. Using the finite element method, the temperature stress distributions are found for high-intensity pulsed heating of a nonlocal half-space.

1. Introduction

The need for non-local thermodynamics in natural sciences and technology dates to the middle of the last century. Several experimental observations of the temperature field of metal interfaces, which occurs due to changes in the conductivity parameters in the local thermostated regions. It shows the temperature gradients aggregation near the boundaries [1]. Nowadays, the similar phenomena are observed in heat transfer molecular dynamics simulation in nanowires. The modeling demonstrates thermostated regions including phonon–phonon scattering which causes some changes in conductivity of materials [2, 3].

A rigorous and complete descriptions of the natural material mechanical behavior is a difficult problem. To solve pragmatic problems, we assume property simplification of the natural material behavior. The basis for constructing a natural material mathematical model is the replacement of the natural material with an artificial one. Almost all natural and artificially created materials have a complex internal structure, which does not remain constant but changes under external influences. It significantly affects the behavior of samples and products made of such materials.

Different models can be used depending on both the material type and study level. The deeper the consideration is the less demanding the model for describing the structure inhomogeneity. At low levels, models must describe structure details and simultaneously provide a hierarchical description, i.e. the connection of low levels to the higher ones.

The development of modern science and technology requires modeling the behavior of materials with fine peculiarities in their structural organization. Such materials are obtained by powder compaction, structural decomposition, and amorphous alloy crystallization [1]. Such materials acquire structural inhomogeneity, not allowing them to be described in the terms of classical mechanics of continuous media. These materials are called "structure–sensitive" due to their structural peculiarities. The research into structure-sensitive materials becomes possible in the terms of general continuum mechanics.

The concept of a microstructure medium for structure-sensitive materials is introduced in [2, 3]. The continuous medium is defined as the medium in which each infinitely small 'point' carries information on the material structure. Comparing it with the classical continuous medium, some additional forces are often applied to these 'points' which influence the nearest points.

Such models are being developed now. They can accurately describe the thermomechanical behavior of structure-sensitive materials under a highly intensive external load. A. K. Eringen's approach is the simplest and most popular one nowadays [4–6]. It states that the nonlocal influence function is used to consider the structure elements influence on a given point. Our research is based on this approach.

We have already studied the propagation of heat in nonlocal media [7–11]. The studies show that considering the nonlocality effect in heat propagation is a difficult task. Mathematical models of nonlocal media based on Eringen's approach use integral differential equations that describe the thermomechanical state of the medium. An analytical solution to such equations is impossible. In [9], algorithms were proposed for the numerical solution to the heat propagation problem in a nonlocal medium. The importance of considering non-local effects was shown in [9–12].

At sufficiently high temperatures and stresses, most materials move from the elastic stage to the viscous flow stage. In this case, the behavior of the material is usually called viscoelastic, since it simultaneously exhibits elastic and viscoelastic properties. Therefore, consideration of rheological effects is of great importance in the design of structural elements exposed to high temperatures.

With an increase in absolute temperature, the material of a solid can acquire viscous properties, while retaining the property of elasticity. When the body is loaded, the so-called viscoelastic deformation occurs, which changes over time even with a constant load, and after removing the load, it gradually disappears. The properties of such materials, which include many polymers, are described by mathematical models of a thermoviscoelastic continuous medium [6].

2. Mathematical model

An equation describing the distribution of stresses along the depth of a viscoelastic layer of thickness in a dimensionless form:

$$
\begin{aligned}
R^2\frac{\partial^2\sigma}{\partial t^2} + R^2\frac{\partial^2\theta}{\partial t^2} &= (1-\xi)\left[p_1\frac{\partial^2\sigma}{\partial z^2} + p_2\int_{z-r}^{z+r}\varphi(|z-z'|)\frac{\partial^2\sigma}{\partial (z')^2}dz'\right] \\
&+ \xi p_1\int_0^t \exp\left(-\frac{t-t'}{D_\sigma^2}\right)\frac{\partial}{\partial t'}\left(\frac{\partial^2\sigma}{\partial z^2}\right)dt' \\
&+ \xi p_2\int_0^t \exp\left(-\frac{t-t'}{D_\sigma^2}\right)\frac{\partial}{\partial t'}\left(\int_{z-r}^{z+r}\varphi(|z-z'|)\frac{\partial^2\sigma}{\partial (z')^2}dz'\right)dt', \qquad (1)
\end{aligned}
$$

where σ is the dimensionless stress, R^2 is a dimensionless parameter inversely proportional to the square of the speed of sound, θ is dimensionless temperature, z is spatial coordinate, t is time, ξ is coefficient of the ratio of viscous and elastic mechanical properties, D_σ^2 is dimensionless relaxation time of the internal parameter of the thermodynamic state, r is radius of the influence function.

The effect of nonlocality of the medium is that the physical characteristics of the elements of the structure-sensitive material are influenced by other surrounding elements of the structure. In this regard, expression (1) includes non-local influence weights p_1 and p_2 ($p_1 + p_2 = 1$) and function of nonlocality influence $\varphi(|x' - x|)$. Moreover: $\int_V \varphi(|x' - x|)dx' = 1$.

The function of the influence of nonlocality in this work is taken in the form [7]:

$$\varphi(|z' - z|) = \frac{1}{2r} \exp\left(-\frac{|z' - z|}{r}\right), \quad |z' - z| < r. \tag{2}$$

The initial and boundary conditions for stresses have the form:

$$\begin{aligned} &\sigma|_{t=0} = 0, \quad \left.\frac{\partial \sigma}{\partial t}\right|_{t=0} = 0, \\ &\sigma|_{z=0} = \sigma|_{t=L} = 0. \end{aligned} \tag{3}$$

To find the temperature distribution in the viscoelastic layer, we use the heat conduction equation [6]:

$$\frac{\partial \theta}{\partial t} = \frac{\partial^2 \theta}{\partial z^2} - \int_0^t \exp\left(-\frac{t - t'}{D_T^2}\right) \frac{\partial}{\partial t'} \left(\frac{\partial^2 \theta}{\partial z^2}\right) dt', \tag{4}$$

where D_T^2 is dimensionless relaxation time of the heat flux density.

We write the initial and boundary conditions for equation (4) in the form:

$$\begin{aligned} &\theta(z, t)|_{t=0} = 0, \\ &z = 0: \quad -\frac{\partial \theta}{\partial z} + \int_0^t \exp\left(-\frac{t - t'}{D_T^2}\right) \frac{\partial}{\partial t'} \frac{\partial \theta}{\partial z} dt' = Q_1, \\ &z = L: \quad \frac{\partial \theta}{\partial z} - \int_0^t \exp\left(-\frac{t - t'}{D_T^2}\right) \frac{\partial}{\partial t'} \frac{\partial \theta}{\partial z} dt' = Q_2, \end{aligned} \tag{5}$$

where Q_1 and Q_2 are dimensionless heat flux densities.

3. Results

To find the temperature stresses from equation (1), the temperature distribution in the layer was previously found. The solution of equation (4) with boundary conditions of the second kind, where the thermal insulation condition was specified at one end of the layer, and a high-intensity impulse action at the other end, was found using the finite element method (FEM) on a uniform grid [8]. To find the temperature stresses from equation (1), the temperature distribution in the layer was previously found. The solution of equation (4) with boundary conditions of the second kind, where the thermal insulation condition was specified at one end of the layer, and a high-intensity impulse action at the other end, was found using the finite element method (FEM) on a uniform grid [8]. In figure 1 shows the temperature distributions at different times in a layer with a thickness $L = 10$ and $D_T^2 = 10$.

Based on the obtained temperature distributions, using the FEM, a solution to equation (1) with initial and boundary conditions (3) was found.

Let us consider the influence of the contribution of the local component p_1 and the parameter D_σ^2 on the solution of equation (1). Let us take the parameters $R^2 = 1$, $\xi = 0.05$, $r = 3$ and obtain graphs of the distribution of temperature stresses at different points in time (figure 2).

We see that a decrease in the contribution of the local component leads to an increase in the stress wave amplitude and a decrease in the depth of penetration into the material layer, which indicates the possibility of using such materials as protective coatings under high-intensity loading [9, 10].

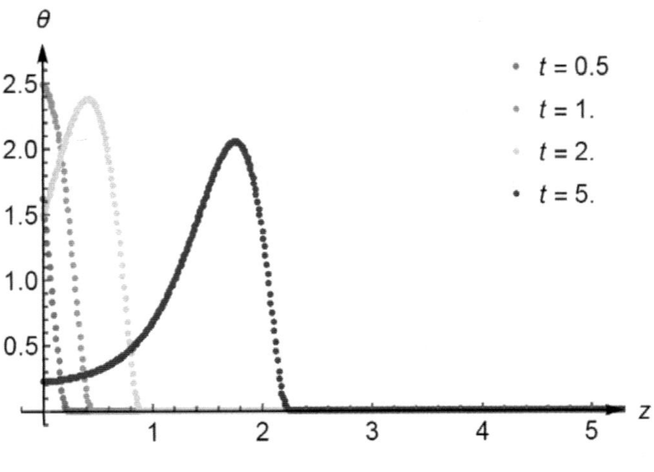

Figure 1. Solution of the hyperbolic heat equation at different times.

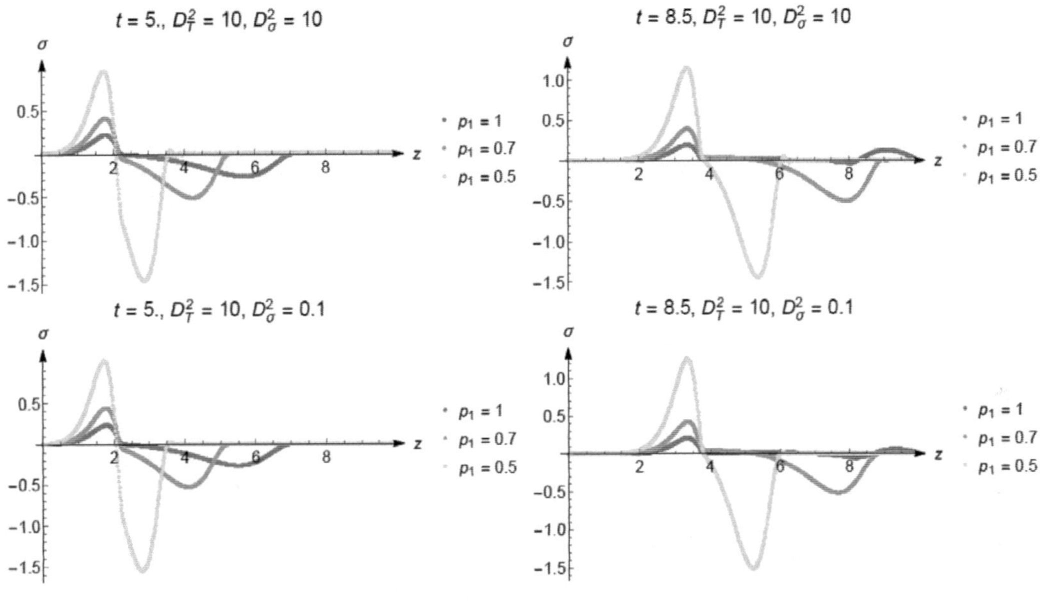

Figure 2. Distribution of nonlocal temperature stresses in the layer with different contributions of the local component.

Acknowledgments

This work was supported by the Ministry of Science and Higher Education (project No. 0705-2020-0032)

References

[1] Andrievskii R A and Ragulya A V 2005 *Nanostructural Materials* (Moscow: Akademiya) [in Russian]
[2] Alymov M I and Zelensky V A 2007 *Methods of Production and Physical and Mechanical Properties of Bulk Nanocrystalline Materials* (Moscow: ELIZ) [in Russian]
[3] Onami M, Iwashimizu S, Hank K, et al. 1987 *Introduction to Micromechanics* (Moscow: Metallurgiya) [in Russian]
[4] Eringen A C 2002 *Nonlocal Continuum Field Teories* (New York-Berlin-Heidelberg: Springer-Verlag)

[5] Eringen A C 2006 Nonlocal continuum mechanics based on distributions *International Journal of Engineering Science* **44** (3-4) 141–7 doi: 10.1016/j.ijengsci.2005.11.002
[6] Pisano A A and Fuschi P 2013 Closed form solution for a nonlocal elastic bar in tension *International journal of Solids and Structures* **40** (1) 13–23 doi: 10.1016/S0020-7683(02)00547-4
[7] Zarubin V S and Kuvyrkin G N 2008 *Mathematical Models of Continuum Mechanics and Electrodynamics* (Moscow: Izdat. MGTU im. N.E. Baumana) [in Russian]
[8] Kuvyrkin G N, Savelyeva I Y, and Kuvshynnikova D A 2018 Mathematical model of the heat transfer process taking into account the consequences of nonlocality in structurally sensitive materials *Journal of Physics: Conference Series* **991** 012050 doi: 10.1088/1742-6596/991/1/012050
[9] Kuvyrkin G N, Savel'eva I Yu, and Kuvshinnikova D A 2019 Nonstationary heat conduction in a curvilinear plate with account for spatial nonlocality *Journal of Engineering Physics and Thermophysics* **92** (3) 608–13 doi: 10.1007/s10891-019-01969-9
[10] Kuvyrkin G N, Savelyeva I Yu, and Kuvshynnikova D A 2021 Nonlocal dynamic temperature stress simulation *Journal of Physics: Conference Series* **1902** 012015 doi: 10.1088/1742-6596/1902/1/012015
[11] Kuvyrkin G N, Savelyeva I Yu, and Kuvshynnikova D A 2019 One mathematical model of thermal conductivity for materials with a granular structure *Thermal Science* **23** (4) S1273–S1280 doi: 10.2298/TSCI19S4273K
[12] Kuvyrkin G N, Savelyeva I Y, and Zhuravsky A V 2018 Numerical simulation of vapor-phase epitaxy with allowance for diffusion processes *Mathematical Models and Computer Simulations* **10** (3) 299–307 doi: 10.1134/S2070048218030080

Examples of exact solutions to boundary value problems of the theory of elasticity for a truncated wedge

M D Kovalenko[1], I V Menshova[2,3], A P Kerzhaev[2,*], and G Yu[4]

[1]Institute of Applied Mechanics of Russian Academy of Sciences, Moscow, Russia
[2]Institute of Earthquake Prediction Theory and Mathematical Geophysics, Russian Academy of Sciences, Moscow, Russia
[3]Bauman Moscow State Technical University, Moscow, Russia
[4]School of Civil Engineering, Qingdao University of Science and Technology, Qingdao, China

E-mail: *alex_kerg@mail.ru

Abstract. The paper presents examples of exact solutions to boundary value problems of the theory of elasticity for a truncated wedge. The straight sides of the wedge are free, and normal stresses are given on its circular "end" (the tangential stresses are considered equal to zero). The solutions are sought in the form of expansions in Papkovich–Fadle eigenfunctions. The coefficients of the expansions are determined in a simple explicit form.

1. Introduction

In the articles [1–3] and others, the theory of expansions in Papkovich–Fadle eigenfunctions in the Cartesian coordinate system was constructed and, on its basis, various problems for the half-strip and rectangle were solved. In these solutions, stresses and displacements are represented by series the coefficients of which are determined by using functions biorthogonal to Papkovich–Fadle eigenfunctions, i.e. in a simple explicit form, as in the known periodic solutions in trigonometric Fourier series [4]. At the end of the half-strip (rectangle), series in Papkovich–Fadle eigenfunctions converge in the same way as trigonometric functions do. This is due to the fact that Papkovich–Fadle eigenfunctions are entire functions of exponential type, and eigenvalues are zeros of entire functions of exponential type [5], the form of which depends on the type of boundary conditions on the long sides of the half-strip. In fact, these two circumstances underlie the possibility of constructing an exact solution to the biharmonic problem in the Cartesian coordinate system.

Solutions in the polar coordinate system for a truncated wedge or part of a ring (a circular rectangle) are also represented by series in Papkovich–Fadle eigenfunctions, which have the same properties as their analogues in the Cartesian coordinate system, i.e. eigenfunctions are entire functions of exponential type, and eigenvalues are zeros of entire functions of exponential type, the form of which is determined by what boundary conditions are given on the straight sides of the wedge. Therefore, the fundamental possibility of constructing a theory similar to the theory in the Cartesian coordinate system exists, although the mathematical means used in this case can sometimes be completely different.

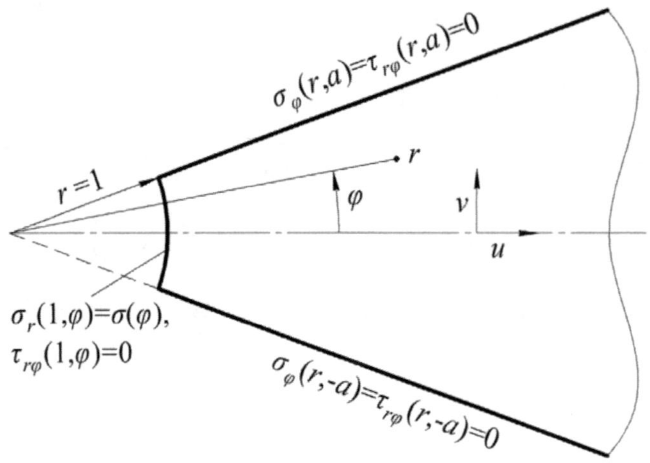

Figure 1. A truncated wedge.

In this short note, of course, there is no way to present at least part of the theory. Therefore, we will restrict ourselves to the most important results and examples of solutions to the simplest and, at the same time, well-known problems of practical importance.

In the theory of solving boundary value problems in the Cartesian coordinate system, the coefficients of expansions in Papkovich–Fadle eigenfunctions can be determined in two ways: directly from equations for determining biorthogonal functions or by using the so-called "finite parts of biorthogonal functions". The first way is used when boundary functions given at the end of the half-strip are entire analytical functions. This way one can find many different solutions (the technique of their construction is described in our works). If expanded functions are not entire (for example, they are piecewise continuous), then one should use the finite parts of biorthogonal functions. Here we will consider only the case when entire analytical functions are given at the circular end of a truncated wedge. In this case it is not necessary to construct the finite parts of biorthogonal functions. It is impossible to solve the problem this way when, for example, at the circular end, normal stresses are represented by a concentrated force. However, this is not required for demonstrating exact solutions yet.

Unfortunately, there are very few analytical works devoted to problems for the truncated wedge. In particular, it is difficult to find articles that provide examples of Papkovich–Fadle eigenfunctions and their basic properties. The results of this article are not the result of other authors' research. They should be viewed as the development of the theory developed in the Cartesian coordinate system. Therefore, we will restrict ourselves to only a few results of the previous research [6–8], which are closest to the ones presented.

2. Statement of the boundary value problem

In the polar coordinate system $\{r \geq 1, |\varphi| \leq a \leq \pi\}$, consider an even-symmetric deformation of the truncated wedge schematically shown in figure 1.

The long sides of the wedge are free, i.e.

$$\sigma_\varphi(r, \pm a) = \tau_{r\varphi}(r, \pm a) = 0, \tag{1}$$

and normal and zero tangential stresses are given at its "end" for $r = 1$:

$$\sigma_r(1, \varphi) = \sigma(\varphi), \quad \tau_{r\varphi}(1, \varphi) = 0. \tag{2}$$

The function $\sigma(\varphi)$ is even and self-balanced, i.e.

$$\int_{-a}^{a} \sigma(\varphi)\, d\varphi = 0. \tag{3}$$

Following [9], we introduce a new independent variable related to the variable r by the formulas

$$r = e^s, \quad \frac{\partial}{\partial r} = \frac{1}{e^s}\frac{\partial}{\partial s}, \quad \frac{\partial}{\partial s} = r\frac{\partial}{\partial r}. \tag{4}$$

Then the equilibrium equations and the equations of Hooke's law can be written as follows:

$$\frac{\partial S(s,\varphi)}{\partial s} - \Phi(s,\varphi) + \frac{\partial R(s,\varphi)}{\partial \varphi} = 0, \quad \frac{\partial R(s,\varphi)}{\partial s} + R(s,\varphi) + \frac{\partial \Phi(s,\varphi)}{\partial \varphi} = 0, \tag{5}$$

$$\begin{aligned} S(s,\varphi) &= \frac{2}{1-\nu}\left\{\frac{\partial U(s,\varphi)}{\partial s} + \nu\left[\frac{\partial V(s,\varphi)}{\partial \varphi} + U(s,\varphi)\right]\right\}, \\ \Phi(s,\varphi) &= \frac{2}{1-\nu}\left[\frac{\partial V(s,\varphi)}{\partial \varphi} + U(s,\varphi) + \nu\frac{\partial U(s,\varphi)}{\partial s}\right], \\ R(s,\varphi) &= \left[\frac{\partial U(s,\varphi)}{\partial \varphi} + \frac{\partial V(s,\varphi)}{\partial s} - V(s,\varphi)\right]. \end{aligned} \tag{6}$$

In these formulas, the following notation is introduced: ν is Poisson's ratio; $U(s,\varphi)$ and $V(s,\varphi)$ are, respectively, the displacements along the coordinate lines r and φ multiplied by the shear modulus G; $S(r,\varphi) = r\sigma_r(r,\varphi)$, $\Phi(r,\varphi) = r\sigma_\varphi(r,\varphi)$, $R(r,\varphi) = r\tau_{r\varphi}(r,\varphi)$.

3. Method of initial functions and Papkovich–Fadle eigenfunctions

The solution to the boundary value problem (1), (2), (5), (6) is best sought by the symbolic method of initial functions in the space of the variables (s,φ) [9]. As a result, we arrive at two ordinary differential equations of infinite order with constant coefficients with respect to two unknown initial functions defined at $\varphi = 0$. Solving it, we obtain the characteristic equation of the problem

$$L(\lambda) = \sin(2\lambda a) + \lambda \sin(2a) = 0. \tag{7}$$

The entire function of exponential type (7) has two zero roots which correspond to the elementary solution to the problem for an infinite wedge and an infinite set of complex roots $\pm\lambda_k$, $\pm\bar{\lambda}_k$. The functions generating systems of eigenfunctions at $\lambda = \lambda_k$ have the form

$$\begin{aligned} S(\lambda,\varphi) &= \frac{1+\nu}{2}\lambda\{(\lambda-1)\sin[a(\lambda-1)]\cos[\varphi(\lambda+1)] - (\lambda-3)\sin[a(\lambda+1)]\cos[\varphi(\lambda-1)]\}, \\ \Phi(\lambda,\varphi) &= \frac{1+\nu}{2}\lambda\{(\lambda+1)\sin[a(\lambda+1)]\cos[\varphi(\lambda-1)] - (\lambda-1)\sin[a(\lambda-1)]\cos[\varphi(\lambda+1)]\}, \\ R(\lambda,\varphi) &= -\frac{1+\nu}{2}\lambda(\lambda-1)\{\sin[a(\lambda-1)]\sin[\varphi(\lambda+1)] - \sin[a(\lambda+1)]\sin[\varphi(\lambda-1)]\}, \\ U(\lambda,\varphi) &= \frac{1+\nu}{4}(\lambda-1)\sin[a(\lambda-1)]\cos[\varphi(\lambda+1)] - \frac{\nu-3+\lambda(1+\nu)}{4}\sin[a(\lambda+1)]\cos[\varphi(\lambda-1)], \\ V(\lambda,\varphi) &= \left(\frac{1+\nu}{4}\lambda + \frac{3-\nu}{4}\right)\sin[a(\lambda+1)]\sin[\varphi(\lambda-1)] - \frac{1+\nu}{4}(\lambda-1)\sin[a(\lambda-1)]\sin[\varphi(\lambda+1)]. \end{aligned} \tag{8}$$

4. Equations for determining biorthogonal functions

Let us give the equations necessary to determine the functions biorthogonal to the Papkovich–Fadle eigenfunctions (λ is a real parameter):

$$\begin{aligned}\int_{-\infty}^{\infty} S^+(\lambda,\varphi)S_k^+(\varphi)\,d\varphi &= \frac{\lambda L(\lambda)(\lambda^2-1)}{\lambda_k(\lambda^2-\lambda_k^2)(\lambda_k^2-1)},\\ \int_{-\infty}^{\infty} R^-(\lambda,\varphi)R_k^-(\varphi)\,d\varphi &= \frac{\lambda^2 L(\lambda)(\lambda^2-1)}{\lambda_k^2(\lambda^2-\lambda_k^2)(\lambda_k^2-1)},\\ \int_{-\infty}^{\infty} \Phi(\lambda,\varphi)\Phi_k(\varphi)\,d\varphi &= \frac{\lambda L(\lambda)(\lambda^2-1)^2}{\lambda_k(\lambda^2-\lambda_k^2)(\lambda_k^2-1)^2}.\end{aligned} \tag{9}$$

In the case when, at the circular end of the wedge, only normal stresses are given and the tangential stresses are zero, it is convenient to introduce the functions $S^+(\lambda,\varphi)=\frac{1}{2}[S(\lambda,\varphi)+S(-\lambda,\varphi)]$ and $R^-(\lambda,\varphi)=\frac{1}{2}[R(\lambda,\varphi)-R(-\lambda,\varphi)]$ even and odd in the parameter λ, respectively. The unknown biorthogonal functions are denoted by $S_k^+(\varphi)$ and $R_k^-(\varphi)$ (their existence and uniqueness follow from the results of the article [1]). The right-hand sides of equations (9) are chosen due to the same reasons as in [1] and, additionally, by taking into account the fact that the generating functions $S^+(\lambda,\varphi)$ and $R^-(\lambda,\varphi)$ have first-order zeros at $\lambda=\pm 1$. At the same points, the function $\Phi(\lambda,\varphi)$ has second-order zeros. At $\lambda=0$, the even functions $S^+(\lambda,\varphi)$ and $\Phi(\lambda,\varphi)$ have second-order zeros, and the function $R^-(\lambda,\varphi)$ odd in λ has a third-order zero.

5. Examples of solutions to boundary value problems

The final formulas that give the solution to the problem for the truncated wedge, regardless of what boundary conditions are given at the circular end, will actually be equivalent to the corresponding formulas for the half-strip. In particular, in the case when only the normal self-balanced stresses (2) are given at the "end" of the wedge, they will be as follows (the formulas for the displacements are omitted):

$$\begin{aligned}\sigma_r(r,\varphi) &= -\frac{1}{r}\sum_{k=1}^{\infty} 2\mathrm{Re}\left[s_k\frac{S^+(\lambda_k,\varphi)}{M_k}\right]\frac{\mathrm{Im}(\bar{\lambda}_k r^{-\lambda_k})}{\mathrm{Im}(\lambda_k)},\\ \sigma_\varphi(r,\varphi) &= -\frac{1}{r}\sum_{k=1}^{\infty} 2\mathrm{Re}\left[s_k\frac{\Phi(\lambda_k,\varphi)}{\lambda_k^2 M_k}\right]\lambda_k\bar{\lambda}_k\frac{\mathrm{Im}(\lambda_k r^{-\lambda_k})}{\mathrm{Im}(\lambda_k)},\\ \tau_{r\varphi}(r,\varphi) &= -\frac{1}{r}\sum_{k=1}^{\infty} 2\mathrm{Re}\left[s_k\frac{R^-(\lambda_k,\varphi)}{\lambda_k M_k}\right]\lambda_k\bar{\lambda}_k\frac{\mathrm{Im}(r^{-\lambda_k})}{\mathrm{Im}(\lambda_k)}.\end{aligned} \tag{10}$$

Here $M_k = L'(\lambda_k)/2\lambda_k$, $\lambda_k M_k$, $\lambda_k^2 M_k$ are normalizing factors for the corresponding eigenfunctions, and s_k is a Lagrange coefficient [1, 2], $\mathrm{Re}\lambda_k>0$.

If in the first equation of (9) we pass to the limit when $\lambda\to 0$, then, under the integral on the left instead of $S^+(\lambda,\varphi)$, we obtain the expanded function

$$\sigma(\varphi) = (1+\nu)\frac{\varphi\sin\varphi\sin a - (\sin a - a\cos a)\cos\varphi}{2a+\sin(2a)}, \tag{11}$$

and, on the right, we obtain the expression for the Lagrange coefficient

$$s_k = \frac{1}{\lambda_k^3(\lambda_k^2-1)} \quad (k=1,2,\ldots). \tag{12}$$

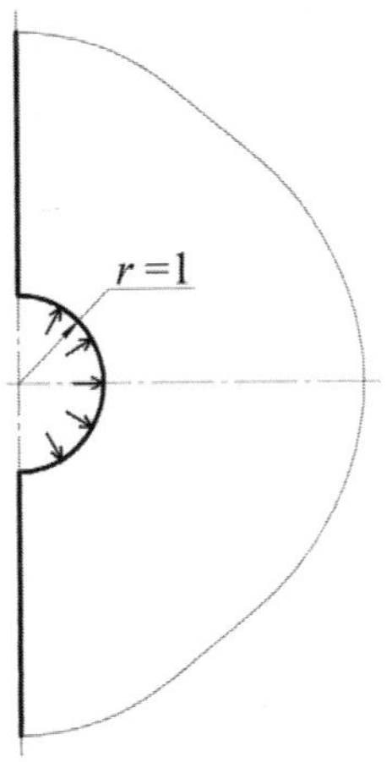

Figure 2. A half-plane with a circular notch.

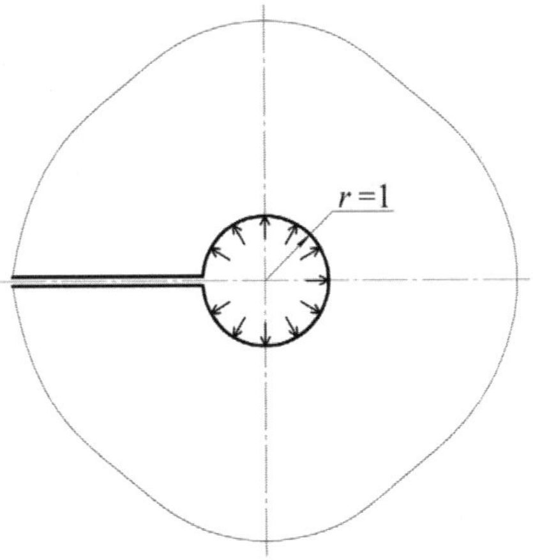

Figure 3. A plane with a semi-infinite cut at the tip of which there is a circular hole.

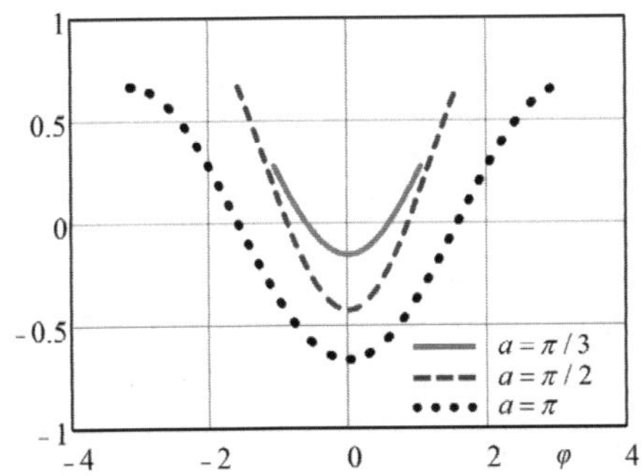

Figure 4. The distribution of the normal stresses $\sigma_r(1, \varphi)$.

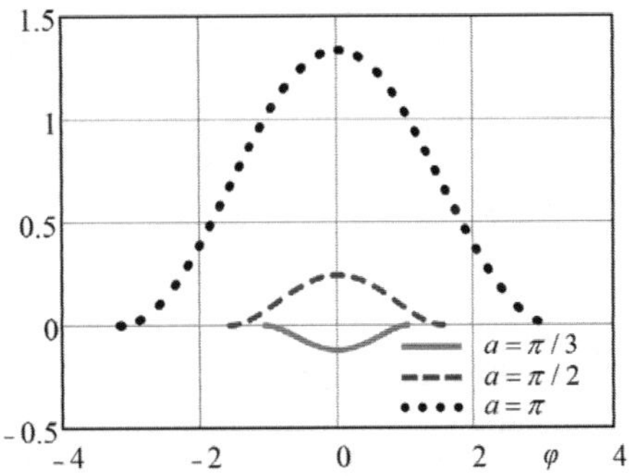

Figure 5. The distribution of the normal stresses $\sigma_\varphi(1, \varphi)$.

In the papers devoted to problems for the half-strip, it was noted that in this way one can obtain formulas for Lagrange coefficients for a very wide class of entire analytical expanded functions $\sigma(\varphi)$.

The calculations were performed for three values of the half-opening angle of the wedge: $a = \pi/3$, $a = \pi/2$ – a half-plane with a circular notch (figure 2) and $a = \pi$ – an infinite plane with a semi-infinite cut at the tip of which there is a circular hole (figure 3). Poisson's ratio ν is equal to 1/3.

Figures 4–9 below show the graphs of the stresses.

6. Conclusions

1. Solutions for all types of loads can be obtained only by using the finite parts of biorthogonal functions.
2. If in formulas (10) we assume that $r \in [0, 1]$, $\mathrm{Re}\lambda_k < 0$, then we obtain solutions: (a) for the vertex of a wedge ($a = \pi/3$); (b) for a half-disc with a free straight boundary and the

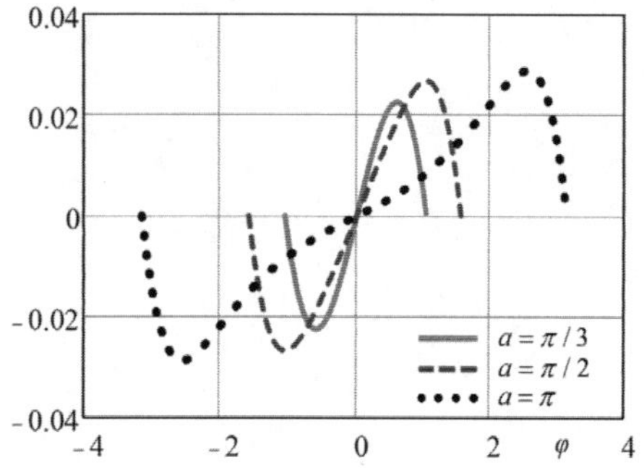

Figure 6. The distribution of the tangential stresses $\tau_{r\varphi}(2, \varphi)$.

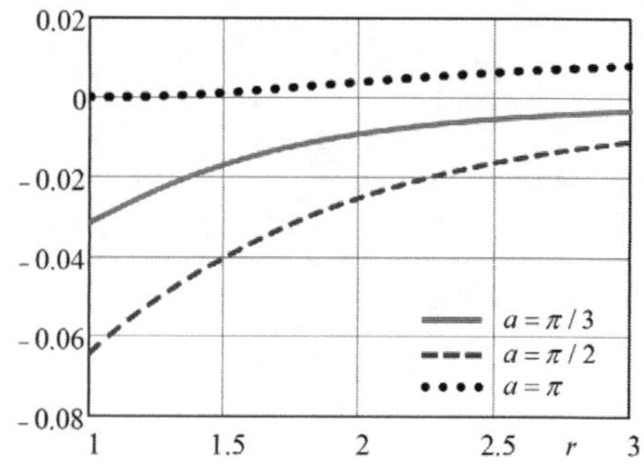

Figure 7. The distribution of the normal stresses $\sigma_r(r, \frac{a}{2})$.

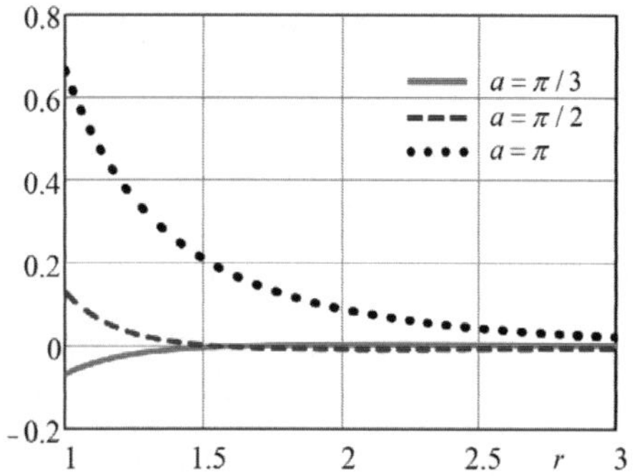

Figure 8. The distribution of the normal stresses $\sigma_\varphi(r, \frac{a}{2})$.

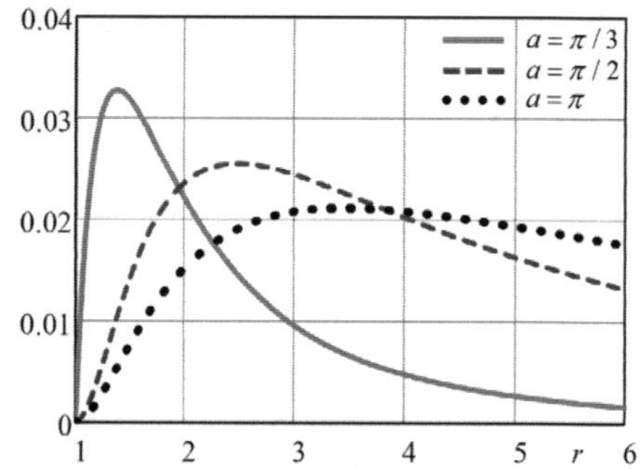

Figure 9. The distribution of the tangential stresses $\tau_{r\varphi}(r, \frac{a}{2})$.

normal stresses (11) given on the arc of the circle ($a = \pi/2$); (c) for a disc cut along a radius ($a = \pi$). The sides of the cut are free, and the normal stresses (11) act on the arc of the circle.

3. Consider an infinite wedge with a vertex, on the sides of which normal and tangential stresses act. This problem is solved by standard means: in general, by using the Fourier transform in the coordinates (s, φ), and even easier for power loads [9]. Let, for example, only normal stresses equal to 1 act on the sides of the wedge. We construct a solution to the problem, remove the vertex and add to this solution a solution for a truncated wedge the straight sides of which are free and at the circular end of which stresses, equal in magnitude and opposite in sign to the corresponding stresses for the wedge with a vertex, act. As a result, we obtain solutions for three domains when $r \geq 1$ and for three domains when $r \leq 1$, in which the arc boundaries are free and normal stresses equal to 1 act on the straight boundaries.

Acknowledgments
The research of MK, IM and AK was funded by RFBR and NSFC according to the research project No. 20-51-53021. The research of GY was funded by NSFC and RFBR (No. 5191101589).

References

[1] Kovalenko M D and Shulyakovskaya T D 2011 Expansions in Fadle-Papkovich functions in a strip. Theory foundations *Mechanics of Solids* **46** (5) 721–38 doi: 10.3103/S0025654411050074

[2] Kovalenko M D, Men'shova I V, and Shulyakovskaya T D 2013 Expansions in Fadle-Papkovich functions: Examples of solutions in a half-strip *Mechanics of Solids* **48** (5) 584–602 doi: 10.3103/S0025654413050154

[3] Kovalenko M D, Menshova I V, Kerzhaev A P, and Yu G 2020 A boundary value problem in the theory of elasticity for a rectangle: Exact solutions *Zeitschrift für angewandte Mathematik und Physik* **71** (6) 199 doi: 10.1007/s00033-020-01425-2

[4] Timoshenko S P and Goodier J N 1951 *Theory of Elasticity* (New-York: McGraw-Hill)

[5] Leontiev A F 1976 *Series of Exponentials* (Moscow: Nauka) [in Russian]

[6] Buchwald V T 1965 Eigenfunctions of plane elastostatics III the Wedge *Journal of the Australian Mathematical Society* **5** (2) 241–57 doi: 10.1017/S1446788700026811

[7] Little R W and Thompson T R 1972 End effects in truncated semi-infinite wedge *Journal of the Engineering Mechanics Division* **98** (5) 1239–52 doi: 10.1061/JMCEA3.0001665

[8] Grinberg G A 1953 On a method applied by P.F. Papkovich for the solution of the plane problem of the theory of elasticity for a rectangular region and of problem of bending of a rectangular thin plate with two clamped edges, and on some of its generalizations *Prikladnaya Matematika i Mekhanika* **17** 211–28

[9] Vlasov V V 1975 *Method of Initial Functions in Problems of the Theory of Elasticity and Structural Mechanics* (Moscow: Stroiizdat) [in Russian]

Reinforcement of polymer composite materials by titanium dioxide nanoparticles synthesized in plasma discharge under ultrasonic cavitation

S A Kolesnik[*] and N A Bulychev[**]
Moscow Aviation Institute (National Research University), Moscow, Russia
E-mail: *ksa@mai.ru, **nbulychev@mail.ru

Abstract. In this work, the synthesis of titanium dioxide nanoparticles in plasma discharge was carried out under ultrasonic cavitation. It was found that applied acoustoplasma method allows to synthesize nanosized particles of titanium dioxide with the controlled shape and narrow size distribution. Polymer composite materials based on synthesized nanoparticles of titanium dioxide and epoxy polymer have been obtained by the in situ method by curing with preliminary ultrasonic dispersing. The composition and structure of the obtained TiO2 polymer composite materials have been studied by scanning electron microscopy. The mechanical properties of the obtained film polymer composite materials were investigated to determine the tensile strength, elastic modulus and relative deformation. It was found that nanoparticles in optimal concentration (3%) lead to the reinforcement of obtained composite materials.

1. Introduction

Epoxy resins are often used as polymer matrices in functional and structural composites due to their good heat resistance, durability, as well as good electrical, chemical and mechanical properties [1–6]. However, the area of application of such materials is often limited due to their fragility and poor toughness. One of the ways to overcome these problems is to improve the properties of epoxy matrices by introducing a solid filler into the polymer matrix [3–5]. A new approach in the field of nanotechnology proposes the use of fillers in the nanometer size range, since nanofillers have a large surface area, which makes them chemically very reactive and helps them to better bond with the matrix. Nanoparticles embedded in a polymer matrix are attracting increasing interest due to the unique mechanical, optical, electrical, and magnetic properties exhibited by such polymer composite materials and can significantly improve them. These property improvements are dependent on the volumetric content of the filler as well as the type and characteristics of the nanofillers.

Nanofillers of a metallic or other inorganic type can be considered as excellent candidates for hardening polymer matrices. Due to their chemical inertness, low toxicity, photocatalytic activity, high refractive index, and other beneficial properties, titanium dioxide (TiO2) nanoparticles have attracted a great deal of attention of many researchers and are used in the food, paint and varnish industry, etc. [5, 6]. A number of studies have shown that the introduction of TiO2 nanoparticles improves some properties of polymer materials [3, 4]. However, the process of interaction and the mechanism of hardening of epoxy resin using nanoparticles is still obscure and not fully understood. There are few works on epoxy polymer composite materials with TiO2 nanoparticles; therefore, it is prospective to obtain new examples of such polymer composite materials and study their physical and mechanical properties, since due to the presence of TiO2, it is possible to use such polymer composite materials in biomedicine, as bactericidal and photocatalytic surfaces.

In a range of previous studies, the possibility to initiate the specific type of plasma discharge in a cavitating liquid has been shown as well as the application of such discharge for the synthesis of nanomaterials with controlled properties [7–9].

Previously, dispersed systems obtained as a result of intensive action of various types were studied in detail [10]. It was found that mechanical, especially ultrasonic activation significantly affects the properties of dispersed systems, significantly increasing their stability and dispersity of the solid phase, and also affects the properties of the particle surface, causing the formation of active centers that can participate in the interaction with (bio)-organic and polymer molecules, matrixes etc. It has been shown that such surface changes depend on the parameters of the intensive action and thus the properties of the particle surface can be controlled.

The aim of this work was to create an epoxy polymer composite materials based on specially synthesized TiO2 nanoparticles and to study their physical and mechanical properties in depending on the concentration of embedded nanoparticles.

2. Experimental

Samples of TiO2 nanoparticles were synthesized in a plasma discharge under the action of ultrasound as described earlier [8]. Titanium wire (99.99%) of 2 mm diameter was used as electrodes. The synthesis was carried out in distilled water. After synthesis, a part of the suspension of nanoparticles was sedimented with a centrifuge at 6000 rpm for 10 min, then dried in air at 150°C. Another part was analyzed as a suspension.

Particle size and ξ-potential value were measured using a Zetasizer Nano instrument (Malvern Instruments). SEM images of nanoparticles were obtained on a Carl Zeiss Evo 50 device. Silicon dioxide plates were used as a substrate. Nanoparticles were applied to substrates by centrifuging the suspension.

Polymer composite materials based on TiO2 nanoparticles with epoxy polymer were obtained by the in situ method at the stage of curing an epoxy resin with pre-dispersed titanium dioxide nanoparticles in an epoxy resin matrix. To improve the dispersion of the introduced nanoparticles in a viscous epoxy resin, we used the method of ultrasonic treatment at a frequency of 40 kHz for 10 min. The curing process was carried out in a two-stage temperature conditions at 90°C for 3 h and then at 160°C for 3 h.

3. Results and Discussion

Since it was found that the acoustoplasma method allows to synthesize nanosized particles of metal oxides with the controlled shape and narrow size distribution, the aim of this work was to obtain and study the properties of aggregation and sedimentation stable dispersed systems of titanium dioxide nanoparticles with an average particle size less than 100 nm, with the same shape and narrow particle size distribution.

The size distribution of TiO2 nanoparticles synthesized in plasma discharge under ultrasonic cavitation is shown in figure 1, and the SEM image of the particles is shown in figure 2.

Figure 1 shows that the particle size distribution is quite narrow, the particles have size in the range from 30 to 50 nm with a distribution peak around 40 nm. These data are confirmed by the results of scanning electron microscopy – the particles have characteristic spherical shape with a size about 30–50 nm (figure 2).

Further observations of nanoparticles suspensions showed that they have high aggregative and sedimentation stability and do not grow larger over time, but can form composite associates when applied to a substrate.

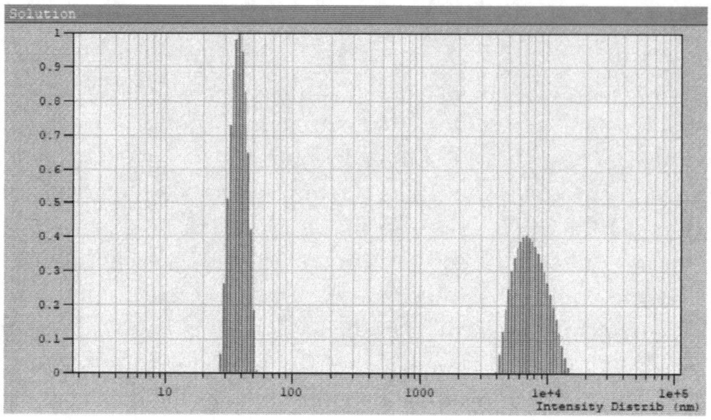

Figure 1. Particle size distribution of TiO2 nanoparticles synthesized with ultrasonic cavitation.

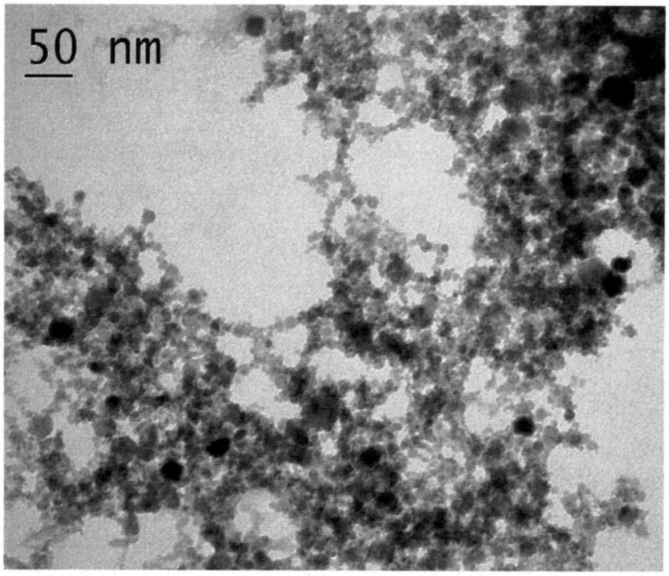

Figure 2. SEM image of TiO2 nanoparticles synthesized with ultrasonic cavitation.

As a result of the further experiments, epoxy polymer composite materials based on titanium dioxide nanoparticles in the form of opaque blue-gray films were obtained. The composition and structure of epoxy polymer composite materials were studied by scanning electron microscopy. In a polymer composite material containing 2% TiO2 nanoparticles, agglomerations of particles are observed, and therefore the average diameter increases from 30–50 nm to 70–90 nm (figure 3). The picture shows single sphere-shaped particles and gradual distribution of particles.

The properties of the obtained film polymer composite materials were investigated to determine the glass transition temperature, and the mechanical properties of the films were tested in order to determine the tensile strength, elastic modulus and relative deformation.

The results of the differential scanning calorimetry studies with an increase in temperature from 20 to 200°C and a heating rate of 5°C/min in a nitrogen atmosphere are shown in figure 4. Figure 4 shows that the glass transition temperature of the obtained epoxy polymer composite materials weakly depends on the concentration of TiO2 nanoparticles, however, one can notice the tendency to insignificant change: at concentrations of nanoparticles less than 1 wt.% the

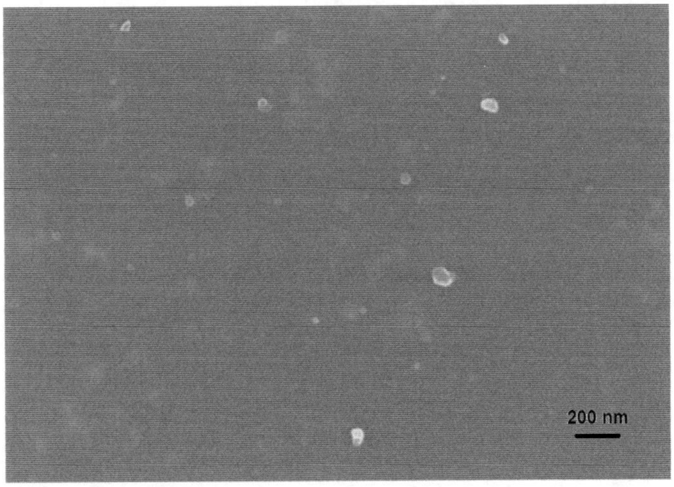

Figure 3. SEM image of the particles distribution in a polymer composite materials containing 2 wt.% of TiO2 nanoparticles.

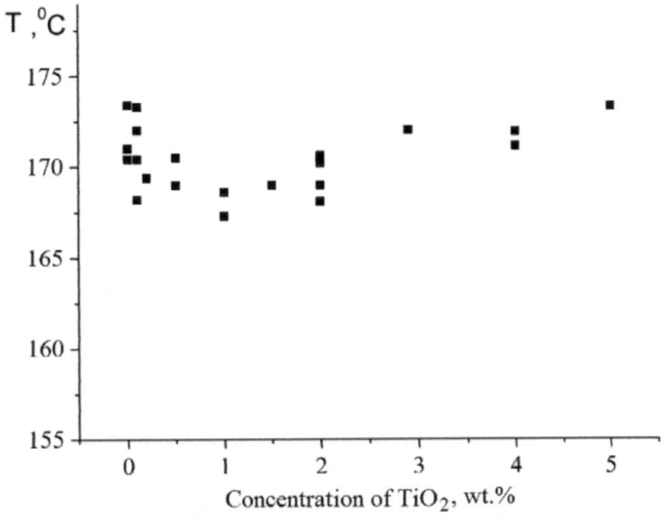

Figure 4. Dependence of the glass transition temperature of TiO2/EP polymer composite materials on the TiO2 nanoparticles concentration.

glass transition temperature decreases to 167°C, then at concentrations up to 5 wt.% increases and reaches 173°C. Such a decrease in the glass transition temperature may indicate a change in the viscosity and elasticity of the system, an increase in the flexibility of molecules and their mobility; while a further increase in the concentration of nanoparticles leads to an increase in the rigidity and, consequently, to an increase in the glass transition temperature of the system.

Based on the results obtained, it can be concluded that the presence of titanium dioxide nanoparticles in the system leads to a slight decrease in the glass transition temperature at a concentration of 1% TiO2.

In following physical and mechanical tensile tests, the efficiency of filling the polymer matrix with TiO2 nanoparticles was monitored according to the following parameters: tensile strength

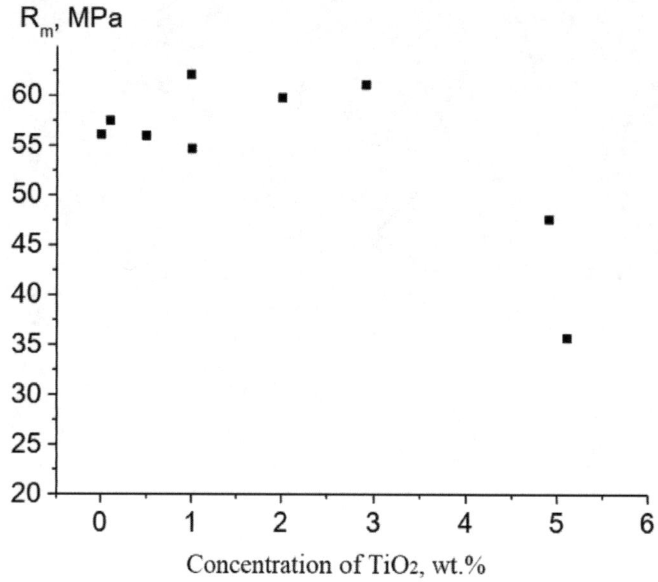

Figure 5. Dependence of the tensile strength of TiO2/EP polymer composite materials on the TiO2 nanoparticles concentration.

(R_m, MPa), elastic modulus (E, GPa), and relative deformation (ε, %). The test included control samples not modified with titanium dioxide and samples of polymer composite materials with different contents of TiO2 nanoparticles. The size of the samples was 10×90 mm, the thickness was 0.08 mm. The tensile test speed was 2 mm/min. The samples were securely fixed in rubberized clamps and a load was applied until complete destruction. Fracture occurred in the middle of the sample. After carrying out all the necessary tests and calculations, diagrams of the dependence of strength, modulus, and deformation on the concentration of TiO2 nanoparticles were obtained. As shown in figure 5, the tensile strength drops sharply at concentrations of TiO2 nanoparticles higher than 3 wt.%, which is possibly associated with secondary processes of agglomeration of TiO2 nanoparticles. Therefore, working concentrations of TiO2 nanoparticles should not exceed 3 wt.%.

The dependence of the elastic modulus on the concentration of TiO2 nanoparticles is given in figure 6. At concentrations above 1 wt.%, the increase of the elastic modulus is 20–25%, which indicates the effect of antiplasticization of the polymer composite materials.

The change in the relative deformation with varying nanofiller concentration is complex: the deformation increases at TiO2 concentrations less than 1 wt.%, and decreases at TiO2 concentrations above 1 wt.%.

Thus, it was demonstrated that mechanical properties depend on the concentration of TiO2 nanoparticles. The values of the elastic modulus and tensile strength are maximum at a nanoparticle concentration of 3 wt.%. The relative deformation increases with increasing concentration up to 1 wt.%.

4. Conclusions

Polymer composite materials based on epoxy resin and titanium dioxide nanoparticles have been obtained by introducing synthesized nanoparticles into epoxy resin at the stage of its curing, and the effect of nanoparticle concentration on physical and mechanical properties of obtained polymer composite materials has been investigated. The physical and mechanical properties of

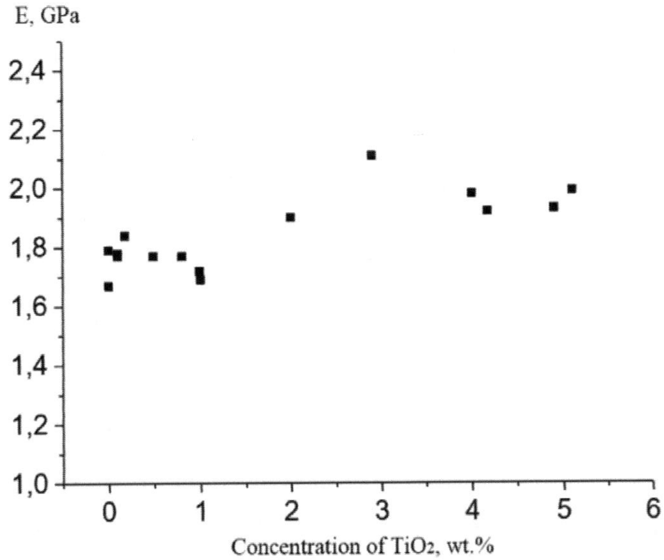

Figure 6. Dependence of the elastic modulus of TiO2/EP polymer composite materials on the concentration of TiO2 nanoparticles.

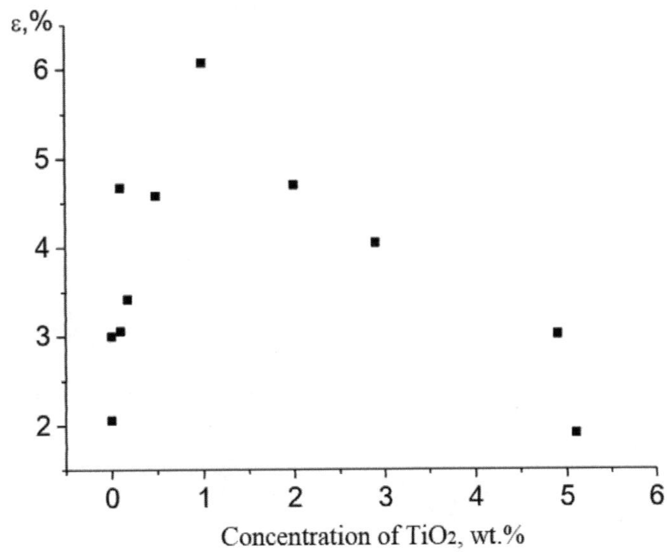

Figure 7. Dependence of the relative deformation of the TiO2/EP polymer composite materials on the TiO2 nanoparticles concentration.

epoxy polymer composite materials have been investigated. It is shown that the introduction of 1 wt.% of TiO2 nanoparticles leads to an increase in the relative deformation by 35–40%. The maximum tensile strength values are achieved at a concentration of 3 wt.% TiO2. The increase in the elastic modulus is 20–25% at TiO2 nanoparticles concentrations was observed above 1 wt.%. Improved mechanical properties indicate the formation of strong interfacial interactions.

Acknowledgments

This work has been carried out with the financial support of Russian Science Foundation, project No. 20-19-00395.

References

[1] Jin F L, Li X, and Park S J 2015 Synthesis and application of epoxy resins: A review *Journal of Industrial and Engineering Chemistry* **29** 1–11 doi: 10.1016/j.jiec.2015.03.026

[2] Jin F L, Ma C J, and Park S J 2011 Thermal and mechanical interfacial properties of epoxy composites based on functionalized carbon nanotubes *Materials Science and Engineering: A* **528** (29-30) 8517–22 doi: 10.1016/j.msea.2011.08.054

[3] Jin F L and Park S J 2006 Thermal properties and toughness performance of hyperbranched–polyimide–modified epoxy resins *Journal of Polymer Science Part B: Polymer Physics* **44** (23) 3348–56 doi: 10.1002/polb.20990

[4] Ahmad Z, Ansell M P, and Smedley D 2010 Effect of nano- and microparticle additions on moisture absorption in thixotropic room temperature cure epoxy-based adhesives for bonded-in timber connections International journal of adhesion and adhesives **30** (6) 448–55 doi: 10.1016/j.ijadhadh.2010.04.001

[5] Tucker S J, Fu B, Kar S, et al. 2010 Ambient cure POSS–epoxy matrices for marine composites *Composites Part A: Applied Science and Manufacturing* **41** (10) 1441–6 doi: 10.1016/j.compositesa.2010.06.005

[6] Yang C and Yang Z G 2013 Synthesis of low viscosity, fast UV curing solder resist based on epoxy resin for ink-jet printing *Journal of Applied Polymer Science* **129** (1) 187–92 doi: 10.1002/app.38738

[7] Bulychev N A, Kazaryan M A, Gridneva E S, et al. 2012 Plasma discharge with bulk glow in the liquid phase exposed to ultrasound *Bulletin of the Lebedev Physical Institute* **39** (7) 214–20 doi: 10.3103/S1068335612070056

[8] Bulychev N A, Kazaryan M A, Chaikov L L, et al. 2014 Nanoscale metal oxide particles produced in the plasma discharge in the liquid phase upon exposure to ultrasonic cavitation. 1. Method for producing particles *Bulletin of the Lebedev Physical Institute* **41** (9) 264–8 doi: 10.3103/S106833561409005X

[9] Klassen N V, Krivko O A, Kedrov V V, et al. 2010 Laser and electric ar synthesis of nanocrystalline scintillators *IEEE Transactions on Nuclear Science* **57** (3) 1377–81 doi: 10.1109/TNS.2010.2040037

[10] Ganiev R F, Bulychev N A, Fomin V N, et al. 2006 Effect of mechanical activation on surface modification in aqueous pigment disperse systems *Doklady Chemistry* **407** (2) 54–6 doi: 10.1134/S0012500806040045

On the influence of an elastic film to the attenuation of wave motion between viscous and moving ideal fluid

D Belonozhko[1*] **and A Ochirov**[1,2**]
[1]P.G. Demidov Yaroslavl State University, Yaroslavl, Russia
[2]Ishlinsky Institute for Problems in Mechanics RAS, Moscow, Russia
E-mail: *belonozhko@mail.ru, **otchirov@mail.ru

Abstract. The regularities of wave motion along the interface of viscous and ideal liquids along which the surfactant is distributed are investigated in theory. The relationship between the decay decrements of capillary–gravitaty waves and the nature of maxima and minima surfactant concentration redistribution is established. The phenomenon has been analized in depend on the film elasticity in the presence of tangential velocity discontinuity at the boundary of the media. Influence of the tangential velocity gap on the relationship is established.

1. Introduction

Surfactants can significantly affect the wave motion realized on the free surface of a liquid or at the interface of two liquids. It is well known about the damping effect of the surfactant film on capillary-gravitaty waves. The theoretical study of this phenomenon began more than half a century ago with the pioneering works of V. G. Levich [1, 2] and was soon continued by other authors [3, 4]. It turned out that in the presence of a surfactant film on the liquid surface there are two types of wave movements: the well-known capillary-gravitaty waves and Marangoni waves caused by alternating longitudinal compressions and stretches of the elastic film [3, 4]. It is established [3, 4] that capillary-gravity waves are most strongly attenuated only at a certain value of the elasticity of the surfactant film. At this value of elasticity the values of the frequency of capillary-gravitaty waves and Marangoni waves coincide.

In the last decade flow properties have been established that are directly responsible for the intensification of dissipative processes in the system at the most optimal values of the elasticity of the surfactant film [5]; the effect of amplification of long waves caused by the presence of a surfactant film has been discovered [6]; the dispersion relations of wave motion have been studied in detail [7]. In the study [8], the regularities of a water thin layer evaporation in the presence of surfactants are considered. In the last few years the vertical structure of the velocity field flow was analyzed in the near-surface boundary layer adjacent to the film PAV [9]; surface wave transport of pollution [10] was concidered; stability of flow in liquid films [11] was investigated; the problem of correctly accounting for viscous surface tension in the presence of the surfactant film was discussed [12].

In a recent paper [13] the relationship between the redistribution of the surfactants surface density and the capillary-gravitaty waves decay related to the presence of the film was considered in detail. Present study is a development of the work [13]. The aim of the study is to analyze the regularities of the redistribution of surfactants along the two environment interface depending on the magnitude of the tangential discontinuity of the velocity field.

2. Problem statement

Let us consider a system of two incompressible fluids in a Cartesian coordinate system. The denser lower fluid has a density ρ and a kinematic viscosity ν. The upper one is ideal with a density of $\rho' < \rho$. The Oz axis is directed vertically upwards against the direction of the gravity field $\mathbf{g}$. The Oxy plane sets the position of the equilibrium interface. An insoluble surfactant film has an equilibrium surface concentration Γ_0. The film is distributed over the interface. A traveling sinusoidal wave has a wave number k, an amplitude ζ and a circular frequency ω. The wave propagates along the positive direction of the Ox axis. It causes a redistribution of the surface concentration $\Gamma \equiv \Gamma(x,t)$. The liquid surface and the surfactant film are in thermodynamic equilibrium. Local changes in the surface concentration of surfactants cause instantaneous local changes in the surface tension coefficient in accordance with the isotherm $\gamma = \gamma(\Gamma)$ supposed to be known [4]. The surface tension coefficient changes along the disturbed surface and thereby generates a surface distribution of tangential elastic forces effected on the dynamics of wave motion. In the case of small amplitude waves the elastic properties of the surfactant film are directly depend on the tangent of the slope angle $\chi = (d\gamma/d\Gamma)_0$ of the isotherm $\gamma(\Gamma)$ at the point $\Gamma = \Gamma_0$. For surfactants one has: $\chi < 0$. In addition we assume that the upper fluid translationally moves relative to the lower one in the positive direction of the Ox axis with a constant velocity of U_0. Let us consider the problem of calculating the effect of the fluids relative motion on the surfactant redistribution and on the attenuation of capillary-gravitaty waves depending on the elastic properties of the surfactant film. The mathematical formulation of the problem of determining the hydrodynamic potential of the upper fluid φ' and the velocity field of the lower fluid $\mathbf{U} = u\mathbf{e}_x + v\mathbf{e}_z$ is as follows:

$$z > \xi: \quad \Delta\varphi' = 0, \quad p' = p_0 - \rho' g z - \rho' \partial_t \varphi' - \frac{\rho'}{2}[(\partial_x \varphi' + U_0)^2 + (\partial_z \varphi')^2], \tag{1}$$

$$z < \xi: \quad \partial_t \mathbf{U} + (\mathbf{U}\cdot\nabla)\mathbf{U} = -\frac{1}{\rho}\nabla p + \nu\Delta\mathbf{U} + \mathbf{g}, \quad \nabla\cdot\mathbf{U} = 0, \tag{2}$$

$$z = \xi: \quad \partial_t \xi + u\partial_x \xi = v, \quad \partial_t \xi + \partial_x \varphi' \partial_x \xi = \partial_z \varphi', \tag{3}$$

$$p - p' - 2\rho\nu(\mathbf{n}\cdot(\mathbf{n}\cdot\nabla)\mathbf{U}) = -\frac{\gamma}{[1+(\partial_x\xi)^2]^{3/2}}\partial_{xx}\xi, \tag{4}$$

$$-\rho\nu[(\boldsymbol{\tau}\cdot(\mathbf{n}\cdot\nabla)\mathbf{U}) + (\mathbf{n}\cdot(\boldsymbol{\tau}\cdot\nabla)\mathbf{U})] + \frac{1}{\sqrt{1+(\partial_x\xi)^2}}\partial_x\gamma = 0, \tag{5}$$

$$\partial_t\Gamma + \frac{1}{1+(\partial_x\xi)^2}\{\partial_x(\Gamma u) + [\Gamma\partial_z u + \partial_x(v\Gamma)]\partial_x\xi + \Gamma(\partial_x\xi)^2\partial_z v\} = 0, \tag{6}$$

$$z \to -\infty: \quad u \to 0, \quad v \to 0, \qquad z \to \infty: \quad \nabla\varphi' \to 0, \tag{7}$$

$$\mathbf{n} = (-\partial_x\xi\mathbf{e}_x + \mathbf{e}_z)[1+(\partial_x\xi)^2]^{-1/2}, \quad \boldsymbol{\tau} = (\mathbf{e}_x + \partial_x\xi\mathbf{e}_z)[1+(\partial_x\xi)^2]^{-1/2}.$$

Here v and u are the vertical and horizontal components of the velocity of the lower fluid; $\mathbf{e}_x$ and $\mathbf{e}_z$ are orts of the axes Ox and Oz respectively; p and p' are hydrodynamic pressures in the lower and upper fluids; $\mathbf{n}$ and $\boldsymbol{\tau}$ are orts of the external normal to the interface and tangent to it.

The mathematical formulation includes standard kinematic (3) and dynamic (4) boundary conditions on the interface. The expression (6) is the continuity equation for the surface redistribution of the surfactant film. Its derivation can be found in [14].

In assuming the ratio of the wave amplitude to its length is a small parameter the expressions (1)–(7) transforms to the problem of calculating quantities of the first order of smallness (here and further they are marked with the subscript '1'):

$$\begin{aligned} &z>0: \quad \Delta\varphi_1' = 0, \\ &z<0: \quad \partial_t \mathbf{U}_1 + \frac{1}{\rho}\nabla p_1 - \nu\Delta\mathbf{U}_1 = 0, \quad \nabla\cdot\mathbf{U}_1 = 0, \end{aligned} \tag{8}$$

$$\begin{aligned} z=0: \quad &\partial_t\xi_1 - v_1 = 0, \quad \partial_t\xi_1 + U_0\partial_x\xi_1 - \partial_z\varphi_1' = 0, \\ &-\rho g\xi_1 + \rho' g\xi_1 + \rho'\partial_t\varphi' + \rho'\partial_x\varphi' U_0 + p_1 - 2\rho\nu\partial_z v_1 + \gamma_0\partial_{xx}\xi_1 = 0, \\ &-\rho\nu(\partial_z u_1 + \partial_x v_1) + \chi\partial_x\Gamma_1 = 0, \quad \partial_t\Gamma_1 + \Gamma_0\partial_x u_1 = 0, \end{aligned} \tag{9}$$

$$\begin{aligned} &z\to-\infty: \quad u_1\to 0, \quad v_1\to 0, \\ &z\to\infty: \quad \nabla\varphi_1'\to 0. \end{aligned} \tag{10}$$

3. Solving the problem of the first order of smallness

The solution of the problem of the first order of smallness (8)–(10) is derived by standard methods and is a superposition of traveling waves with different wave numbers k:

$$\begin{pmatrix} \xi_1 \\ u_1 \\ v_1 \\ \Gamma_1 \\ \varphi_1' \end{pmatrix} = \frac{\zeta}{2}\begin{pmatrix} 1 \\ A\exp(kz) + B\exp(qz) \\ i[A\exp(kz) + Bk\exp(qz)/q] \\ ik\Gamma_0(A+B)/S \\ (iU_0 - S/k)\exp(-kz) \end{pmatrix}\exp(St - ikx) + \text{C.C.}, \tag{11}$$

$$A = iS\left[\frac{q}{k-q} - \frac{kS\nu\rho}{k^2\Pi - (k+q)S\nu\rho}\right], \quad B = -\frac{ikqS(k\Pi - 2S\nu\rho)}{(k-q)[k^2\Pi - (k+q)S\nu\rho]},$$

$$q = \sqrt{k^2 + \frac{S}{\nu}}, \quad \Pi = \Gamma_0\chi, \quad \chi = \left(\frac{d\gamma}{d\Gamma}\right)_{\Gamma=\Gamma_0}.$$

Here, the symbol C.C. denotes complex conjugate terms. The complex frequency of the wave motion S is determined from the dispersion relation:

$$\begin{aligned} &(S + 2k^2\nu)^2 + \omega_0^2 + \frac{k^3\Pi\omega_0^2}{S^2\rho} + (S - ikU_0)^2\frac{\rho'(k^3\Pi + S^2\rho)}{S^2\rho^2} + \frac{gk\rho'(k^3\Pi - S^2\rho)}{S^2\rho^2} \\ &= q\left\{4k^3\nu^2 + \frac{k^2\Pi}{\rho}\left(1 + \frac{\omega_0^2}{S^2}\right) - \frac{k^2\Pi\rho'[kg - (S - ikU_0)^2]}{S^2\rho^2}\right\}, \quad \omega_0^2 = kg\left(1 + k^2\frac{\gamma_0}{\rho g}\right). \end{aligned} \tag{12}$$

The real part of the complex frequency $\text{Re}(S)$ is negative and describes the wave decrement. The imaginary part $\text{Im}(S)$ is responsible for the frequency of the wave motion. The auxiliary parameter ω_0^2 has the meaning of the circular frequency square of capillary-gravity waves on the surface of an infinitely deep ideal fluid with a capillary length $\alpha = \sqrt{\gamma_0/(\rho g)}$. The value of the surface tension coefficient $\gamma_0 = \gamma(\Gamma_0)$ corresponds to the equilibrium concentration of surfactants. The parameter q is included in the multiplier $\exp(qz)$. It responcible for the vortex component of the velocity of the lower fluid.

The dispersion relation (12) is irrational with respect to the complex frequency S. Irrationality can be eliminated by squaring both parts of (12). The new equation may contain "extra" roots. For all the roots of this equation it is possible to derive a rather cumbersome analytical expression. The selection of physically correct roots have to carried out as follows. The parameter q is expressed from the relation (12). Result is auxiliary expression for q. The verifiable root is substituted into the auxiliary expression. If the real part of the auxiliary expression turned out to be positive the root is valid (for it, the expression $\exp(qz)$ attenuates at depth at $z \to -\infty$). Otherwise the root is discarded.

The parameter Π characterizes the elastic properties of the film and for surfactants is negative as well as the coefficient χ. It should be noted that the solution of the problem depends on the product of $\Gamma_0\chi$ but not on each of the parameters separately. A particular type of dependency $\gamma = \gamma(\Gamma)$ (for determination $\gamma_0 = \gamma(\Gamma_0)$ and $\chi = (d\gamma/d\Gamma)_0$) is rather cumberstone question. It is quite difficult to specify the type of this dependence without going into some purely special cases (see [15, 16] and the literature cited there). Meanwhile, various sources emphasize some common features for most surfactants. Increase the surfactant concentration leads to a change in $|\Pi|$ from zero (in the absence of surfactants) to values that are usually several times higher than the so-called surface pressure $\pi = \gamma_f - \gamma_0$. The surface tension coefficient of a liquid without surfactants is denoted by γ_f (see [17–20] and the literature cited there). In fact π is a decrease in surface tension due to the presence of surfactants. In connection with above it was decided to analyze the present problem in terms of the dependence of the solution on the most general parameter equal to $|\Pi|/\gamma_0$. The parameter varies from zero to at least one. In the next paragraph the value of this parameter will be called the dimensionless value of the elastic modulus of the surfactant film.

4. Analysis of the solution

The solution of the dispersion relation (12) describes capillary-gravity waves and Marangoni waves. All the calculations below are performed in dimensionless variables. The fluid density, the equilibrium value of the surface tension coefficient and the acceleration of the gravity field are taken as basic parameters of the dimensionless system. In the system of dimensionless variables we have $\rho = g = \gamma_0 = 1$ and other physical quantities keep be denoted by the former symbols but are measured in units of proper characteristic scales:

$$k_{\text{scl}} = \sqrt{\frac{\rho g}{\gamma_0}}, \quad S_{\text{scl}} = \sqrt[4]{\frac{\rho g^3}{\gamma_0}}, \quad \nu_{\text{scl}} = \sqrt[4]{\frac{\gamma_0^3}{\rho^3 g}}, \quad U_{0\text{scl}} = \sqrt[4]{\frac{\gamma_0 g}{\rho}}, \quad \rho'_{\text{scl}} = \rho, \quad \Pi_{\text{scl}} = \gamma_0.$$

The dimensionless viscosity is taken to be $\nu = 0.002$. In dimension variables it corresponds to viscosity of water. The lower environment is assumed to be 1000 times denser than upper one: $\rho' = 0.001$ (that is same as in the case of water and air).

For the integrity and visibility of the narrative the fitting structure of analysis was chosen. Preliminary the attention is taken for one known but not obviously phenomenon. The phenomenon is illustrated by figures 1 and 2. Additionally figure 3 describes one recently discovered property of the phenomenon. Figures 1–3 are performed at $U_0 = 0$. They are necessary foundation for the transition to the formulation of our new results illustrating by figures 4–6. The new results relate to behavior if system at different values $U_0 > 0$. All figures are plotted as result of calculations by means of dispersion relation (12) with using the chosen dimensionless system of variable.

Figure 1 shows the dependence of the dimensionless frequency $\omega \equiv \text{Im}(S)$ on the dimensionless modulus of elasticity of the film $|\Pi|$. The plot is performed for capillary-gravity waves (denoted by C-G) and Marangoni waves (denoted by M). The dependencies are plotted for different values of the dimensionless wavenumber k. It can be seen that for each value of the wavenumber k there is a certain special value of the elastic modulus of surfactant film at which the frequency of the capillary-gravity waves is equal to the frequency of the Maragoni waves (see points A, B, C). The special values of elastic modulus are denoted at the figure by symbols $|\Pi_A^*|$, $|\Pi_B^*|$, $|\Pi_C^*|$. It is known that if the value of the elasticity modulus is special (in the above-described sense) the decrement of capillary-gravity waves comes to the peak value. The fact is illustrated by plots at the figure 2. Here one can see the dependence of the dimensionless decrement $r \equiv |\text{Re}(S)|$ on

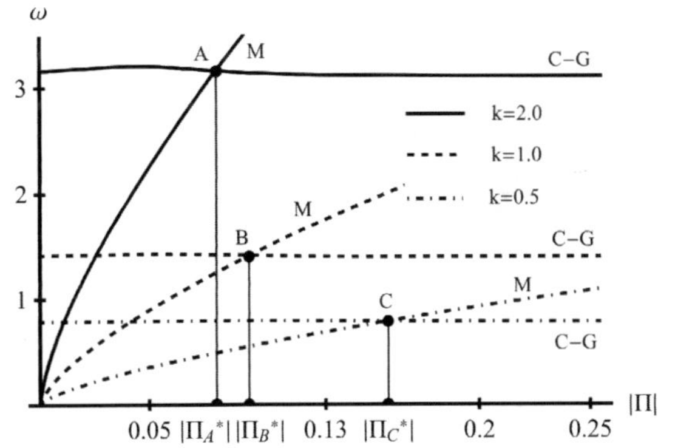

Figure 1. Dependence of dimensionless frequencies of capillary–gravity waves and Marangoni waves on the elasticity of the surfactant film for waves with different dimensionless wave numbers in the absence of tangential velocity gap.

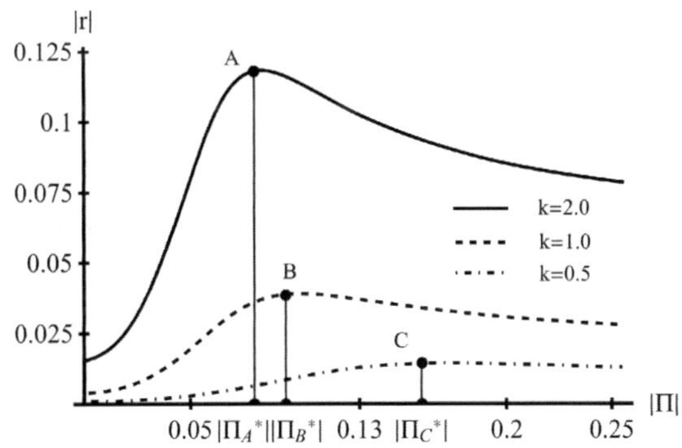

Figure 2. Dependence of dimensionless decay decrements of capillary–gravitaty waves on the elasticity of the surfactant film for waves with different dimensionless wave numbers in the absence of tangential velocity gap.

the dimensionless modulus of elasticity of the film $|\Pi|$. Only the decrements of capillary-gravity waves are shown since decrements of Marangoni waves increase extremely and merge with a vertical axis on the scale of figure 2. Though the phenomenon of reinforcement of capillary-gravity waves damping at special value of surfactant elasticity is known and repeatedly have discussed some new features of the phenomena are discovering so far still. In a recent study [13] it was shown that with the of maximum damping of capillary-gravity waves the surfactant concentration maximum shifts to the position at the middle of the front slope of the wave (in the direction of the wave propagation). The figure 3 illustrates foresaid. The horizontal axe is the dimensionless modulus of elasticity of the film $|\Pi|$ and vertical one is phase shift Φ between the wave of surfactant concentration thickening Γ_1 and wave deviations of interface form equilibrium horizontal level. It can be seen that if the elasticity of the surfactant film takes its special significance the value Φ comes to $\pi/2$. This means the surfactant concentration maximum turns up at the middle of the wave slope.

Now let proceed to the analysis of the upper medium velocity of the horizontal motion influence on the phenomena described above. It is known that the presence of a tangential discontinuity of the velocity field on the interface between liquids causes instability of wave disturbances. In presented formulation the effect of both gravitational forces and surface tension forces is taken into account. Under these conditions the instability of the tangential discontinuity of the velocity field has a threshold. The instability occurs only if the velocity of the upper environment U_0 becomes greater than the critical value $U_0 > W$. To estimate the threshold value W it is expedient to turn to the classical model of instability of the velocity field tagential discontinuity at the horizontal interface between two ideal incompressible fluids without any film of surfactant. The model with taken into account the gravity field and the surface tension force gives the known expression for W_k. The index k means that the expression is individual for each value of wavenumber. In the system of our dimensionless variables we have the expression:

$$W_k = \sqrt{\frac{k^2(1+\rho') + (1-\rho'^2)}{k}}.$$

Analyzing the above expression one can establish the notable fact. For all values of wavenumbers the inequality is valid $W_k \geq W$, where $W = \min_k W_k = W_{k=K}$ with $K = 1 - \rho'$. The "capital"

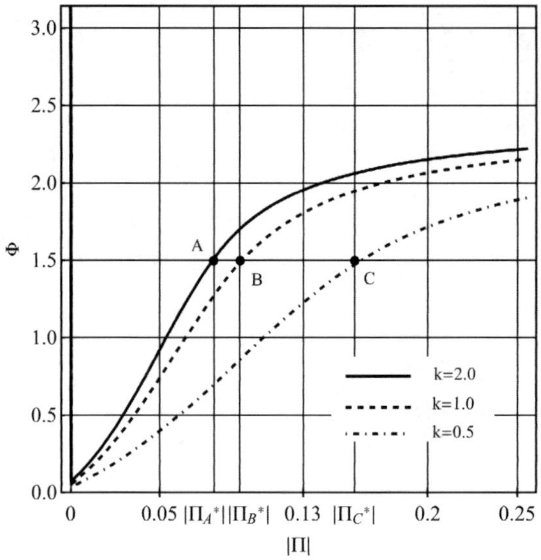

Figure 3. The dependence of the phase shift between the position of the maximum concentration of surfactants and the crest of the wave on the elasticity of the surfactant film for waves with different dimensionless wave numbers in the absence of a tangential velocity gap.

wavenumber $K = 1 - \rho'$ is most sensible to the instability since corresponding value $W = W_{k=K}$ (the instability occurs if upper environment exceed this threshold of velocity) is minimum among all possible values. In our case $\rho' = 0.001$. We have $K = 1 - \rho' = 1 - 0.001 \simeq 1$ and $W = W_{k=1} \simeq 45$. The object of our research is waves with an indestructible shape. Therefore in all our calculations the velocity value U_0 varies in the range $0 \leq U_0 < W = 45$. The main attention is taken for the value of wavenumber $k = K \simeq 1$ as corresponding to wave being most sensible to changing velocity U_0. If lower environment is water and upper one is air then the dimensionless value $W \simeq 45$ corresponds dimensional value 7.3 m/s. Wherein the dimensionless value $K = 1$ corresponds dimensional value of wavelength 1.7 cm.

Analysis of the dispersion relation discovered that the capillary-gravity wave frequency decreases markedly with an increase in the magnitude of the tangential velocity fields gap at the interface. At the same condition the frequency of the Marangoni waves changes rather weakly. This fact is illustrated by figure 4. The figure shows the dependences similar to figure 1. It is plotted for a waves with a dimensionless wavenumber $k = 1$ at various dimensionless values of the horizontal velocity of upper environment $0 < U_0 < W$. The dependences corresponding to the frequency of Marangoni waves merge in one curve. It is clear that the greater the velocity of the environments relative motion the lower the special value of elasticity (the value necessary for coinciding of frequencies of two types of wave motions). Recall that the special value of elasticity sets the conditions for the strongest damping of capillary-gravity waves. It turns out that the larger the velocity U_0 the lower the elasticity providing the most efficient damping of capillary-gravity waves.

Examples of decrements calculating are presented at figure 5. The dependencies are similar with those that are shown by solid line at figure 2, but it is plotted at various dimensionless values of the horizontal velocity U_0. One can see that if velocity U_0 exceed a slight decrease in the damping decrement is observed. It is important to understand that the elasticity module of the surfactant film significantly increases with the equilibrium surface concentration Γ_0 of surfactant increasing. For example let consider the system water-air. If wind appears above the surface of water covered with an elastic surfactant film the maximum effect of centimeter waves

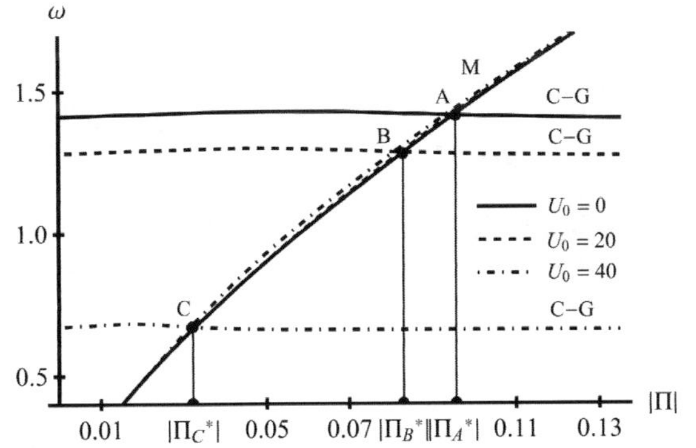

Figure 4. The dependence of dimensionless frequencies of capillary–gravity waves and Marangoni waves with a dimensionless wavenumber $k = 1$ on the elasticity of the surfactant film for different dimensionless values U_0.

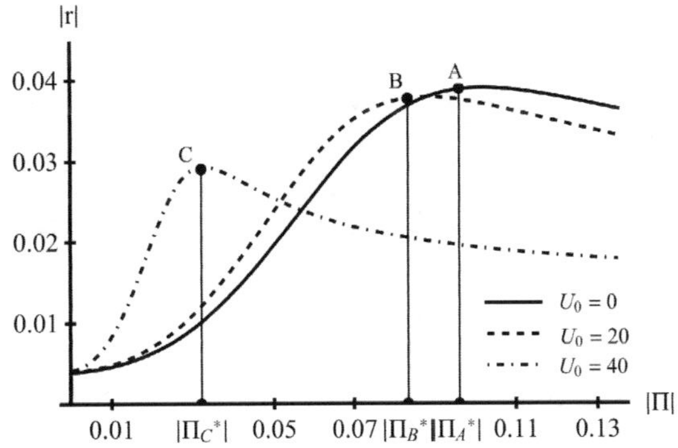

Figure 5. Dependence of dimensionless decay decrements of capillary–gravity waves with a dimensionless wavenumber $k = 1$ on the elasticity of the surfactant film for different dimensionless values U_0.

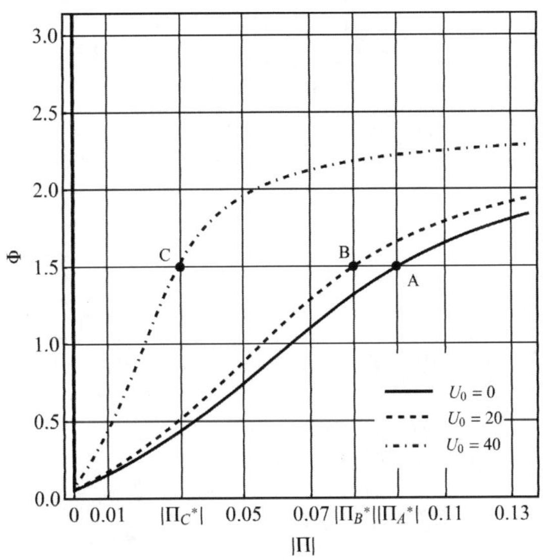

Figure 6. The dependence of the phase shift between the position of the maximum concentration of surfactants and the crest of a wave with a dimensionless wavenumber $k = 1$ on the elasticity of the surfactant film at different dimensionless values U_0.

damping is observed at greatly lower surface surfactant concentrations than in the absence of wind. Of course the decrement is also reduced but not too significant compared to damping without wind.

It is interesting again to turn to behavior of phase shift Φ between the wave of surfactant concentration thickening Γ_1 and wave deviations of interface from horizontal level on depending on of the film elasticity module $|\Pi|$. The example of the dependencies plots at figure 6. Ut illustrates that for all considered velocity values U_0 the phase shift $\Phi = \pi/2$ corresponds to the special value of the elasticity module. In a qualitative sense the plots is repeated the figure 3. Thus when with wind as well as without it the position of the maximum surfactant concentration indicate of damping efficiency of capillary-gravity waves at set equilibrium surface concentration Γ_0. If the maximum concentration of surfactant is close to the top of the wave the damping effect can be enhanced by increasing the surface concentration of the surfactant. If the maximum concentration of the surfactant is located closer to the trough of the wave to enhance the effect of damping it is necessary to decrease the concentration of the surfactant. If the maximum concentration is near the middle the front slope of the wave the maximum damping effect for this type of surfactant has already been achieved and it can be enhanced only if the surfactant itself is replaced.

5. Conclusion

It is known that in presence of a surfactant film on the horizontal surface of a liquid it is possible to achieve a film elasticity value at which the effect of capillary-gravity waves damping by this film becomes the most significant. At the same time the maxima of the surfactant concentration and the positions of the wave humps alternate in a strictly defined way. At each spatial period the maximum concentration of surfactant is located in the center of the front slope of the wave (in the direction of propagation). In the present study it has been shown that this pattern persists in a more complex situation when an upper less dense liquid moves above the surfactant film-coated lower liquid. It turned out that the effect of maximum capillary-gravity waves damping with a given wavelength depend on horizontal velocity of the upper liquid (without reaching the values beyond the limit in the sense of the beginning of the tangential discontinuity instability). The lager the horizontal velocity of the upper liquid the lower value of the elasticity of the surfactant film is required for most efficient damping of capillary gravity waves. Wherein the value of the decrement turns out to be somewhat less than those one without moving upper environment and the maximum concentration of surfactant is always located at middle of the front slope of the wave. In laboratory studies it can serve as an indicator of achieving the maximum effect of damping waves for a given type of surfactant.

Acknowledgments

The study was supported by the Government program (contract No. AAAA-A20-120011690131-7).

References

[1] Levich V G 1941 Damping of waves by surface-active substances. I *Acta Physicochimica U.R.S.S.* **14** 307–28
[2] Levich V G 1962 *Physicochemical Hydrodynamics* (New Jersey: Prentice-Hall, Englewood Cliffs)
[3] Lucassen J 1968 Longitudinal capillary waves. Part 1. – Theory *Transactions of the Faraday Society* **64** 2221–29 doi: 10.1039/TF9686402221
[4] Lucassen–Reynders E N and Lucassen J 1969 Properties of capillary waves *Advances in Colloid and Interface Science* **2** (4) 347–95 doi: 10.1016/0001-8686(70)80001-X
[5] Ermakov S A 2003 Resonance damping of gravity-capillary waves on the water surface covered with a surface-active film *Izvestiya, Atmospheric and Oceanic Physics* **39** (5) 624–28
[6] Ermakov S A 2010 On the intensification of decimeter-range wind waves in film slicks *Izvestiya, Atmospheric and Oceanic Physics* **46** (2) 208–13 doi: 10.1134/S0001433810020088
[7] Averbukh E L, Kurkin A A, Stepanyants Y A, and Talipova T G 2014 Dispersion properties of waves on the surface of a viscous fluid covered by an elastic film *Fluid Dynamics* **49** (6) 796–803 doi: 10.1134/S0015462814060118
[8] Gordeeva V Yu and Lyushnin A V 2014 Peculiarities of evaporation of a thin water layer in the presence of a solvable surfactant *Technical Physics* **59** (5) 656–62 doi: 10.1134/S1063784214050090
[9] Averbukh E L, Talipova T G, and Kurkin A A 2015 Vertical structure of the velocity field in the waves on the surface of a viscous liquid covered by an elastic film *Fluid Dynamics* **50** (3) 442–52 doi: 10.1134/S0015462815030138
[10] Strickland S L, Shearer M, and Daniels K E 2015 Spatiotemporal measurement of surfactant distribution on gravitycapillary waves *Journal of Fluid Mechanics* **777** 523–43 doi: 10.1017/jfm.2015.352

[11] Karapetsas G and Bontozoglou V 2014 The role of surfactants on the mechanism of the long-wave instability in liquid film flows *Journal of Fluid Mechanics* **741** 139–55 doi: 10.1017/jfm.2013.670

[12] Elfring G J, Leal L G, and Squires T M 2016 Surface viscosity and Marangoni stresses at surfactant laden interfaces *Journal of Fluid Mechanics* **792** 712–39 doi: 10.1017/jfm.2016.96

[13] Belonozhko D F and Ochirov A A 2018 Mutual influence of wave motion and patterns of surfactant distribution *Bulletin of the Russian Academy of Sciences: Physics* **82** (1) 40–4 doi: 10.3103/S1062873818010045

[14] Belonozhko D F and Grigor'ev A I 2004 On the correct writing of the law of conservation of amount of substance at the moving fluid-fluid interface *Technical Physics* **49** (11) 1415–21 doi: 10.1134/1.1826184

[15] Fainerman V B, Lucassen-Reynders E H, and Miller R 1998 Adsorption of surfactants and proteins at fluid interfaces *Colloids and Surfaces A: Physicochemical and Engineering Aspects* **143** (2-3) 141–65 doi: 10.1016/S0927-7757(98)00585-8

[16] Guzmán E, Llamas S, Maestro A, et al. 2016 Polymersurfactant systems in bulk and at fluid interfaces *Advances in Colloid and Interface Science* **233** 38–64 doi: 10.1016/j.cis.2015.11.001

[17] Zhou Z-H, Ma D-S, Zhang Q, et al. 2018 Surface dilational rheology of betaine surfactants: Effect of molecular structures *Colloids and Surfaces A: Physicochemical and Engineering Aspects* **538** 739–47 doi: 10.1016/j.colsurfa.2017.11.064

[18] Lucassen-Reynders E H, Cagna A, and Lucassen J 2001 Gibbs elasticity, surface dilational modulus and diffusional relaxation in nonionic surfactant monolayers *Colloids and Surfaces A: Physicochemical and Engineering Aspects* **186** (1-2) 63–72 doi: 10.1016/S0927-7757(01)00483-6

[19] Ravera F, Ferrari M, Santini E, and Liggieri L 2005 Influence of surface processes on the dilational visco-elasticity of surfactant solutions *Advances in Colloid and Interface Science* **117** (1-3) 75–100 doi: 10.1016/j.cis.2005.06.002

[20] Rippner Blomqvist B, Ridout M J, Mackie A R, et al. 2004 Disruption of viscoelastic β-lactoglobulin surface layers at the airwater interface by nonionic polymeric surfactants *Langmuir* **20** (23) 10150–58 doi: 10.1021/la0485475

On a pseudo-periodicity in the elastodynamics

S P Kurdina[1*] **and H A Gevorgyan**[2]
[1]Bauman Moscow State Technical University, Moscow, Russia
[2]Institute of Mechanics of the National Academy of Sciences of Armenia, Yerevan, Republic of Armenia

E-mail: *kurdinasp@bmstu.ru

Abstract. The periodic laws of motion of a real mechanism imply the periodicity of all dynamic characteristics of the mechanism during such a movement, including the elastic displacements of its constituent links. This statement covers the most difficult problem of the dynamics of an elastically deformable body. However the existing publications devoted to the study of flexible mechanisms do not mention the periodicity of elastic displacements of the links when the mechanisms move according to periodic laws. This article discusses the periodicity and pseudo-periodicity of the elastic displacements of the links using examples of calculations of the motion of a double pendulum and a slider–crank mechanism without taking into account dissipative forces.

1. Introduction

In recent decades, in a variety of publications, progressive formalisms have appeared that allow efficient numerical solution (with characteristic complexity $O(n)$ of direct and inverse problems of the dynamics of systems with rigid links without inverting their mass matrices [1, 2]. The strategy for efficient numerical calculation of multilink dynamical systems of a simple open kinematic chain is called Newton–Euler method. Special attention has been paid to the study of the deformations of the executive links of real mechanisms. Generalizations of effective formalisms for the case of elastic links of the kinematic chain are given in [3–6]. Recursive symbolic model for the numerical solution of the first problem of manipulators dynamics with elastic links is given in [4]. Such a model for the case of a mixed dynamics problem for the same simplest class of multilink systems is given in [5]. The Newton–Euler method, applied and generalized for this case, was called generalized Newton–Euler method [4]. In work [6], the generalized Newton–Euler method was applied for numerically solve the mixed problem of the dynamics of tree-like multilink systems without external connections.

One of the most famous dynamical systems are pendulums and, in particular, double pendulum. Studies of pendulum oscillations are investigated, for example, in works [7, 8]. At the same time, one of the most common mechanisms in technology is the slider–crank mechanism. Naturally, much attention is paid to these mechanisms in the scientific literature [5, 6, 9–13]. These article shows the periodicity of the change in the elastic displacement function of the second link of pendulum in time and the pseudo-periodicity of the change in the elastic displacement function of the connecting rod in time in the case when the links of the corresponding rigid mechanisms move periodically. The second example shows that the period of pseudo-oscillations of the elastic displacement function can differ significantly from the period of revolution of the mechanism itself [14].

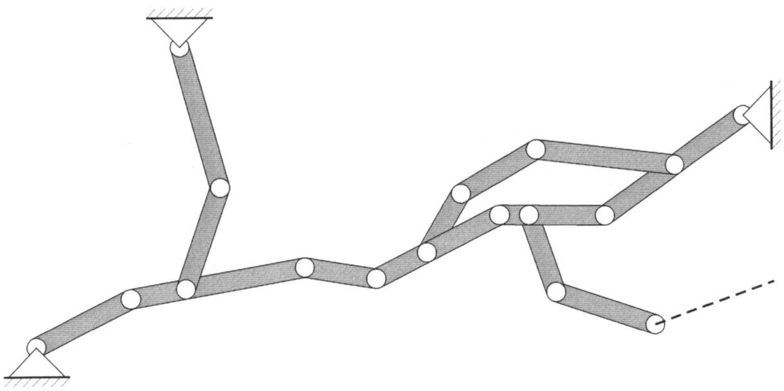

Figure 1. Scheme of the tree-like dynamic system with closed kinematic chains.

2. Traditional formulation of the problem

As is known, the process of dynamic analysis of elastic tree-like dynamic systems (figure 1) is reduced to a numerical solution of the system of nonlinear differential equations [9] with holonomic constraints (algebraic equations), relative to the generalized vector argument $\mathbf{x}$:

$$f_i(\mathbf{x}, \dot{\mathbf{x}}, \ddot{\mathbf{x}}, \boldsymbol{\lambda}, t) = 0, \quad g_j(\mathbf{x}, t) = 0, \quad i = 1, 2, \ldots, p, \quad p = \sum_{j=1}^{n} N_j + n, \quad j = 1, 2, \ldots, m, \tag{1}$$

where n is the number of generalized coordinates, m is number of additional links (holonomic constraints), and N_j is the number of elastic degrees of freedom of the jth link associated with neighboring links by means of rotational or translational hinges. In these equations generalized vector argument $\mathbf{x}$ and vector of indefinite Lagrange multipliers $\boldsymbol{\lambda}$ defined as follows

$$\mathbf{x} = \begin{pmatrix} \mathbf{x}_1 \\ \vdots \\ \mathbf{x}_n \end{pmatrix}, \quad \boldsymbol{\lambda} = \begin{pmatrix} \lambda_1 \\ \vdots \\ \boldsymbol{\lambda}_m \end{pmatrix}, \quad \mathbf{x}_j = \begin{pmatrix} q_j^{\mathrm{r}} \\ \mathbf{q}_j^{\mathrm{e}} \end{pmatrix}, \quad \mathbf{q}_j^{\mathrm{e}} = \begin{pmatrix} q_{j,1}^{\mathrm{e}} \\ \vdots \\ q_{j,n}^{\mathrm{e}} \end{pmatrix}, \quad j = 1, 2, \ldots, n. \tag{2}$$

Here q_j^{r} and $q_{j,i}^{\mathrm{e}}$ $(i, j = 1, 2, \ldots, n)$ are generalized coordinates and the symbols 'r' and 'e' express abbreviations for the words rigid and elastic. Examples of the implementation of dynamic analysis for other systems can be found, for example, in [15–17].

The article discusses the results of solving two problems: double pendulum elastodynamics problem and slider–crank mechanism elastodynamics problem. These problems are special cases of problem (1), and their solutions are built on the basis of the approach detailed in [14]. Article [14] presents a technique for the numerical dynamic analysis of elastic tree-like multilink systems, which combines the procedure for inverting the mass matrix with the procedure for effective kinematic calculation, developed on the basis of the generalized Newton–Euler method.

3. Numerical simulation of the double pendulum elastodynamics problem

Let us consider a numerical simulation of the double pendulum with a flexible second link (figure 2) performing free oscillations under the action of gravity. The beams has equal lengths $L_1 = L_2 = 1\,\mathrm{m}$, equal mass densities $\rho_1 = \rho_2 = 7800\,\mathrm{kg/m^3}$, and equal masses $m_1 = m_2 = 0.3\,\mathrm{kg}$. Young's modulus of second beam is $E_2 = 200 \times 10^9\,\mathrm{Pa}$. Initial deviations of links are $\varphi_1(0) = 0$, $\varphi_2(0) = -45°$, initial angular velocities of links are equal to zero.

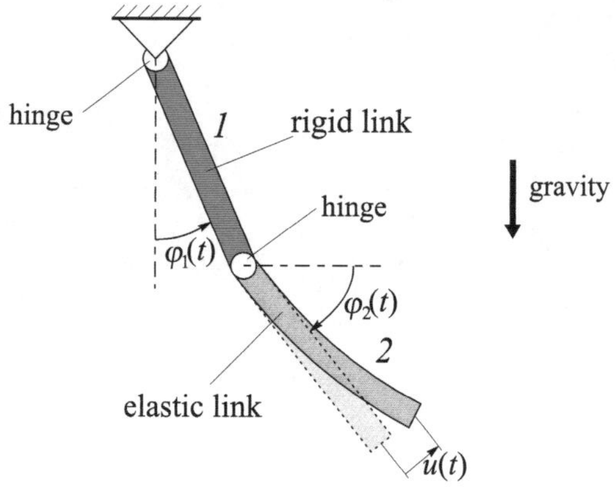

Figure 2. Scheme of the double pendulum problem.

Mathematical model for the double pendulum with rigid links is nonlinear and has a form

$$\begin{cases} [I_1 + I_2 + m_2 l_1^2 + m_2 l_1 l_2 \cos\varphi_2(t)]\ddot{\varphi}_1(t) + [I_2 + \frac{1}{2} m_2 l_1 l_2 \cos\varphi_2(t)]\ddot{\varphi}_2(t) \\ \quad -\frac{1}{2} m_2 l_1 l_2 [2\dot{\varphi}_1(t)\dot{\varphi}_2(t) + \dot{\varphi}_2^2(t)] \sin\varphi_2(t) + g(\frac{1}{2} m_1 l_1 + m_2 l_1)\cos\varphi_1(t) \\ \quad +\frac{1}{2} g m_2 l_2 \cos[\varphi_1(t) + \varphi_2(t)] = 0, \\ [I_2 + \frac{1}{2} m_2 l_1 l_2 \cos\varphi_2(t)]\ddot{\varphi}_1(t) + I_2\ddot{\varphi}_2(t) + \frac{1}{2} m_2 l_2 \{l_1 \dot{\varphi}_1^2(t) \sin\varphi_2(t) + g\cos[\varphi_1(t) + \varphi_2(t)]\} = 0, \end{cases} \tag{3}$$

where I_1 and I_2 are moments of inertia of the rods. Numerical calculations for this problem with parameters described above show that the oscillations of a double pendulum with rigid links have pseudo periodicity (figure 3*a*). A numerical solution for this is not difficult and can be obtained, for example, in any system of computer algebra. If the second link is elastic, then the solution can be obtained numerically only using special approaches. Note that the mathematical model in this case also becomes more complicated. To obtain a numerical solution for this problem, the effective approach described in paper [14] was used. Results of calculations presented in figure 3*b*. It can be notice that the movement of its end point will also oscillate with the same pseudo period.

4. Numerical solution for the slider–crank mechanism elastodynamics problem

Consider a numerical solution of the mixed problem of the dynamics of elastic mechanisms with external holonomic constraints using the dynamic simulation of a slider–crank mechanism with a rigid crank A_3A_4 and an elastic rod A_2A_3 without taking into account dissipative forces (figure 4). It is 4 generalized coordinates: $q_1^{\mathrm{r}}(t)$, $q_2^{\mathrm{r}}(t)$, $q_3^{\mathrm{r}}(t)$, and $q^{\mathrm{e}}(t)$. Let us make calculations of the mechanism functioning with following periodic law of slider motion [9]

$$q_1^{\mathrm{r}}(t) = 0.8 l_3 \sin(100t). \tag{4}$$

Obviously, the scleronomic geometric constraints have the form

$$\begin{cases} (0.35 - q_1^{\mathrm{r}})\sin(q_2^{\mathrm{r}} + q_3^{\mathrm{r}}) - l_2 \sin q_3^{\mathrm{r}} = 0, \\ l_3 \sin(q_2^{\mathrm{r}} + q_3^{\mathrm{r}}) + l_2 \sin q_2^{\mathrm{r}} = 0. \end{cases} \tag{5}$$

We will assume that geometrical and physical-mechanical characteristics of the links (subscripts 2 and 3) and slider (subscript 1) have following meanings: radii $r_2 = r_3 = 3.0 \times 10^{-3}$ m, lengths $l_2 = 0.35$ m and $l_3 = 0.15$ m, densities $\rho_2 = \rho_3 = 7800\,\mathrm{kg/m^3}$, Young's modulus $E_2 = 200$ MPa,

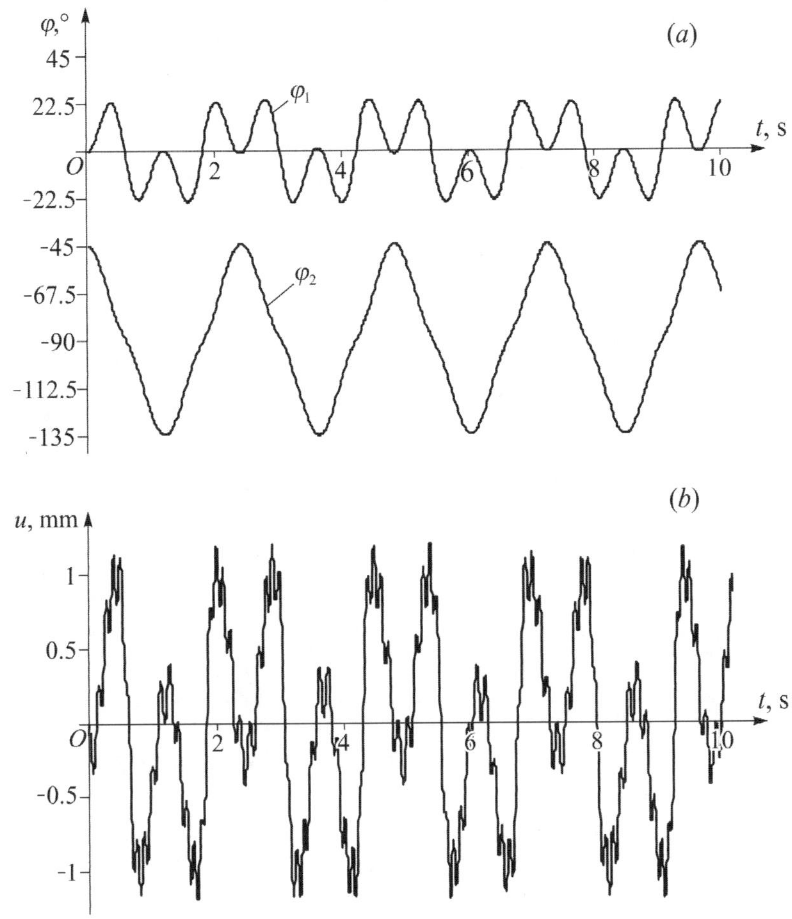

Figure 3. Functions of generalized coordinates $\varphi_1(t)$ and $\varphi_2(t)$ (a) and elastic displacement $u(t)$ (b).

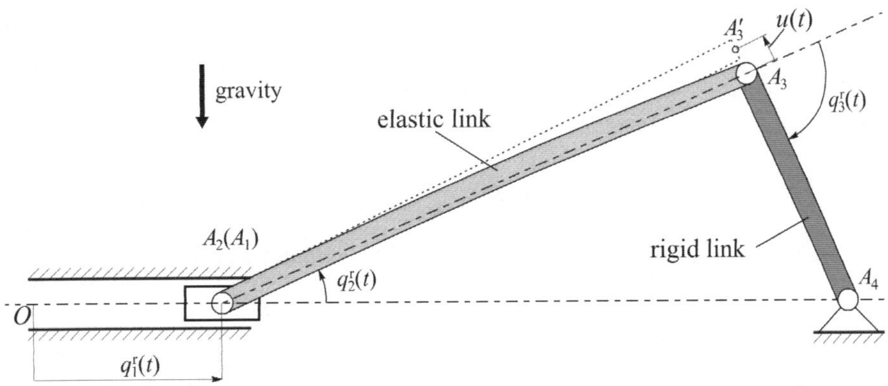

Figure 4. Slider–crank mechanism with elastic connecting rod.

mass $m_1 = 0.1\,\mathrm{kg}$ in these numerical experiments. We will also assume that initial conditions for generalized coordinates and elastic variable of mechanism movement and their velocities are $q_1^{\mathrm{r}}(0) = 0$, $q_2^{\mathrm{r}}(0) = 0.432\,\mathrm{rad}$, $q_3^{\mathrm{r}}(0) = -1.787\,\mathrm{rad}$, $\dot{q}_1^{\mathrm{r}}(0) = 80l_3$, $\dot{q}_2^{\mathrm{r}}(0) = \dot{q}_3^{\mathrm{r}}(0) = 0$, and

$q_2^e(0) = \dot{q}_2^e(0) = 0$. Note that $N_2 = 1$ for this problem.

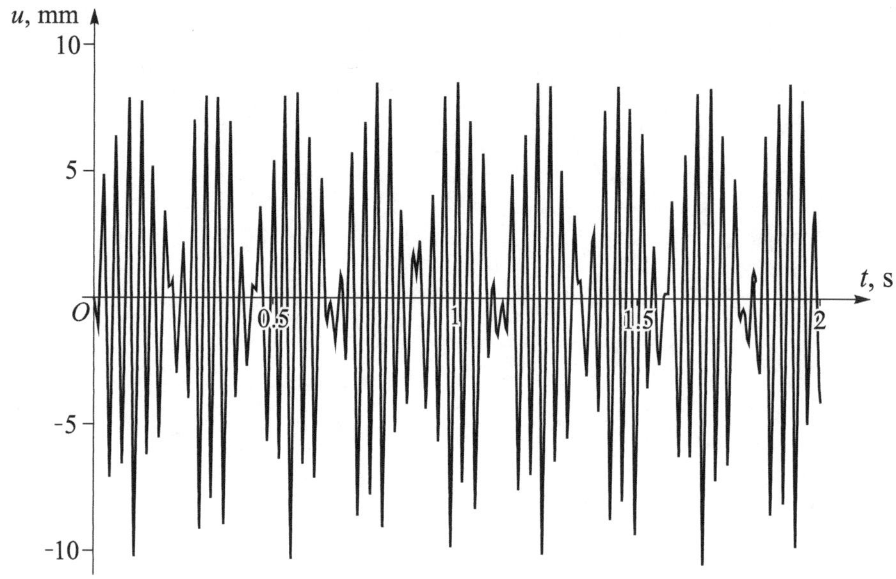

Figure 5. Pseudo-periodic function of elastic displacement $u(t)$ of the connecting rod.

The results of numerical modeling for this problem for 27 rotation cycles of the crank-sliding mechanism are shown in figure 5. To obtain the results was used algorithm of numerical calculations, developed and described in detail in [14]. It is seen that the function of elastic displacement $u(t)$ of the point A_3 (distance between points A_3 and A_3', where point A_3' is on the tangent at point A_2 to the curved profile of the elastic link A_2A_3 at a distance l_2 from A_2) have a pseudo period (about 1.14 s) that differs from the period of the connecting rod $2\pi/100\,\text{s} \approx 0.063\,\text{s}$, that is, they differ by 18 times.

5. Conclusions

The result obtained is of importance in the theory of elastic dynamical systems. This article shows that in a number of cases, when the initial systems oscillate periodically, then the oscillations of individual elastic links can have pseudo periodicity, and the value of such a pseudo period can differ from the oscillation period of the system links. No such effects were found in earlier works.

Most of the results were obtained using an efficient technique for the numerical dynamic analysis of elastic tree-like multilink systems, which combines the procedure for inverting the mass matrix with the procedure for effective kinematic calculation, developed on the basis of the generalized Newton–Euler method, described in [14].

References

[1] Featherstone R 1983 The calculation of robot dynamics using articulated-body inertias *The International Journal of Robotics Research* **2** (1) 13–30 doi: 10.1177/027836498300200102
[2] Dombre E and Khalil W 1999 *Modelisation, Identification et Commande des Robots* (Paris: Hermes)
[3] Boyer F and Coiffet P 1996 Symbolic modeling of a flexible manipulator via assembling of its generalized Newton Euler model *Mechanism and Machine Theory* **31** (1) 45–56 doi: 10.1016/0094-114X(95)00038-Z
[4] Boyer F and Khalil W 1998 An efficient calculation of flexible manipulator inverse dynamics *The International Journal of Robotics Research* **17** (3) 282–293 doi: 10.1177/027836499801700305

[5] Sargsyan Yu L, Stepanyan K G, Azuz N, and Gevorgyan H A 2004 Dynamic analysis of flexible manipulators using the generalized Newton–Euler method *Proceedings of NAS RA and NPUA: Series of Technical Sciences* **57** (1) 3–10

[6] Sargsyan Yu L, Stepanyan K G, and Gevorgyan H A 2006 Dynamic analysis of flexible tree-like mechanical systems without external constraints *Proceedings of NAS RA and NPUA: Series of Technical Sciences* **59** (1) 3–9

[7] Gribkov V A and Khokhlov A O 2017 Experimental study of inverted regulable pendulum stability *Herald of the Bauman Moscow State Technical University. Series Natural Sciences* No. 2 (71) 22–39 doi: 10.18698/1812-3368-2017-2-22-39

[8] Krasnova S A and Antipov A S 2019 Block Synthesis of Tracking System for Inverted Pendulum via Sigmoidal Feedbacks *2019 International Russian Automation Conference (RusAutoCon), Sochi, Russia, 2019* (IEEE) 1–5 doi: 10.1109/RUSAUTOCON.2019.8867695

[9] Gofron M and Shabana A A 1993 Control structure interaction in the nonlinear analysis of flexible mechanical systems *Nonlinear Dynamics* **4** (2) 183–206 doi: 10.1007/BF00045253

[10] Pascal M and Gagarina T 1999 A pseudo-rigid model for the dynamical simulation of flexible mechanisms *Multibody System Dynamics* **3** 303–31 doi: 10.1023/A:1009872629790

[11] Azouz N 1994 *Modélisation des Sructures Souples Polyarticulées. Application à la Simulation des Robots. Doctoral Thesis* (University of Paris VI)

[12] Verlinden O 1994 *Simulation du Comportement Dynamique de Systèmes Multicorps Flexibles Comportant des Membrures de Forme Complexe. Doctoral Thesis* (Polytechnic Faculty of Mons)

[13] Gevorgyan H A 2004 Dynamic analysis of flexible tree-like mechanical systems in the presence of external holonomic constrtaints *Information Technologies and Management. – Encyclopedia-Armenica.* No. 4 36–43

[14] Gevorkian H A 2020 Simulation of elastic tree-like dynamic systems in presence of external holonomic constraints *Herald of the Bauman Moscow State Technical University. Series Natural Sciences* No. 2 (89) 4–24 doi: 10.18698/1812-3368-2020-2-4-24

[15] Nosova N Y, Kheilo S V, Glazunov V A, and Tsar'kov A V 2018 Dynamic analysis of the spherical part of the parallel manipulator taking into account the control law *Journal of Machinery Manufacture and Reliability* **47** 205–212 doi: 10.3103/S1052618818030111

[16] Malakhov A A and Fashchevskiy N N 2019 MATLAB and SIMULINK educational potentiality in analysis of dynamic systems *AIP Conference Proceedings* **2195** 020069 doi: 10.1063/1.5140169

[17] Zolin A and Udovik I 2019 Dynamic analysis of non-linear processes in super heavy-lift launch vehicle erector and polyspast lifting system *AIP Conference Proceedings* **2171** 120010 doi: 10.1063/1.5133266

The dependence of the deformation behavior of fiberglass pipes on the initial disorientation of reinforcement subjected to repeated-static internal hydrostatic pressure and axial tension

K A Karapetyan[1*], S Sh Valesyan[1,2], and N S Muradyan[1]
[1]Institute of Mechanics of the National Academy of Sciences of Armenia, Yerevan, Republic of Armenia
[2]National University of Architecture and Construction of Armenia, Yerevan, Republic of Armenia

E-mail: *koryan@mechins.sci.am

Abstract. The problems of influence of the technological reinforcement disorientation on the deformation behavior of composite tubular elements subjected to repeated-static internal hydrostatic pressure and axial tension, as well as the resistance to fracture of these elements subjected to the above-mentioned loading are discussed.

The experimental part of research is carried out applying thin-walled tubular elements manufactured from the fiberglass fabric prepreg based on the modified epoxy resin.

It is established that a possible deviation of the reinforcement angle from 6o to 8o taking place during the processing of material into goods has little effect on the fracture resistance both subjected to the internal hydrostatic pressure and axial tension of fiberglass pipes manufactured so that the directions of reinforcing fiberglass fabric warp and ones axis are coincided.

Meanwhile, the following was stated that the accompanied shearing deformations are arising besides the basic cyclic deformations (noted in the direction of force application) for initially disorientation fiberglass pipes in difference from non-disoriented ones at the condition of pulsating (irregular) loading by the above-mentioned load-bearing factors.

Calculations are proved that the reinforcement disorientation direction in the above-underlined limits is brought to the significant decrease of the energy dissipation coefficient of fiberglass pipes.

The statement aiming the constructional solution of the problem is formulated in a sense of negative influence of the technological reinforcement disorientation on the composite tubular elements functioning at the conditions of cyclic axial tension and internal hydrostatic pressure.

1. Introduction

Disorientation of the reinforcement (deviation of the direction of the reinforcing fibers in the layers of the material from the projected) is one of the common defects in the macrostructure of reinforced composite materials [1–3]. This phenomenon, associated mainly with the imperfection of the technological process of processing the composite into a product, can cause a large scatter in the indices of the like mechanical characteristics defined for the same class of composite material [2, 4, 5]. It can also lead to the appearance of a significant discrepancy between the values of the identical mechanical characteristics of composites into products and those obtained as a result of testing standard specimens of a rectangular cross section in the majority of cases [6]. The probability of a disorientation of the reinforcement is usually random in nature, however turns out to be higher in the case of the manufacture of spatial products from composites, including thin-walled tubular elements of construction [7, 8].

According to the experimentally established data, a possible technological deviation of the reinforcement angle from 6° to 8° from its zero value, having little effect on the fracture resistance, leads to a significant increase in the ductility relative to the deformation of fiberglass pipes subjected to static axial tension, as well as internal hydrostatic pressure [9].

The results of a study of the influence of technological disorientation of reinforcement on the deformation behavior and fracture resistance of fabric fiberglass tubular elements subjected to repeated-static axial tension, or internal hydrostatic pressure (tensile in the annular direction) are presented and discussed in this paper.

2. Research methodology

The phenomenon of significant technological reinforcement disorientation was particularly discovered in the manufacture of experimental tubular specimens based on fiberglass thus the direction of the warp of the fabric knitting coincided with the direction of the axis of the pipes (the value of the angle of reinforcement $\varphi = 0°$). However, according to the carried out measurements, a part of the pipes (about 8% of the total number) had a deviation of the reinforcement angle from its zero value within $6 - 8°$ [8].

In order to define the problems stated in this paper, for experimental research carried out here, the following research methodology was developed. For this reason, two batches of tubular specimens were manufactured with an inner diameter of 38 mm, a wall thickness of 2.25 mm and a length of 285 mm applying technology [10] for a fiberglass prepreg based on a modified epoxy resin. For one batch of pipes, the angle between the directions of the fiberglass warp and their axis was $\varphi = 0°$, and for the other batch was $\varphi = 6 - 8°$. The plain weave fiberglass was used with the main overlap [11] of the disorientation density (number of threads per 1 cm^2 of fabric) 36:20 (warp:weft), produced by the Sevan factory "Glass insulation" (Republic of Armenia). The fiberglass reinforcement coefficient is $\mu = 0.45$ ($\mu_{\mathrm{warp}} = 0.29$, $\mu_{\mathrm{weft}} = 0.16$).

Preliminarily, on one part of the total number of manufactured experimental tubular specimens, the limits of fracture resistance under axial tension $\sigma_{zz}^{\mathrm{ult}}$ and internal hydrostatic pressure $\sigma_{\theta\theta}^{\mathrm{ult}}$ were determined.

The remaining part of the above-mentioned two batches of fiberglass tubular specimens was divided into 2 groups. Pipes of the first group were tested for repeated-static axial tension, and ones of the second group-for repeated-static internal hydrostatic pressure.

During cyclic tests of fiberglass pipes for axial tension (average rate of change of axial tensile stress σ_{zz} is approximately 16.2 MPa/min) and for internal hydrostatic pressure (average rate of change of internal pressure is about 26.7 atm/min, this corresponds to rate of stress change in the circumferential direction of the pipes $\sigma_{\theta\theta} = 25.8$ MPa/min) the value of the amplitude stress was applied equal to $0.6\sigma_{zz}^{\mathrm{ult}}$ and $0.42\sigma_{\theta\theta}^{\mathrm{ult}}$, respectively.

Regarding the choice of the value of the amplitude stress $\sigma_{\theta\theta} = 0.42\sigma_{\theta\theta}^{\mathrm{ult}}$, it was noted, based on direct measurements, obeying the condition $\sigma_{\theta\theta} \leq 0.45\sigma_{\theta\theta}^{\mathrm{ult}}$, fiberglass pipes with $\varphi = 0°$ during increasing internal pressure do not change the direction of deformation and experience only shortening [12].

As preliminary carried out tests showed, almost the same above-mentioned phenomenon is observed for fiberglass pipes with an angle of reinforcement $\varphi = 6 - 8°$.

The basic number of the loading-unloading cycles for pipes subjected to axial tension was equal to 12, and for pipes subjected to internal hydrostatic pressure was equal to 6.

Let's note that the cyclic tensile tests were carried out after the manufacture of fiberglass pipes 9 years later, and the internal hydrostatic pressure-after 30 years. During the indicated time periods, the pipes were stored in a laboratory room at an ambient temperature of $20 \pm 6°$C and a relative humidity of $60 \pm 8\%$.

It is known that in order to determine some mechanical characteristics of an element deformable in the loading-unloading mode, including the scattering or dissipation of energy coefficient, it is advisable to proceed from the relationship between the intensities of the shear stress T and the shear strain Γ [13], the expressions of those in the cylindrical frame are as follows [14]:

$$\mathrm{T} = \frac{1}{\sqrt{6}}\sqrt{(\sigma_z - \sigma_r)^2 + (\sigma_r - \sigma_\theta)^2 + (\sigma_\theta - \sigma_z)^2 + 6(\tau_{zr}^2 + \tau_{r\theta}^2 + \tau_{\theta z}^2)}, \quad (1)$$

$$\Gamma = \sqrt{\frac{2}{3}}\sqrt{(\varepsilon_z - \varepsilon_r)^2 + (\varepsilon_r - \varepsilon_\theta)^2 + (\varepsilon_\theta - \varepsilon_z)^2 + \frac{3}{2}(\gamma_{zr}^2 + \gamma_{r\theta}^2 + \gamma_{\theta z}^2)}. \quad (2)$$

In the case of axial tension and internal hydrostatic pressure of fiberglass pipes for shear stress intensities both with the reinforcement angle $\varphi = 0°$ and $\varphi = 6 - 8°$, respectively, we have

$$\mathrm{T} = \frac{1}{\sqrt{3}}|\sigma_z|, \quad (3)$$

$$T = \frac{1}{\sqrt{3}}|\sigma_\theta|. \quad (4)$$

The expressions of the intensities of shear deformations for these two types of tests take the following form:

for pipes with $\varphi = 0°$

$$\Gamma = \sqrt{\frac{2}{3}}\sqrt{(\varepsilon_z - \varepsilon_r)^2 + (\varepsilon_r - \varepsilon_\theta)^2 + (\varepsilon_\theta - \varepsilon_z)^2}, \quad (5)$$

for pipes with $\varphi = 6 - 8°$

$$\Gamma = \sqrt{\frac{2}{3}}\sqrt{(\varepsilon_z - \varepsilon_r)^2 + (\varepsilon_r - \varepsilon_\theta)^2 + (\varepsilon_\theta - \varepsilon_z)^2 + \frac{3}{2}\gamma_{\theta z}^2}. \quad (6)$$

In the case of the fiberglass pipes loading by tension in the direction of the z-axis in equations (5) and (6) $\varepsilon_r = -\nu_1\varepsilon_z$, $\varepsilon_\theta = -\nu_2\varepsilon_z$, where ν_1 and ν_2 are the Poisson's ratios in the radial and in the circumferential directions, respectively. Moreover, for fabric fiberglass with the both values of the reinforcement angles considered here, the values of ν_1 and ν_2 can be taken equal to approximately 0.2 and 0.16, respectively [6]. In the case of the fiberglass pipes loading by internal hydrostatic pressure (tension in the circumferential direction) can be assumed $\varepsilon_r = -\mu_1\varepsilon_\theta$, where the allowed measure of inaccuracy (error) is too small.

An analytical description of the experimentally obtained dependences between T and Γ in the sections of the ascending ($\rightarrow$) and descending ($\leftarrow$) branches of the hysteresis loop of the cycle for the considered cases of testing tubular specimens was carried out by a linear-fractional approximation of the following form:

$$\overset{\rightleftarrows}{\mathrm{T}} = \frac{\mathrm{T}_{\lim}}{\overset{\rightleftarrows}{a}} \cdot \frac{\overset{\rightleftarrows}{\Gamma} - \Gamma_{\mathrm{perm}}}{1 \pm \frac{\overset{\rightleftarrows}{b}}{\overset{\rightleftarrows}{a}}(\overset{\rightleftarrows}{\Gamma} - \Gamma_{\mathrm{perm}})}, \quad (7)$$

where $\mathrm{T}_{\lim}$ is the limiting value of T determined by formulas (3) and (4) applying the data obtained from the corresponding pipe tests, Γ_{perm} is the permanent strain intensity of the cycle shear, a and b are the experimental approximation parameters.

In the denominator of expression (7), the sign (+) is selected for the convex deformation curve, and the sign (−) is for the concave deformation curve.

The characteristic of scattering or dissipation of the energy of an element deformed in the loading-unloading mode is its relative value, expressed by the dependence [15]

$$\psi = \frac{\Delta W}{W}, \tag{8}$$

where ΔW is the value of the energy dissipated per cycle, and W is the value of the strain energy.

Taking into account that ΔW and W into equation (8) are the area of the hysteresis loop and the area bounded by the ascending branch of the hysteresis loop and the Γ-axis, respectively, applying (7) and (8) and denoting $\frac{\overset{\rightleftarrows}{a}}{\overset{\rightleftarrows}{b}} = \overset{\rightleftarrows}{X}$ after some calculations made up of the corresponding integral expressions for ψ, the following equation can be obtained [8]

$$\psi = 1 - \frac{\overrightarrow{b}\left[-(\Gamma_{\text{ampl}} - \Gamma_{\text{perm}}) - \overleftarrow{X}\ln\left|1 - \frac{1}{\overleftarrow{X}}(\overleftarrow{\Gamma} - \Gamma_{perm})\right|\right]}{\overleftarrow{b}\left[\Gamma_{\text{ampl}} - \overrightarrow{X}\ln\left(1 + \frac{1}{\overrightarrow{X}}\Gamma_{\text{ampl}}\right)\right]}, \tag{9}$$

where Γ_{ampl} is the value of the intensity of the amplitude deformations of the cycle shear.

3. Results and discussion

3.1. Cyclical axial tension of fiberglass pipes

Before proceeding to the consideration of the issue presented in this section, should be noted that the results of a study of the influence of reinforcement disorientation in the above-mentioned limits on the deformation behavior and fracture resistance of fiberglass pipes subjected to repeated-static simple torsion, as well as axial tension, were detailed discussed in [8]. Some basic results are brought in this work related to the cyclic tension of pipes in order to formulate general conclusions.

So, it was found by short-term tests that the deviation of the reinforcement angle value within $6 - 8°$ from its zero value practically does not affect the fracture resistance value of fiberglass pipes subjected to axial tension. The average value of this characteristic for the above-mentioned batches of pipes can be taken equal to 147.1 MPa.

Values of corresponding deformations were defined and approximations of these curves were plotted for all carried experimental cycles. Based on data shown in figure 1, it is possible to form a fair statement on the deformation behavior of the fiberglass pipes of the considered groups (pipes with $\varphi = 0°$ and pipes with $\varphi = 6 - 8°$) subjected to repeated-static axial tension.

From the data of figure 1 notes that at the condition of cyclic tension for disoriented pipes (figure 1*b*) in contrast to non-disoriented (figure 1*a*), the cyclic shear deformations accompanying cyclic basic longitudinal deformations are appeared.

The comparison of the obtained data brought on figures 1*a* and 1*b* shows that the ascending and descending branches of the hysteresis loop of the tubular specimens $\varphi = 0°$ longitudinal deformations and both longitudinal and shear deformations of the tubular specimens $\varphi = 6 - 8°$ approach each other due to the increasing test cycle number n maintaining the direction of convexity or concavity.

The changes in the amplitude (fixed at $\sigma_{z\,\max}$) and permanent (after complete unloading) strains within the cycle depending on the number of the test cycle n of tubes were compared. This comparison showed that the values of the longitudinal amplitude deformations of the cycle for both batches of fiberglass pipes and shear deformations for pipes $\varphi = 6 - 8°$ decrease significantly after the first loading-unloading cycle, and subsequently, with an increase in the number of cycles n, their monotonic increase with a small intensity can take place. This phenomenon is relatively more sensitivity in case of longitudinal deformations.

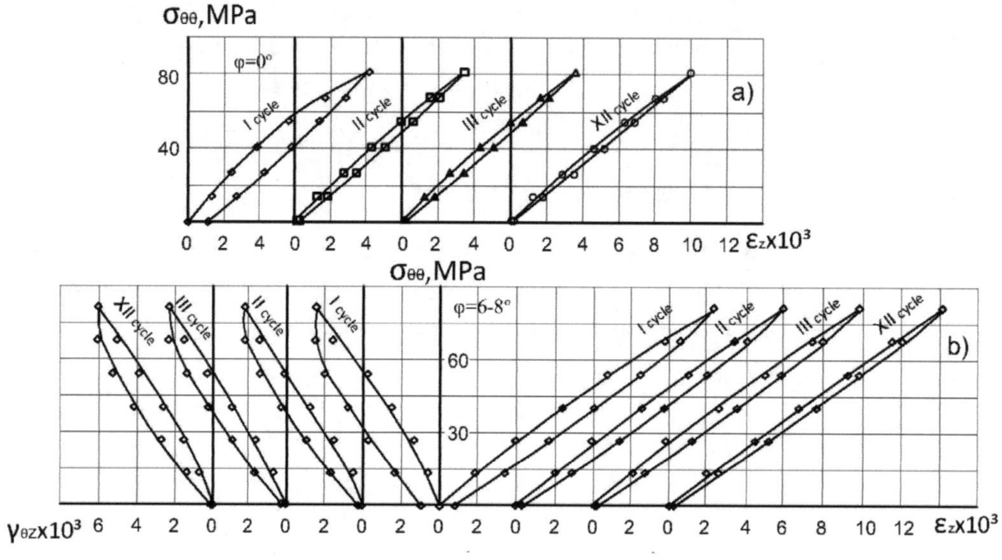

Figure 1. Deformation diagrams of pipes $\varphi = 0°$ (a) and $\varphi = 6$–$8°$ (b) within the cycle in loading-unloading mode subjected to axial tension.

The permanent longitudinal deformations of the cycle for both tubular specimens $\varphi = 0°$ and $\varphi = 6 - 8°$, and the shear permanent deformations of the cycle for pipes $\varphi = 6 - 8°$ with an increase in the number of cycles n to 6 monotonically decrease and become steadily smaller than their corresponding values determined during cycle I. There is a slight upward trend in the values of these strains with a further increase of the number of loading and unloading cycles.

The experimental dependence of the amplitude and permanent shear deformations intensities of the cycle on the cycle number of the testing tubular elements with the disorientated and non-disorientated direction of reinforcement was stated.

A comparison of the amplitude and permanent strains intensities recorded at the same number of test cycle shows that the appearance of a closed hysteresis loop for both batches of tubular specimens of the considered test case occurs after $5 - 6$ loading-unloading cycles.

According to the data calculated applying expression (9), the value of the relative energy dissipation coefficient ψ during the second loading-unloading cycle for glass fabric pipes $\varphi = 0°$ is $2.1 - 2.3$ times less than the value of the same coefficient determined for test cycle I. A further increase of the number of test cycles to 12 has little effect on the value of the coefficient ψ.

Almost the same above-mentioned phenomenon is observed for fiberglass pipes with the reinforcement angle $\varphi = 6 - 8°$.

Calculations show that the ratio of the energy dissipation coefficients of fiberglass pipes $\varphi = 6 - 8°$ and $\varphi = 0°$ determined for the same number of loading-unloading cycles does not depend much on the number of test cycles. The average value of this ratio can be taken equal to 0.9.

3.2. Cyclic internal hydrostatic pressure of fiberglass pipes

According to the experimental data obtained as a result of short-term tests, the deviation of the reinforcement angle within $6 - 8°$ from its zero value, as one was observed in the case of axial tension, has little effect on the fracture resistance in the circumferential direction of fabric glass tubular specimens subjected to internal hydrostatic pressure. The average value of this characteristic can be taken equal to $\sigma_{\theta\theta}^{\text{ult}} = 424.9\,\text{MPa}$.

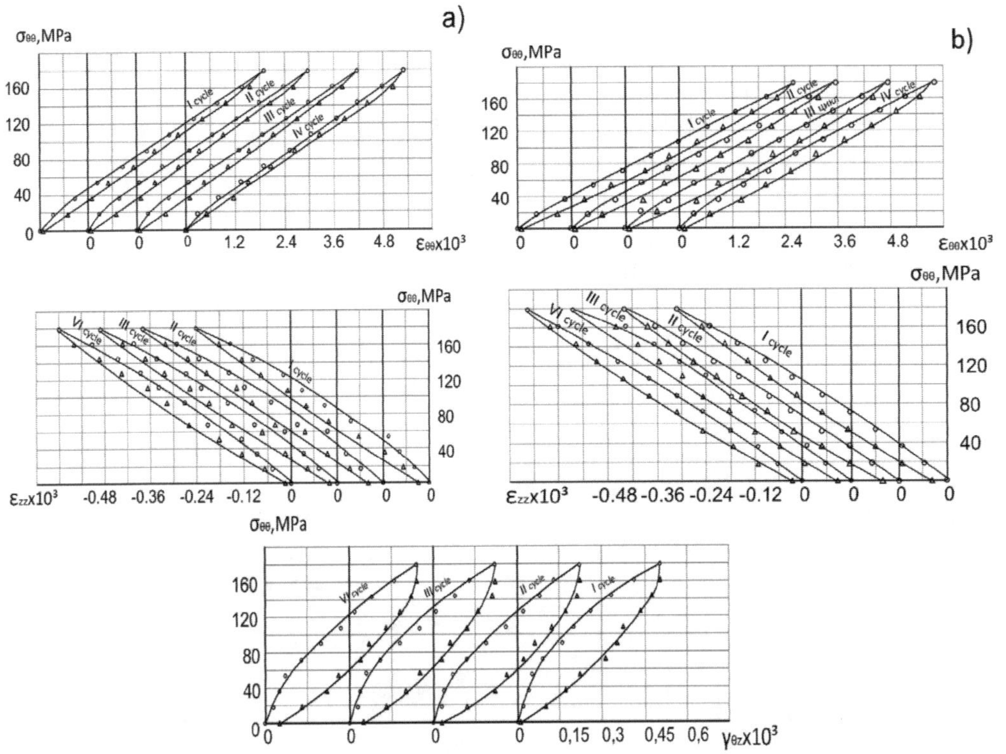

Figure 2. Deformation diagrams of pipes $\varphi = 0°$ (a) and $\varphi = 6$–$8°$ (b) within the cycle in loading-unloading mode subjected to internal hydrostatic pressure.

Deformation data were determined and approximating curves were plotted both for the above-mentioned cases of testing and for the testing above-underlined two groups of fabric fiberglass pipes subjected to repeated-static internal hydrostatic pressure for all cycles. These results obtained for I, II, III, and VI testing cycles are brought on figure 2.

From data of this figure 2 can be noted that for fabric fiberglass pipes with the reinforcement angle $\varphi = 0°$ subjected to repeated-static internal hydrostatic pressure besides main circumferential (annular) cyclic strains $\varepsilon_{\theta\theta}$, the accompanying to main the longitudinal cyclic strains ε_{zz} are arising (figure 2a), and for fabric fiberglass pipes with the reinforcement angle $\varphi = 6 - 8°$ — the accompanying cyclic shear strains $\gamma_{\theta z}$ as well (figure 2b).

The characteristic of the fabric fiberglass pipes cyclic deformation process is that the branches of a hysteresis loop keep the direction of convexity and concavity during the step-by-step increase of internal hydrostatic pressure or its decrease for all above-mentioned cases. With an increase in the number of cycles n, the gradual reduction of discrepancy between the sections of the ascending ($\rightarrow$) and descending ($\leftarrow$) branches of the hysteresis loop of one-valued deformations for considering here reinforcement angles is observed.

Let's discuss the behaviors of main amplitude and permanent, as well as accompanying to main deformations of the cycle of fabric fiberglass pipes with the non- disoriented $\varphi = 0°$ and disoriented $\varphi = 6 - 8°$ reinforcement subjected to repeated-static internal hydrostatic pressure.

For the case of amplitude deformations, the following behavior is common independent of the tubular element reinforcement orientation: due to the initial increasing test loading-unloading cycle number n to 3, some decrease of these deformations is taken place, and due to further increasing test loading-unloading cycle number n to 6, the slight tendency of increase of the

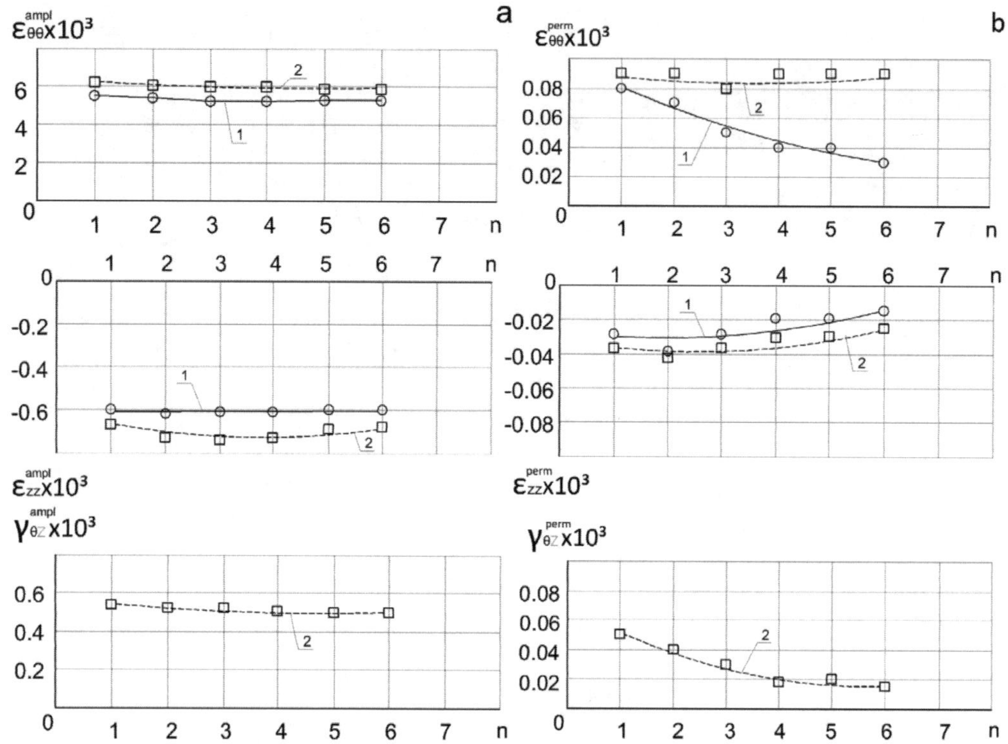

Figure 3. Curves of the alteration of amplitude (*a*) and permanent (*b*) main ($\varepsilon_{\theta\theta}$) and accompanying to main (ε_{zz} and $\gamma_{\theta z}$) deformations per cycle from the testing cycle number for tubular elements with $\varphi = 0°$ (curve *1*) and $\varphi = 6$–$8°$ (curve *2*) subjected to repeated-static internal hydrostatic pressure.

main amplitude and accompanying to main amplitude deformations values is observed (figure 3*a*).

Similar to above-mentioned behaviors due to cyclic internal pressure are appeared for the permanent main cycle strain $\varepsilon_{\theta\theta}^{\text{perm}}$ of the tubular specimens with the reinforcement angle $\varphi = 6 - 8°$, and for the accompanying to main permanent longitudinal cycle strain $\varepsilon_{zz}^{\text{perm}}$ of both $\varphi = 0°$ and $\varphi = 6 - 8°$ pipes (figure 3*b*).

The characteristics of main permanent cycle strain $\varepsilon_{\theta\theta}^{\text{perm}}$ for pipes with $\varphi = 0°$ and the accompanying to main permanent shear cycle strains $\gamma_{\theta z}$ for pipes with $\varphi = 6-8°$ in dependence on the testing cycle number n are similar. This similarity is set that at the initial increase of n to 4, the monotonous decrease at the high rate of underlined characteristics values is taken place, and due to the further increase of n to 6, the gradual drop of this rate is observed (figure 3*b*).

Comparing the values of permanent and amplitude deformations taken from the same testing cycle number, should be taken into account that the close loop of hysteresis of the main strains ($\varepsilon_{\theta\theta}$) of fabric fiberglass pipes with the above-mentioned reinforcement angles values is occurred at the first cycle. For the case of accompanying to main strains (ε_{zz} and $\gamma_{\theta z}$), the underlined phenomenon is observed after the third cycle of loading-unloading of these pipes by the internal pressure.

The experimental dependences of the intensity of cycle shear deformation Γ upon the intensity of tangent deformation T for all test cycles of pipes with considered reinforcement angles were

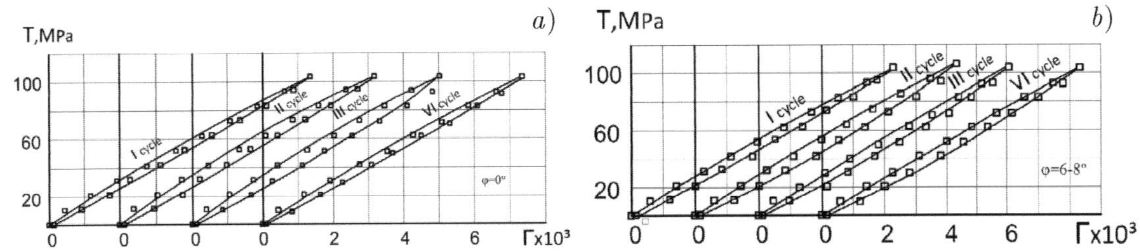

Figure 4. Curves of the alteration of amplitude and permanent shear deformation of cycle from the testing cycle number n for tubular elements with reinforcement angles $\varphi = 0°$ (a) and $\varphi = 6$–$8°$ (b) subjected to repeated-static internal hydrostatic pressure.

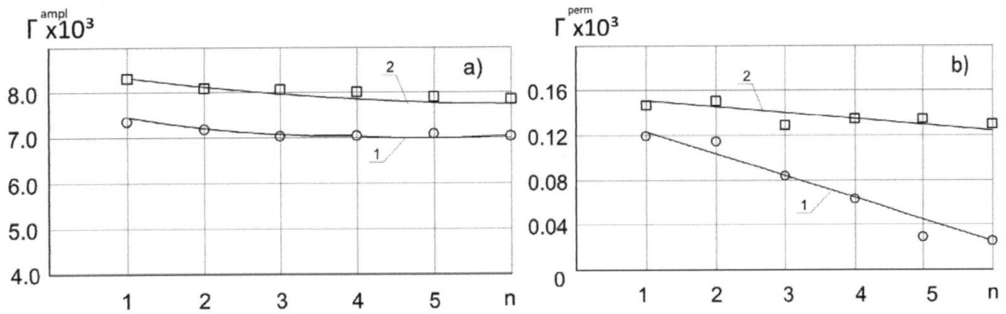

Figure 5. Curves of the alteration of amplitude (a) and permanent (b) shear deformation of cycle from the testing cycle number n for fabric fiberglass pipes with reinforcement angles $\varphi = 0°$ (curve *1*) and $\varphi = 6$–$8°$ (curve *2*).

defined by the application of (3)–(6). These results obtained for I, II, III, and VI testing cycles are shown by marks.

The comparison of above-mentioned results is shown that for the case of shear deformation intensity, the direction of convexity and concavity of the branches of a hysteresis loop is the same during the testing process and the similar to the above-considered cases. So that, the increase of testing cycles number leads to the decrease of discrepancy between the branches of a hysteresis loop, and this phenomenon is more significant for fabric fiberglass tubular specimens with the reinforcement angle $\varphi = 0°$.

Approximating curves of experimental dependences $\Gamma - \mathrm{T}$ plotted according to (7) are brought on figure 4. Based on the data of this figure, the application of (7) for these dependences can be acceptable.

The curves of dependences of the intensity of amplitude and permanent shear deformations of the cycle on the testing cycle number n for above-mentioned two types of fabric fiberglass pipes are brought on the figure 5.

The comparison of this figure data is shown that the values of intensity both amplitude Γ^{ampl} and permanent Γ^{perm} shear deformations of the cycle for fabric fiberglass pipes with the disoriented reinforcement are greater than for fabric fiberglass pipes with the non-disoriented reinforcement for the same loading-unloading cycle number n. The decrease of these values is observed with the testing cycle number increase. Meanwhile, defined for the same cycle number n, the ration of values Γ^{ampl} of pipes with $\varphi = 6 - 8°$ and $\varphi = 0°$ depends a little upon the quantity of testing cycles and varies in limits $1.1 - 1.2$, whereas, the ration of values Γ^{perm} increases from 1.2 to 5.2 with the increase of testing cycles number n to 6. Based on the

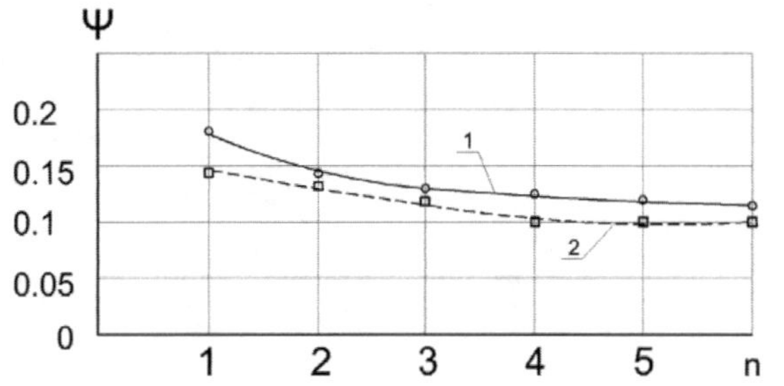

Figure 6. Curves of the alteration of the coefficient ψ from the testing cycle n for fabric fiberglass pipes with reinforcement angles $\varphi = 0°$ (curve *1*) and $\varphi = 6$–$8°$ (curve *2*) subjected to repeated-static internal hydrostatic pressure.

comparison of amplitude and permanent shear deformation intensity values of cycle determined for the same testing cycle number is followed that the formation of closed hysteresis loop is occurred at $2 - 3$ loading-unloading cycles at the internal pressure of fabric fiberglass pipes with considered reinforcement angles.

The curves of the alteration of the energy dissipation coefficient ψ defined by (9) in dependence upon the cycle number n of the repeated-static internal pressure testing of above-mentioned two types fabric fiberglass pipes are brought on figure 6.

From the data of this figure can be concluded that the initial increase of cycle number n to 4 leads to the monotonous decrease with the diminishing rate of coefficient ψ value of tubular elements both $\varphi = 0°$ and $\varphi = 6 - 8°$ reinforcement angles. The alteration of the coefficient ψ value isn't observed during the further increase of cycle number n to the end of cyclic testing of the underlined types of fabric fiberglass pipes. Determined for the same testing cycle number, according to the carried calculations, the ratio of energy dissipation coefficient values for pipes with $\varphi = 6 - 8°$ and $\varphi = 0°$ reinforcement angles after the deformation stabilization process doesn't change and it's 0.83 approximately.

4. Conclusions

Experimental determined common behaviors of the deformation resistance and failure of composite thin-walled tubular elements with the initially disoriented and non-disoriented reinforcement subjected to repeated-static, as well as axial static tension by the internal hydrostatic pressure (tension in circumferential direction) are brought here. Stated some considerations are supporting to the optimal design of these elements assigned for the functioning at the pulsating conditions of underlined loading.

The possible technological deviation of reinforcement angle in limit $6 - 8°$ of the fabric fiberglass pipes manufactured so that the fabric warp direction coincides with the pipe axis direction (reinforcement angle $\varphi = 0°$) doesn't influence on the failure resistance practically, and leads to the slenderness essential increase with respect to the deformation of pipes subjected to static and repeated-static axial tension, as well as internal hydrostatic pressure.

Fabric fiberglass pipes subjected to both repeated-static axial tension and internal hydrostatic pressure with initial disoriented reinforcement in difference from non- disoriented one, the accompanying to main cyclic shear deformations are arising except cyclic main ones (occurred in the direction of load application). The following can be stated with high level of probability that the above-mentioned phenomenon can be occurred for tubular elements manufactured by the cross spinning for the case of spinning irregularity with respect to the pipe axis.

The value of energy dissipation coefficient is practically constant for disoriented tubular elements; however, it is less than the value of energy dissipation coefficient – for non- disoriented ones after the deformation stabilization process. For the case of loading of the pipes by the repeated-static internal pressure, the mentioned difference is significantly greater that for the case of loading of the pipes by the cyclic axial tension.

So that, the possible reinforcement angle deviation from projected during the processing from material to goods can significantly influence on the expected deformation behavior of the spatial elements for reinforced composites. Particularly, the reinforcement symmetry breakdown with respect to composite tubular element axis subjected to pulsating axial tension and internal hydrostatic pressure can cause the appearance of unpredicted cyclic shear deformations. This means that, for both cases of operation, the composite material matrix, in view of one is the vulnerable link, is under the condition of cyclic changing complex stress. Meanwhile, based on the practice, it is known that the matrix failure or the surface between fiber and matrix destruction rather than the reinforcement fiber is the main reason of composite goods destruction for the majority of cases [16].

The minimization problem of the internal unpredicted stresses formation during the operation caused by the technological reinforcement disorientation preventing numerical reasons leading to the composite structure destruction should be defined by the constructive way. Particularly, contributing to the slenderness rate decrease about deformation, stiffening ribs optimal oriented symmetrically with respect to axis are necessary to design for thin-walled composite tubular elements.

References

[1] Tarnopol'skii Yu M, Rose A V, and Portnov G G 1969 Some negative characteristics of fiber-reinforced materials *Polymer Mechanics* **5** (1) 115–23 doi: 10.1007/BF00859033

[2] Tarnopol'skii Yu M, Rose A V, Zhigun I G, and Gunyaev G M 1971 Structural characteristics of materials reinforced with high-modulus fibers *Polymer Mechanics* **7** (4) 600–9 doi: 10.1007/BF00855201

[3] Ermolenko S F and Ermolenko A F 1991 Strength of laminate with random deviations in reinforcement directions *Mechanics of Composite Materials* **27** (1) 29–33 doi: 10.1007/BF00630715

[4] Gol'dman A Ya and Savel'eva N F 1967 The stressed state and certain fracture characteristics of glass-reinforced plastics loaded in tension at an angle to the direction of reinforcement *Polymer Mechanics* **3** (6) 1030–9 doi: 10.1007/BF00858740

[5] Chiao C C and Chiao T T 1977 *Aramid fibers and composites. Preprint UCRL-80400* (California: Lawrence Livermore Laboratory)

[6] Tarnopolsky Yu M and Kintsis T Ya 1981 *Methods of Static Testing of Reinforced Plastics* (Moscow: Khimiya) [in Russian]

[7] Zhigun I G and Polyakov V A 1978 *Properties of spatially reinforced plastics* (Riga: Zinatne) [in Russian]

[8] Karapetyan K A 2001 On strength and strain properties of glass-fibre pipes at repeatedly-static loadings depending on deviations of orientation of reforcement *Mechanics — Proceedings of National Academy of Sciences of Armenia* **54** (2) 70–9

[9] Karapetyan K A, Valesyan S Sh, and Muradyan N S 2018 The influence of the reinforcement disorientation on the deformative behavior and resistance to fracture of the glass-plastic tubes subjected to the axial tension and inner hydrostatic pressure *Mechanics — Proceedings of National Academy of Sciences of Armenia* **71** (3) 36–43 doi: 10.33018/71.3.5

[10] Martirosyan M M 1971 Manufacturing pressed thin-walled fiberglass pipes *Industry of Armenia* **10** pp 56-57

[11] Martynova A A, Yatchenko O F, and Vasiliev A V 2007 *Fabric Manufacturing Technology* (Moscow: Akademiya) [in Russian]

[12] Karapetyan K A, Valesyan S Sh, and Muradyan N S 2019 Dependence of deformability and destruction of glass fiber pipes from initial torsion under the action of internal pressure *Mechanics — Proceedings of National Academy of Sciences of Armenia* **72** (1) 49–60 doi: 10.33018/72.1.4

[13] Il'yushin A A 1948 *Plasticity* (Moscow–Leningrad: Gostekhizdat) [in Russian]

[14] Kachanov L M 1969 *Fundamentals of the Theory of Plasticity* (Moscow: Nauka)

[15] Davidenkov N N 1938 Energy dissipation in vibration *Journal of Technical Physics* **8** pp 403–17

[16] Hahn H T 1973 A note on deformation of the shear stress-strain response of unidirectional composites *Journal of Composite Materials* **7** (3) 383–6 doi: 10.1177/002199837300700308

Inhomogeneous transversely isotropic space under influence of concentrated power and temperature sources

O F Kryvyi[1*] and Yu O Morozov[2]**
[1]National University "Odessa Maritime Academy", Odessa, Ukraine
[2]Odessa Polytechnic State University, Institute of Mechanical Engineering, Odessa, Ukraine
E-mail: *krivoy-odessa@ukr.net, **morozovyu@gmail.com

Abstract. The problem of constructing fundamental solutions to the thermoelasticity problem for a piecewise-homogeneous transversely isotropic space is reduced to the matrix Riemann problem in the space of generalized slow growth functions. As a result of the solution of which, were obtained expressions in explicit form for the components of the stress tensor and the displacement vector in plane of connection of transversely isotropic elastic half-spaces containing concentrated stationary heat sources. The temperature distribution is investigated depending on the thermophysical characteristics of the half-space materials.

Introduction

The study of stress concentration in the vicinity of interfacial and internal defects such as cracks or inclusions in thermoelastic fields is of great practical importance. Many works have been devoted to this problem for various environments. In particular, in [1-4], the problems of elasticity and thermoelasticity about interfacial stress concentrators such as cracks or rigid inclusions in piecewise homogeneous isotropic and transversely isotropic spaces are considered, which are reduced to systems of two-dimensional singular integral equations (SIE) and proposed a method for their solution.

In the mathematical formulation and solution of such problems about defects, it is necessary to set the boundary conditions on the defect itself, such as stress on the crack edges or displacement at the inclusion. Since in thermophysical formulation of the problems from determining the stress and displacement fields in the vicinity of the stress concentrators, known the stresses or displacements at the boundary of the region, at some interior points or at infinity (for unbounded bodies), then the determination of the boundary conditions on the defect is a separate problem.

Within the framework of the linear theory of thermoelasticity, to solve this problem, it is necessary to know the distribution of the temperature, stress and displacement fields in the corresponding piecewise homogeneous bodies without defects in the presence of volumetric forces and concentrated heat sources.

In particular, for piecewise homogeneous isotropic and transversally isotropic spaces, such solutions are given, respectively, in [5, 6]. Green's functions for piecewise homogeneous transversally isotropic spaces in the presence of a concentrated heat source and in the absence of thermal diffusion were constructed in [7–10], and in the presence of thermal diffusion — in [11]. In [12], the Green's function was constructed for a continuous transversely isotropic space with allowance for the orientation of the symmetry axes. For piezo and magnetoelectric compound transversely isotropic spaces, the Green's function was constructed in [13–15]. The fundamental solution for porous transversely isotropic materials was constructed in [16, 17], and for functionally graded materials in [18, 19]. In [20–22], Green's functions for a layered thermal environment were constructed.

In [23–25], similar problems for composite materials were solved by numerical methods.

In this work, using the approach of works [26–28], the problem of constructing fundamental solutions for piecewise homogeneous two-dimensional anisotropic media is reduced to the matrix Riemann problem and obtained its exact solution, which made it possible to construct in an explicit form a fundamental solution for a piecewise-homogeneous transversely isotropic space in the presence of concentrated forces and heat sources.

1. Statement of the problem

Let in an inhomogeneous space composed of two different transversally isotopic half-spaces completely linked in the plane $z = 0$, at an arbitrary point $M_0(x_0, y_0, z_0)$ concentrated force $P = (P_1, P_2, P_3)$ and at an arbitrary point $M_1(x_1, y_1, z_1)$ stationary heat sources.

The thermoelastic state of space is described by the vector

$$\mathbf{v} = \{v_k(x, y, z)\}_{k=\overline{1,9}} = \{\sigma_x, \sigma_y, \sigma_z, \tau_{yz}, \tau_{xz}, \tau_{xy}, u, v, w\}. \quad (1)$$

Based on the equilibrium equations and the generalized Hooke's law, and also taking into account the Duhamel–Neumann relation with respect to the components of the vector $\mathbf{v}$, in the space of generalized functions of slow growth $\Im'(\mathbb{R}^3)$ we write the following boundary value problem

$$\mathbf{D}[z, \partial_1, \partial_2, \partial_3]\mathbf{v} = \mathbf{F}, \quad \mathbf{v}, \mathbf{F} \in \Im'(\mathbb{R}^3), \quad (2)$$

$$v_k(x, y, +0) = v_k(x, y, -0), \quad k = \overline{1,9}, \quad k \neq 1, 2, 6, \quad (3)$$

$$v_k(x, y, z)|_{(x,y,z)\to\infty} = 0, \quad k = \overline{1,9}. \quad (4)$$

Here we use the notation

$$\mathbf{D} = \begin{Vmatrix} \mathbf{D}_0 & \mathbf{O}_{3\times3} \\ -\mathbf{S} & \mathbf{D}_0^T \end{Vmatrix}, \quad \mathbf{F}^T = \mathbf{F}_0^T + \mathbf{F}_*^T, \quad \mathbf{F}_0^T = -\delta(x - x_0, x - x_0, x - x_0)\|P_1, P_2, P_3, \mathbf{O}_{1\times6}\|,$$

$$\mathbf{F}_*^T = \|\mathbf{O}_{1\times3}, \beta_1 T, \beta_2 T, \beta_3 T, \mathbf{O}_{1\times3}\|, \quad \mathbf{S} = \begin{Vmatrix} \mathbf{S}_1 & \mathbf{O}_{3\times3} \\ \mathbf{O}_{3\times3} & \mathbf{S}_2 \end{Vmatrix},$$

$$\mathbf{D}_0 = \begin{Vmatrix} \partial_1 & 0 & 0 & 0 & \partial_3 & \partial_2 \\ 0 & \partial_2 & 0 & \partial_3 & 0 & \partial_1 \\ 0 & 0 & \partial_3 & \partial_2 & \partial_1 & 0 \end{Vmatrix}, \quad \mathbf{S}_1 = \begin{Vmatrix} s_{11} & s_{12} & s_{13} \\ s_{21} & s_{11} & s_{13} \\ s_{13} & s_{13} & s_{33} \end{Vmatrix}, \quad \mathbf{S}_2 = \begin{Vmatrix} s_{44} & 0 & 0 \\ 0 & s_{44} & 0 \\ 0 & 0 & s_{66} \end{Vmatrix},$$

$\partial_1 = \frac{\partial}{\partial x}$, $\partial_2 = \frac{\partial}{\partial y}$, $\partial_3 = \frac{\partial}{\partial z}$, $s_{kj} = \theta(z)s_{kj}^+ + \theta(-z)s_{kj}^-$, $s_{kj}^\pm$ — of the generalized Hooke's law, respectively, for the upper $z > 0$ and lower $z < 0$ half-spaces; $O_{n\times m}$ is zero matrix of dimension $n \times m$, $\beta_k = \theta(z)\beta_k^+ + \theta(-z)\beta_k^-$, $\beta_k^\pm$ is thermal expansion coefficients, T is temperature from a concentrated heat source with a capacity Q which was received in [13]

$$T(x, y, z) = \frac{m_{31}}{\sqrt{r_1^2 + (\xi_0|z - z_1|)^2}} + \frac{m_{32}}{\sqrt{r_1^2 + (\xi_0(z + z_1))^2}} + \frac{m_{33}}{\sqrt{r_1^2 + (\hat{\xi}_0 z + \breve{\xi}_0 z_1)^2}},$$

where

$$m_{31} = \theta(z,z_0)m_{31}^{+} + \theta(-z,-z_0)m_{31}^{-}, \quad m_{32} = \theta(-z,-z_0)m_{32}^{-} - \theta(z,z_0)m_{32}^{+},$$
$$m_{33} = \theta(z,-z_0)m_{33}^{+} - \theta(-z,z_0)m_{33}^{-},$$
$$m_{11}^{\pm} = \frac{\lambda_1^{\pm}}{\xi_0^{\pm}}, m_{13}^{\pm} = \lambda_1^{\mp}(\lambda_3^{\pm}m_1^{\mp} \pm \frac{m_2^{\mp}}{\xi_0^{\pm}}), \quad m_{12}^{\pm} = \lambda_1^{\pm}(\lambda_3^{\pm}m_1^{\pm} \pm \frac{m_2^{\pm}}{\xi_0^{\pm}}), \quad m_{21}^{\pm} = 1,$$
$$m_{22}^{\pm} = \frac{\lambda_1^{\pm}\lambda_3^{+}}{\xi_0^{\pm}}m_1^{\pm} \pm \lambda_3^{\pm}m_2^{\pm}, \quad m_{23}^{\pm} = \frac{\lambda_1^{\pm}\lambda_3^{+}}{\xi_0^{\pm}}m_1^{\mp} \pm \lambda_3^{\mp}m_2^{\mp}, \quad m_{31}^{\pm} = \frac{1}{\lambda_3^{\pm}\xi_0^{\pm}},$$
$$m_{32}^{\pm} = \lambda_3^{\pm}m_1^{\pm} \pm \frac{m_2^{\pm}}{\xi_0^{\pm}}, \quad m_{33}^{\pm} = \lambda_3^{\pm}m_1^{\mp} + \frac{m_2^{\mp}}{\xi_0^{\pm}}, \quad \xi_0^{\pm} = \sqrt{\lambda_1^{\pm}/\lambda_3^{\pm}},$$
$$\hat{\xi}_0 = \theta(z,z_0)\xi_0^{+} + \theta(z,-z_0)\xi_0^{+} + \theta(-z,-z_0)\xi_0^{-} + \theta(-z,z_0)\xi_0^{-}, \quad \xi_0 = \theta(z,z_0)\xi_0^{+} + \theta(-z,-z_0)\xi_0^{-},$$
$$\breve{\xi}_0 = \theta(z,z_0)\xi_0^{-} - \theta(z,-z_0)\xi_0^{-} + \theta(-z,-z_0)\xi_0^{+} - \theta(-z,z_0)\xi_0^{+}, \quad Q = \theta(z,z_1)Q^{+} + \theta(-z,-z_1)Q^{-},$$

$\lambda_i = \lambda_i^{+}\theta(x_3) + \lambda_i^{-}\theta(-x_3)$, $i = \overline{1,3}$, $\lambda_i^{\pm}$ is thermal conductivity coefficient for the upper $z > 0$ and lower $z < 0$ half-spaces, $r_1 = \sqrt{(x-x_1)^2 + (y-y_1)^2}$.

2. Construction of the fundamental solution problem of thermoelasticity

We apply to the matrix equation (2) the operator of the three-dimensional Fourier transform F_3 from $\Im'(\mathbb{R}^3)$, given the following representation for vector components $\mathbf{v}$: $v_k = \theta(z)v_k + \theta(-z)v_k = v_k^{+} + v_k^{-}$, where $v_k^{\pm} \in \Im'(\mathbb{R}^3_{\pm})$, $\mathbb{R}^3_{\pm} = \mathbb{R}^2 \times \mathbb{R}_{\pm}$. Then, considering the conditions (3), (4) and the results of [8–14], with respect to $V_k^{\pm}(\alpha_1,\alpha_2,\alpha_3) = F_3[v_k^{\pm}] \in \Im'(\mathbb{R}^3)$, and also that the functions $V_k^{\pm} \in \Im'(\mathbb{R}^3_{\pm})$ admit an analytical representation [8, 10] in the variable α_3 we obtain the following matrix equation:

$$\mathbf{M}_{\pm}\mathbf{V}^{\pm} = \mathbf{F}^{\pm}, \quad \mathbf{W}^{\pm}, \mathbf{F}^{\pm} \in \Im'(\mathbb{R}^3). \tag{5}$$

Here we use the notation

$$\mathbf{M}_{\pm} = \mathbf{D}[\pm 0, -i\alpha_1, -i\alpha_2, -i\alpha_3], \quad \mathbf{F}_j^{\pm} = \{f_k^{\pm}\}_{k=\overline{1,4}}, \quad f_k^{\pm} = e_0^{\pm}P_k \mp \frac{1}{2}\chi_k, \quad k = 1,2,3,$$
$$f_k^{\pm} = \beta_{k-2}^{\pm}T^{\pm}(\alpha_3) \mp \frac{1}{2}\chi_k, \quad k = 4,5,6, \quad f_k^{\pm} = \mp\frac{1}{2}\chi_k, \quad k = 7,8, \quad f_9^{\pm} = \mp\frac{1}{2}\chi_k,$$
$$\boldsymbol{\chi} = \{\chi_k\}_{k=\overline{1,4}} \in \Im'(\mathbb{R}^2), \quad \chi_k = 0, \quad k = 4,5,9, \quad e_0^{\pm} = \theta(\pm z_0)\exp(i\alpha_1 x_0 + i\alpha_2 y_0 + i\alpha_3 z_0),$$
$$T^{\pm}(\alpha_3) = e_0\left\{Q^{\pm}\theta(\pm z_1)\left[\frac{e^{i\alpha_3 z_1}}{\alpha_3^2\lambda_3^{+} + \lambda_1^{+}r^2} + \frac{1}{r}(\frac{(-i\alpha_3)\lambda_3^{\pm}m_1^{\pm} - rm_2^{\pm}}{\alpha_3^2\lambda_3^{\pm} + \lambda_1^{\pm}r^2})e^{\mp\xi_0^{+}rz_1}\right]\right.$$
$$\left. -\theta \mp z_1)Q^{\mp}(\frac{(-i\alpha_3)\lambda_3^{\pm}m_1^{\mp} - rm_2^{\mp}}{\alpha_3^2\lambda_3^{\pm} + \lambda_1^{\pm}r^2})e^{\pm\xi_0^{-}rz_1}\right\}, \quad e_0 = \exp(i\alpha_1 x_1 + i\alpha_2 y_1),$$

$\chi_k(\alpha_1,\alpha_2)$ are unknown functions from $\Im'(\mathbb{R}^2)$ for determine which, we need to use conditions (3) after the Fourier-transformed.

We represent the sought functions as

$$V_7^{\pm} = -(-i\alpha_2)\Psi_1^{\pm} - (-i\alpha_1)\Psi_2^{\pm}, \quad V_8^{\pm} = (-i\alpha_1)\Psi_1^{\pm} - (-i\alpha_2)\Psi_2^{\pm}, \tag{6}$$
$$V_5^{\pm} = -(-i\alpha_2)\Upsilon_1^{\pm} - (-i\alpha_1)\Upsilon_2^{\pm}, \quad V_4^{\pm} = (-i\alpha_1)\Upsilon_1^{\pm} - (-i\alpha_2)\Upsilon_2^{\pm}, \tag{7}$$

where $\Psi_k^{\pm}$, $\Upsilon_2^{\pm}$, $k = 1,2$ are new unknown functions. Then matrix equation (5) can be separated into two independent equations

$$\mathbf{L}_{\pm}\mathbf{U}_1^{\pm} = \mathbf{F}_1^{\pm}, \quad \mathbf{G}_{\pm}\mathbf{U}_2^{\pm} = \mathbf{F}_2^{\pm}, \tag{8}$$

where

$$\mathbf{U}_1^{\pm} = \{U_k^{1,\pm}\}_{k=1,2} = \{\Upsilon_{1j}^{\pm}, \Psi_{1j}^{\pm}\}, \quad \mathbf{U}_2^{\pm} = \{U_k^{2,\pm}\}_{k=1,4} = \{V_3^{\pm}, \Upsilon_2^{\pm}, \Psi_2^{\pm}, V_9^{\pm}\},$$

$$\mathbf{F}_1^{\pm} = \{(-i\alpha_2)f_1^{\pm} - (-i\alpha_1)f_2^{\pm}, (-i\alpha_2)f_7^{\pm} - (-i\alpha_1)f_8^{\pm}\},$$

$$\mathbf{F}_2^{\pm} = \{f_3^{\pm}, (-i\alpha_1)f_1^{\pm} + (-i\alpha_2)f_2^{\pm}, (-i\alpha_2)f_8^{\pm} + (-i\alpha_1)f_7^{\pm}, f_6^{\pm}\},$$

$$\mathbf{G}_{\pm} = \{g_{kj}^{\pm}\}_{k,j=1,\ldots,4}, \quad g_{11}^{\pm} = g_{44}^{\pm} = (-i\alpha_3), \quad g_{12}^{\pm} = r^2, \quad g_{21}^{\pm} = -\frac{c_{13}^{\pm}}{c_{33}^{\pm}} g_{12}^{\pm},$$

$$g_{23}^{\pm} = -\frac{\bar{c}_{13}^{\pm} + c_{13}^{\pm 2}}{c_{33}^{\pm}} r^4, \quad g_{22}^{\pm} = g_{33}^{\pm} = (-i\alpha_3)r^2, \quad g_{kj}^{\pm} = g_{jk}^{\pm} = 0, \quad k = 1, 2, \quad j = 3, 4,$$

$$g_{32}^{\pm} = -\frac{1}{c_{44}^{\pm}} g_{12}^{\pm}, \quad g_{34}^{\pm} = -g_{12}^{\pm}, \quad g_{41}^{\pm} = -\frac{1}{c_{33}^{\pm}}, \quad g_{43}^{\pm} = \frac{c_{13}^{\pm}}{c_{33}^{\pm}} g_{12}^{\pm}, \quad r^2 = \alpha_1^2 + \alpha_2^2,$$

$$\mathbf{L}_{\pm} = \{l_{kj}^{\pm}\}_{k,j=1,2}, \quad l_{11}^{\pm} = (-i\alpha_3)r^{-2}, \quad l_{22}^{\pm} = (-i\alpha_3)r^2, \quad l_{21}^{\pm} = -\frac{r^2}{c_{44}^{\pm}}, \quad l_{21}^{\pm} = -c_{66}r^4,$$

Directly from equations (8) we obtain $\mathbf{U}_1^{\pm} = \mathbf{L}_{\pm}^{-1}\mathbf{F}_1^{\pm}$, $\mathbf{U}_2^{\pm} = \mathbf{G}_{\pm}^{-1}\mathbf{F}_2^{\pm}$, where $\mathbf{L}_{\pm}^{-1} = \{l_{kj}^{*,\pm}\}_{k,j=1,2}$, $\mathbf{G}_{\pm}^{-1} = \{g_{kj}^{*,\pm}\}_{i,j=1,\ldots,4}$. Further, using representations (6), (7) after applying the inverse Fourier transform, we obtain such representations for σ_z, τ_{xz}, τ_{yz}:

$$\sigma_z = -\sum_{n=1}^{3}\left(\frac{R_{1,n}^0}{\sqrt{r_1^2 + (\xi_n|z - z_1|)^2}} - \frac{\omega_{1,n}}{\sqrt{r_1^2 + (\hat{\xi}_n z + \breve{\xi}_0 z_1)^2}}\right) + \sum_{n=1}^{2}\sum_{m=1}^{3}\frac{\alpha_{1,n,m}}{\sqrt{r_1^2 + (\hat{\xi}_n z + \breve{\xi}_m z_1)^2}}$$
$$+ \sum_{j=1}^{2} P_j \vartheta_{2j}\left\{\sum_{n=1}^{2}\frac{R_{1,2,n}^*}{[r_0^2 + (\xi_n|z - z_0|)^2]^{3/2}} + \sum_{n,m=1}^{2}\frac{\beta_{1,n,m}^2}{[r_0^2 + (\hat{\xi}_n z + \breve{\xi}_m z_0)^2]^{3/2}}\right\}$$
$$- P_3\left\{\sum_{n=1}^{2}\frac{|z - z_0|\hat{R}_{1,1,n}^*}{[r_0^2 + (\xi_n|z - z_0|)^2]^{3/2}} - \sum_{n,m=1}^{2}\frac{z\hat{\beta}_{1,n,m}^1 + z_0\breve{\beta}_{1,n,m}^1}{[r_0^2 + (\hat{\xi}_n z + \breve{\xi}_m z_0)^2]^{3/2}}\right\}, \tag{9}$$

$$\tau_{xz} = \vartheta_{11}\left\{\sum_{n=1}^{3}\frac{R_{2,n}^0[r_0^2 + (\xi_n|z - z_0|)^2]^{-1/2}}{\xi_n|z - z_0| + \sqrt{r_0^2 + (\xi_n|z - z_0|)^2}} - \sum_{n=1}^{3}\frac{\omega_{2,n}[r_0^2 + (\hat{\xi}_n z + \breve{\xi}_0 z_0)^2]^{-1/2}}{\hat{\xi}_n|z| + \breve{\xi}_0|z_0| + \sqrt{r_0^2 + (\hat{\xi}_n z + \breve{\xi}_0 z_0)^2}}\right.$$
$$\left. - \sum_{n=1}^{2}\sum_{m=1}^{3}\frac{\alpha_{2,n,m}[r_0^2 + (\hat{\xi}_n z + \breve{\xi}_m z_0)^2]^{-1/2}}{\hat{\xi}_n|z| + \breve{\xi}_m|z_0| + \sqrt{r_0^2 + (\hat{\xi}_n z + \breve{\xi}_m z_0)^2}}\right\}$$
$$+ \sum_{j=1}^{2} P_j\left(\partial_2\vartheta_{1j}\left\{\frac{(-1)^{j-1}S_{11}[r_0^2 + (\xi_0^+|z - z_0|)^2]^{-1/2}}{\xi_0^+|z - z_0| + \sqrt{r_0^2 + (\xi_0^+|z - z_0|)^2}} + \frac{(-1)^{j-1}\tilde{\beta}_1[r_0^2 + (\hat{\xi}_0 z + \breve{\xi}_0 z_0)^2]^{-1/2}}{\hat{\xi}_0 z + \breve{\xi}_0 z_0 + \sqrt{r_0^2 + (\hat{\xi}_0 z + \breve{\xi}_0 z_0)^2}}\right\}\right.$$
$$\left. + \partial_1\vartheta_{2j}\sum_{n,m=1}^{2}\left\{\frac{-R_{2,2,n}^*[r_0^2 + (\xi_n|z - z_0|)^2]^{-1/2}}{|z - z_0|\xi_n^+ + \sqrt{r_0^2 + (\xi_n|z - z_0|)^2}} + \frac{\beta_{2,n,m}^2[r_0^2 + (\hat{\xi}_n z + \breve{\xi}_m z_0)^2]^{-1/2}}{\hat{\xi}_n|z| + \breve{\xi}_m|z_0| + \sqrt{r_0^2 + (\hat{\xi}_n z + \breve{\xi}_m z_0)^2}}\right\}\right)$$
$$- \vartheta_{11}\left\{\sum_{n=1}^{2}\frac{R_{2,1,n}^*}{[r_0^2 + (\xi_n|z - z_0|)^2]^{3/2}} + \sum_{n,m=1}^{2}\frac{\beta_{q,n,m}^1}{[r_0^2 + (\hat{\xi}_n z + \breve{\xi}_m z_0)^2]^{3/2}}\right\}, \tag{10}$$

$$\tau_{yz} = \vartheta_{12}\left\{\sum_{n=1}^{3}\frac{R^0_{2,n}[r_0^2+(\xi_n|z-z_0|)^2]^{-1/2}}{\xi_n|z-z_0|+\sqrt{r_0^2+(\xi_n|z-z_0|)^2}} - \sum_{n=1}^{3}\frac{\omega_{2,n}[r_0^2+(\hat{\xi}_n z+\check{\xi}_0 z_0)^2]^{-1/2}}{\hat{\xi}_n|z|+\check{\xi}_0|z_0|+\sqrt{r_0^2+(\hat{\xi}_n z+\check{\xi}_0 z_0)^2}}\right.$$
$$\left. - \sum_{n=1}^{2}\sum_{m=1}^{3}\frac{\alpha_{2,n,m}[r_0^2+(\hat{\xi}_n z+\check{\xi}_m z_0)^2]^{-1/2}}{\hat{\xi}_n|z|+\check{\xi}_m|z_0|+\sqrt{r_0^2+(\hat{\xi}_n z+\check{\xi}_m z_0)^2}}\right\}$$
$$+\sum_{j=1}^{2}P_j\left(\partial_1\vartheta_{1j}\left\{\frac{(-1)^{j-1}S_{11}[r_0^2+(\xi_0^+|z-z_0|)^2]^{-1/2}}{\xi_0^+|z-z_0|+\sqrt{r_0^2+(\xi_0^+|z-z_0|)^2}} + \frac{(-1)^{j-1}\tilde{\beta}_1[r_0^2+(\hat{\xi}_0 z+\check{\xi}_0 z_0)^2]^{-1/2}}{\hat{\xi}_0 z+\check{\xi}_0 z_0+\sqrt{r_0^2+(\hat{\xi}_0 z+\check{\xi}_0 z_0)^2}}\right\}\right.$$
$$\left.+\partial_2\vartheta_{2j}\sum_{n,m=1}^{2}\left\{\frac{-R^*_{2,2,n}[r_0^2+(\xi_n|z-z_0|)^2]^{-1/2}}{|z-z_0|\xi_n^+ +\sqrt{r_0^2+(\xi_n|z-z_0|)^2}} + \frac{\beta^2_{2,n,m}[r_0^2+(\hat{\xi}_n z+\check{\xi}_m z_0)^2]^{-1/2}}{\hat{\xi}_n|z|+\check{\xi}_m|z_0|+\sqrt{r_0^2+(\hat{\xi}_n z+\check{\xi}_m z_0)^2}}\right\}\right)$$
$$+\vartheta_{12}\left\{\sum_{n=1}^{2}\frac{R^*_{2,1,n}}{[r_0^2+(\xi_n|z-z_0|)^2]^{3/2}} + \sum_{n,m=1}^{2}\frac{\beta^1_{q,n,m}}{[r_0^2+(\hat{\xi}_n z+\check{\xi}_m z_0)^2]^{3/2}}\right\}, \quad (11)$$

where

$$q_j^{\pm}(\alpha_3,r) = g^{*,\pm}_{j2}(\alpha_3,r)\gamma_1^{\pm}r^2 + g^{*,\pm}_{j4}(\alpha_3,r)\gamma_2^{\pm}, \quad j=\overline{1,4},$$
$$R^{0,+}_{1,n} = \frac{q_1^+(-i\xi_n^+,1)\tau^{++}(-i\xi_n^+,1)}{\xi_n^+h_n^+}, \quad \bar{R}^{0,+}_{1,n} = \frac{q_1^+(i\xi_n^+,1)\tau^{++}(i\xi_n^+,1)}{\xi_n^+h_n^+},$$
$$\tau^{++}(\alpha_3,r) = (-i\alpha_3)\lambda_3^+m_1^+ - rm_2^+, \quad \tau^{-+}(\alpha_3,r) = (-i\alpha_3)\lambda_3^-m_1^+ - rm_2^+,$$
$$\tau^{+-}(\alpha_3,r) = (-i\alpha_3)\lambda_3^+m_1^- - rm_2^-, \quad \tilde{\beta}^{-+}_{1,n} = \frac{q_1^-(i\xi_n^{\pm},1)\tau^{-+}(i\xi_n^-,1)}{\xi_n^-h_n^-},$$
$$\tilde{\beta}^{+-}_{1,n} = \frac{q_1^+(-i\xi_n^+,1)\tau^{+-}(-i\xi_n^+,1)}{\xi_n^+h_n^+}, \quad \tilde{\beta}^{++}_{j,n} = \frac{q_j^+(-i\xi_n^+,1)\tau^{++}(-i\xi_n,1)}{\xi_n^+h_n^+}, \quad j=\overline{1,4},$$
$$h_n^{\pm} = \prod_{l=1,l\neq n}^{3}(\xi_n^{\pm})^2-(\xi_l^{\pm})^2, \quad \tau^{--}(\alpha_3,r) = (-i\alpha_3)\lambda_3^-m_1^- - rm_2^-,$$
$$\tilde{\beta}^{--}_{1,n} = \frac{q_1^-(i\xi_n^-,1)\tau^{--}(i\xi_n^-,1)}{\xi_n^-h_n^-}, \quad R^{\pm}_{j,k} = \sum_{n=1}^{2}R^{*,+}_{j,k,n}, \quad \beta_j^{\pm\pm} = \sum_{n=1}^{3}\tilde{\beta}^{\pm\pm}_{j,n}, \quad \beta_j^{\pm\mp} = \sum_{n=1}^{3}\tilde{\beta}^{\pm\mp}_{j,n},$$
$$\mathbf{A}_0^{-1} = \{a^*_{kj}\}_{k,j=\overline{1,4}}, \quad \mathbf{A}_0 = \{a_{k,j}\}_{k,j=\overline{1,4}} = \mathbf{N}^+ + \bar{\mathbf{N}}^-, \quad \mathbf{N}^{\pm} = \{R^{\pm}_{j,k}\}_{k,j=\overline{1,4}},$$
$$\alpha^+_{j,m} = \sum_{k=1}^{4}a^*_{jk}\bar{R}^{0,+}_{k,n}, \quad \alpha^-_{j,m} = \sum_{k=1}^{4}a^*_{jk}R^{0,-}_{k,m}, \quad \mu_j^+ = \sum_{k=1}^{4}a^*_{jk}\beta_k^+, \quad \mu_j^- = \sum_{k=1}^{4}a^*_{jk}\beta_k^-,$$
$$\beta_j^+ = \beta_j^{++}+\beta_j^{-+}, \quad \beta_j^- = \beta_j^{--}+\beta_j^{+-}, \quad \vartheta_{1j} = \frac{(y-y_0)^{2-j}}{(x-x_0)^{1-j}}, \quad \vartheta_{2j} = \frac{(x-x_0)^{2-j}}{(y-y_0)^{1-j}}$$
$$\alpha^{++}_{j,n,m} = \sum_{k=1}^{4}R^{*,+}_{j,k,n}\alpha^+_{k,m}, \quad \alpha^{+-}_{j,n,m} = \sum_{k=1}^{4}R^{*,+}_{j,k,n}\alpha^+_{k,m},$$
$$\alpha^{--}_{j,n,m} = \sum_{k=1}^{4}R^{*,-}_{j,k,n}\alpha^+_{k,m}, \quad \alpha^{-+}_{j,n,m} = \sum_{k=1}^{4}R^{*,-}_{j,k,n}\alpha^+_{k,m},$$
$$\mu_j^{++} = \sum_{k=1}^{4}R^{*,+}_{j,k,n}\mu_j^+, \qquad \mu^{+-}_{j,n} = \sum_{k=1}^{4}R^{*,+}_{j,k,n}\mu_j^-, \quad n=1,2, \qquad \mu^{+-}_{j,n}=0, \quad n=3,$$

$$\mu_{j,n}^{--} = \sum_{k=1}^{4} R_{j,k,n}^{*,-} \mu_j^{-}, \qquad \mu_{j,n}^{-+} = \sum_{k=1}^{4} R_{j,k,n}^{*,-} \mu_j^{+}, \quad n = 1, 2, \qquad \mu_{j,n}^{--} = \mu_{j,n}^{-+} = 0, \quad n = 3,$$
$$\tilde{\omega}_{j,n}^{\pm\pm} = \tilde{\beta}_{j,n}^{\pm\pm} - \mu_{j,n}^{\pm\pm}, \quad \tilde{\omega}_{j,n}^{\pm\mp} = \tilde{\beta}_{j,n}^{\pm\mp} - \mu_{j,n}^{\pm\mp},$$
$$\hat{R}_{j,n}^{0,+} = \theta(z - z_0) R_{j,n}^{0,+} + \theta(z_0 - z) \bar{R}_{j,n}^{0,+}, \quad \breve{R}_{j,n}^{0,-} = \theta(z - z_0) R_{j,n}^{0,-} + \theta(z_0 - z) \bar{R}_{j,n}^{0,-}.$$

3. Stress fields in the plane of connection of half-spaces

Putting $z = 0$ in the obtained expressions (9)–(11), we obtain the distribution of normal and tangential stresses in the plane of connection of half-spaces in the presence of a stationary source of heat of the average force:

$$\sigma_z(x, y) = -\sum_{n=1}^{3} \frac{R_{1,n}^0}{\sqrt{r_0^2 + (\xi_n z_0)^2}} + \frac{\omega_1}{\sqrt{r_0^2 + (\xi_0 z_0)^2}} + \sum_{n=1}^{3} \frac{\alpha_{1,n}}{\sqrt{r_0^2 + (\xi_n z_0)^2}}$$
$$+ \sum_{j=1}^{2} P_j \sum_{n=1}^{2} B_{1,n} \frac{\vartheta_{2j} z_0}{[r_0^2 + (\xi_n z_0)^2]^{3/2}} - P_3 \sum_{n=1}^{2} \frac{A_{1,n} z_0}{[r_0^2 + (\xi_n z_0)^2]^{3/2}},$$
$$\tau_{xz}(x, y) = \vartheta_{11} \left\{ \sum_{n=1}^{3} \frac{R_{2,n}^0 [r_1^2 + (\xi_n z_1)^2]^{-1/2}}{\xi_n |z_1| + \sqrt{r_1^2 + (\xi_n z_1)^2}} - \frac{\omega_2 [r_1^2 + (\xi_1 z_0)^2]^{-1/2}}{\breve{\xi}_0 |z_0| + \sqrt{r_1^2 + (\xi_0 z_1)^2}} \right.$$
$$\left. - \sum_{n=1}^{3} \frac{\alpha_{2,n} [r_1^2 + (\xi_n z_1)^2]^{-1/2}}{[\xi_n |z_1| + \sqrt{r_1^2 + (\xi_n z_1)^2}]} \right\} - \sum_{j=1}^{2} P_j \left\{ \partial_2 \frac{\vartheta_{1j} (-1)^j S_1 [r_0^2 + (\xi z_0)^2]^{-1/2}}{\xi |z_0| + \sqrt{r_0^2 + (\xi z_0)^2}} \right.$$
$$\left. + \partial_1 \sum_{n=1}^{2} \frac{\vartheta_{2j} B_{2,n} [r_0^2 + (\xi_n z_0)^2]^{-1/2}}{\xi_n |z_0| + \sqrt{r_0^2 + (\xi_n z_0)^2}} \right\} + P_3 \sum_{n=1}^{2} \frac{A_{2,n}(x - x_0)}{[r_0^2 + (\xi_n z_0)^2]^{3/2}},$$
$$\tau_{yz}(x, y) = \vartheta_{12} \left\{ \sum_{n=1}^{3} \frac{R_{2,n}^0 [r_0^2 + (\xi_n z_0)^2]^{-1/2}}{\xi_n |z_0| + \sqrt{r_0^2 + (\xi_n z_0)^2}} - \frac{\omega_2 [r_0^2 + (\breve{\xi}_0 z_0)^2]^{-1/2}}{\breve{\xi}_0 |z_0| + \sqrt{r_0^2 + (\breve{\xi}_0 z_0)^2}} \right.$$
$$\left. - \sum_{n=1}^{3} \frac{\alpha_{2,n} [r_0^2 + (\breve{\xi}_n z_0)^2]^{-1/2}}{\breve{\xi}_n |z_0| + \sqrt{r_0^2 + (\breve{\xi}_n z_0)^2}} \right\} - \sum_{j=1}^{2} P_j \left\{ -\partial_1 \frac{\vartheta_{1j} (-1)^{3-j} [r_0^2 + (\xi z_0)^2]^{-1/2} S_1}{\xi |z_0| + \sqrt{r_0^2 + (\xi z_0)^2}} \right.$$
$$\left. + \partial_2 \sum_{n=1}^{2} \frac{\vartheta_{2j} B_{2,n} [r_0^2 + (\xi_n z_0)^2]^{-1/2}}{\xi_n |z_0| + \sqrt{r_0^2 + (\xi_n z_0)^2}} \right\} + P_3 \sum_{n=1}^{2} \frac{A_{2,n}(y - y_0)}{[r_0^2 + (\xi_n z_0)^2]^{3/2}},$$

where

$$S_p = \theta(z_0) S_{p1}^{+} + \theta(-z_0) S_{p1}^{-}, \quad A_{p,n} = \theta(z_0) A_{p,n}^{+} + \theta(-z_0) A_{p,n}^{-}, \quad p = 1, 2,$$
$$B_{p,n} = \theta(z_0) B_{p,n}^{+} + \theta(-z_0) B_{p,n}^{-}, \quad A_{1,n}^{+} = -\hat{R}_n^{+} + \hat{\beta}_n^{++}, \quad A_{1,n}^{-} = -\breve{\beta}_n^{+-},$$
$$\hat{\beta}_n^{\pm\pm} = \sum_{m=1}^{2} \hat{\beta}_{m,n}^{\pm\pm}, \quad \breve{\beta}_n^{\pm\mp} = \sum_{m=1}^{2} \breve{\beta}_{m,n}^{\pm\mp}, \quad \beta_{p,k,n}^{\pm\pm} = \sum_{m=1}^{2} \beta_{k,m,n}^{p,\pm\pm}, \quad \beta_{p,k,n}^{\pm\mp} = \sum_{m=1}^{2} \beta_{k,m,n}^{p,\pm\mp},$$
$$A_{k,n}^{+} = -R_{k,1,n}^{-} + \beta_{1,k,n}^{++}, \quad A_{k,n}^{-} = -\beta_{1,k,n}^{+-}, \quad k = 2, 3, 4,$$
$$B_{k,n}^{+} = -R_{k,2,n}^{+} + \beta_{2,k,n}^{++}, \quad B_{k,n}^{-} = -\beta_{2,k,n}^{+-}, \quad k = 1, \ldots, 4.$$

4. Discussion and numerical results

Numerical studies of the stress distribution were carried out depending on the thermophysical properties and the presence at the points $M_1(1, -1, 1)$ and $M_2(-1, 1, -1)$ stationary concentrated

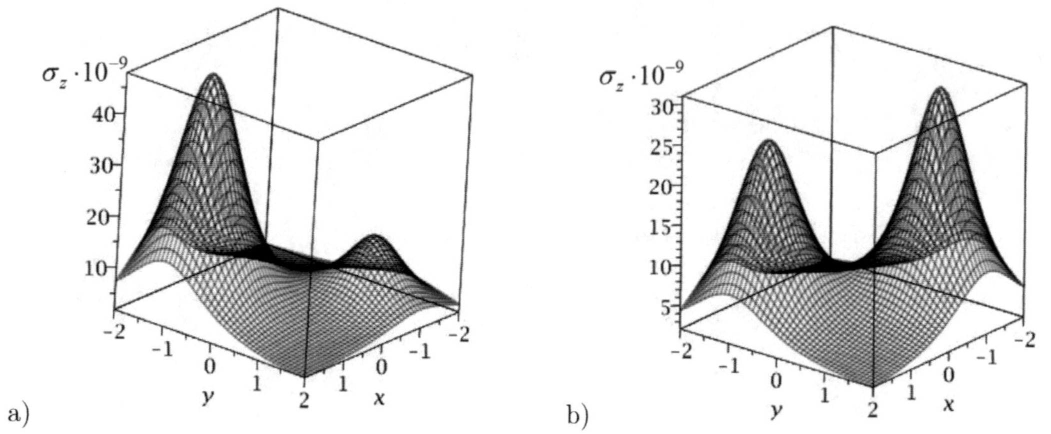

Figure 1. Normal stress distribution for material combination m1–m2.

heat sources with a capacity $Q_1 = 2 \times 10^4$ J and $Q_2 = 10^4$ J. Figures 1 and 2 show the distribution of normal stresses σ_z, in plane $z0y$ for some combinations of transversely isotropic materials. Calculations were performed for the combination of materials: cadmium (material $m1$), magnesium (material $m2$).

Table 1 shows the values of thermoelastic constants of these materials and table 2 shows the attitudes of thermoelastic constants.

Table 1. Material properties of transversely isotropic materials.

	$c_{ij} \times 10^{11}$, N/m^2						$\alpha_j \times 10^{-5}$, (°C)$^{-1}$		λ_j, W/m·°C	
Material	c_{11}	c_{12}	c_{13}	c_{33}	c_{44}	c_{55}	$\alpha_1 = \alpha_2$	α_3	$\lambda_1 = \lambda_2$	λ_3
$m1$	1.08	0.389	0.375	0.46	0.156	0.156	54	20.2	93	94
$m2$	0.5952	0.256	0.214	0.6147	0.1647	0.1647	27.7	20.2	159	160

Table 2. Attitudes of thermoelastic constants.

Material combination	$\zeta_{ij} = c^+_{ij}/c^-_{ij}$						α^+_1/α^-_1	α^+_3/α^-_3	λ^+_1/λ^-_1	λ^+_3/λ^-_3
$m1$–$m2$	1.814	1.519	1.752	0.748	0.947	0.947	1.949	1	0.596	0.587
$m2$–$m1$	0.551	0.658	0.571	1.336	1.055	1.055	0.513	1	1.677	1.702

In figures 1a and 2b show the distribution of normal stresses when the power of the heat source in the upper half-space is greater than in the lowerand in figures 1a and 2b vice versa. As can be seen from the above figures, the ratio of the values of the thermoelastic constants of half-spaces has a significant effect on the distribution of stresses near the inclusion, in particular, the value of the coefficient $\zeta_{33} = A^+_{33}/A^-_{33}$, and attitudes of the thermal conductivity coefficients λ^+_3/λ^-_3 which characterizes the difference between the elastic and thermal properties of half-spaces in the direction of the axis z. This results in both a qualitative and quantitative change in the distribution of the normal stresses, as evidenced, for example, by comparing the graphs of figures.

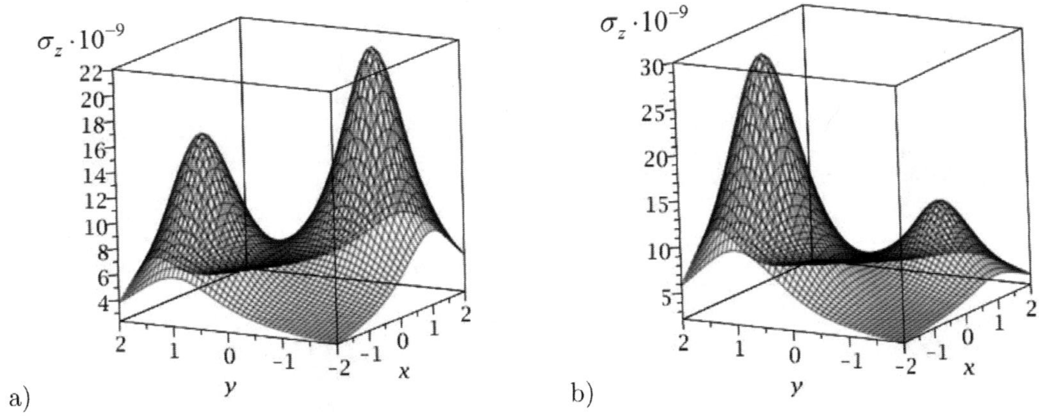

Figure 2. Normal stress distribution for material combination m2–m1.

Conclusion

Built fundamental solution of the thermoelasticity problem for a piecewise-homogeneous transversely isotropic space is constructed in an explicit form, which made it possible to study how temperature affects the stress distribution in the plane of joining materials. In particular, it is shown that the inhomogeneity of the thermoelastic characteristics of the medium significantly affects the stress distribution in space.

References

[1] Kryvyi O F 2012 Interface crack in the inhomogeneous transversely isotropic space *Materials Science* **47** (6) 726-36 doi: 10.1007/s11003-012-9450-9

[2] Kryvyi O F 2014 Delaminated interface inclusion in a piecewise homogeneous transversely isotropic space *Materials Science* **50** (2) 245–53 doi: 10.1007/s11003-014-9714-7

[3] Kit H S and Sushko O P 2011 Problems of stationary heat conduction and thermoelasticity for a body with a heat permeable disk-shaped inclusion (crack) *Journal of Mathematical Sciences* **174** (3) 309 doi: 10.1007/s10958-011-0300-3

[4] Kit H S and Sushko O P 2011 Axially symmetric problems of stationary heat conduction and thermoelasticity for a body with thermally active or thermally insulated disk inclusion (crack) *Journal of Mathematical Sciences* **176** (4) 561-77 doi: 10.1007/s10958-011-0422-7

[5] Li X-F and Fan T-Y 2001 The asymptotic stress field for a ring circular inclusion at the interface of two bonded dissimilar elastic half-space materials *International Journal of Solids and Structures* **38** (44-45) 8019–35 doi: 10.1016/S0020-7683(01)00010-5

[6] Yue Z Q 1995 Elastic fields in two joined transversely isotropic solids due to concentrated forces *International Journal of Engineering Science* **33** (3) 351–69 doi: 10.1016/0020-7225(94)00063-P

[7] Hou P-F, Leung A T Y, and He Y-J 2008 Three-dimensional Green's functions for transversely isotropic thermoelastic bimaterials *International Journal of Solids and Structures* **45** (24) 6100–13 doi: 10.1016/j.ijsolstr.2008.07.022

[8] Hou P-F, Zhao M, and Ju J-W 2013 Three-dimensional Green's functions for transversely isotropic thermoporoelastic biomaterials *Journal of Applied Geophysics* **95** 36–46 doi: 10.1016/j.jappgeo.2013.05.001

[9] Kryvyi O F and Morozov Yu O 2019 Solution of the problem of heat conduction for the transversely isotropic piecewise-homogeneous space with two circular inclusions *Journal of Mathematical Sciences* **243** (1) 162–82 doi: 10.1007/s10958-019-04533-1

[10] Kryvyi O and Morozov Yu 2020 Thermally active interphase inclusion in a smooth contact conditions with transversely isotropic half-spaces *Frattura ed Integrità Strutturale* **14** (52) 33–50 doi: 10.3221/IGF-ESIS.52.04

[11] Kumar R and Gupta V 2014 Green's function for transversely isotropic thermoelastic diffusion bimaterials *Journal of Thermal Stresses* **37** (10) 1201–29 doi: 10.1080/01495739.2014.936248

[12] Mantič V, Távara L, Ortiz J E, and París F 2012 Recent developments in the evaluation of the 3D fundamental solution and its derivatives for transversely isotropic elastic materials *Electronic Journal of Boundary Elements* **10** (1) 1–41 doi: 10.14713/ejbe.v10i1.1116

[13] Houa P-F, Li Z-S, and Zhanga Y 2014 Three-dimensional quasi-static Green's function for an infinite

transversely isotropic pyroelectric material under a step point heat source *Mechanics Research Communications* **62** 66–76 doi: 10.1016/j.mechrescom.2014.08.008

[14] Pan E and Chen W 2015 Green's functions in a transversely isotropic magnetoelectroelastic bimaterial space In *Static Green's Functions in Anisotropic Media* (Cambridge University Press) pp. 220–259 doi: 10.1017/CBO9781139541015.008

[15] Zhao YF, Zhao MH, Pan E, and Fan CY 2015 Green's functions and extended displacement discontinuity method for interfacial cracks in three-dimensional transversely isotropic magneto-electro-elastic bi-materials *International Journal of Solids and Structures* **52** 56–71 doi: 10.1016/j.ijsolstr.2014.09.018

[16] Hou P-F, Zhao M, Tong J, and Fu B 2013 Three-dimensional steady-state Green's functions for fluid-saturated, transversely isotropic, poroelastic biomaterials *Journal of Hydrology* **496** 217–24 doi: 10.1016/j.jhydrol.2013.05.017

[17] Sahebkar K and Eskandari-Ghadi M 2017 Displacement ring load Green's functions for saturated porous transversely isotropic tri-material full-space *International Journal for Numerical and Analytical Methods in Geomechanics* **41** (3) 359–81 doi: 10.1002/nag.2560

[18] Akbari F, Khojasteh A, and Rahimian M 2016 Three-dimensional interfacial Green's function for exponentially graded transversely isotropic bi-materials *Civil Engineering Infrastructures Journal* **49** (1) 71–96 doi: 10.7508/CEIJ.2016.01.006

[19] Zafari Y, Shahmohamadi M, Khojasteh A, and Rahimian M 2019 Asymmetric Green's functions for a functionally graded transversely isotropic tri-material *Applied Mathematical Modelling* **72** 176–201 doi: 10.1016/j.apm.2019.02.038

[20] Kushnir R M and Protsyuk Yu B 2011 Thermoelastic state of layered thermosensitive bodies of revolution for the quadratic dependence of the heat-conduction coefficients *Materials Science* **46** (1) 1–15 doi: 10.1007/s11003-010-9258-4

[21] Kushnir R and Protsiuk B 2009 A method of the Green's functions for quasistatic thermoelasticity problems in layered thermosensitive bodies under complex heat exchange In *Operator Theory: Advances and Applications.* Vol. 191. Adamyan V M, Gohberg I, Kochube A, et al. (Editors) *Modern Analysis and Applications* (Birkhäuser Basel) pp. 143–54 doi: 10.1007/978-3-7643-9921-4_9

[22] Kushnir R M 2010 Thermal stresses Advanced theory and applications *Journal of Thermal Stresses* **33** (1) 76-8 doi: 10.1080/01495730903538421

[23] Tokovyy Yu V and Ma C C 2017 Three-dimensional elastic analysis of transversely-isotropic composites *Journal of Mechanics* **33** (6) 821-30 doi: 10.1017/jmech.2017.91

[24] Tokovyy Y 2019 Direct integration of three-dimensional thermoelasticity equations for a transversely isotropic layer *Journal of Thermal Stresses* **42** (1) 49–64 doi: 10.1080/01495739.2018.1526150

[25] Boiko D S and Tokovyy Y V 2021 Determination of three-dimensional stresses in a semi-infinite elastic transversely isotropic composite *Mechanics of Composite Materials* **57** (4) 481-92 doi: 10.1007/s11029-021-09971-0

[26] Kryvyy O 2009 The discontinuous solution for the piece-homogeneous transversal isotropic medium In *Operator Theory: Advances and Applications.* Vol. 191. Adamyan V M, Gohberg I, Kochube A, et al. (Editors) *Modern Analysis and Applications* (Birkhäuser Basel) pp. 395–406 doi: 10.1007/978-3-7643-9921-4_25

[27] Kryvyy O F 2012 Interface circular inclusion under mixed conditions of interaction with a piecewise homogeneous transversally isotropic space *Journal of Mathematical Sciences* **184** (1) 101–19 doi: 10.1007/s10958-012-0856-6

[28] Kryvyy O 2014 Tunnel internal crack in a piecewise homogeneous anisotropic space *Journal of Mathematical Sciences* **198** (1) 62–74 doi: 10.1007/s10958-014-1773-7

Methods for determining shear resistance by a system of cracks and old slip surfaces with the purpose to evaluate the stability of landslides

S G Hayroyan* **and H S Hayroyan**
Yerevan State University, Yerevan, Armenia
E-mail: *hairoyan@ysu.am

Abstract. On slopes, landslide displacements mainly occur along bedding boundaries, lithogenetic and tectonic fractures, as well as along existing ancient slip surfaces, which are the weak zones and potential slip surfaces for new landslides.

According to experimental studies to determine the shear resistance by the standard and double cut test methods: a) without wetting the sliding surface, and b) by the wetting the sliding surface, we can conclude: when double- cutting the clay soils of a solid, semi-solid and tight-plastic consistency, the cohesion is mainly reduced, but the internal friction factor is slightly changed.

Double-cut tests of soft-plastic and flow-plastic clay soils have shown the cohesion and the internal friction factors are being changed insignificantly.

At the same time, taking into account the fact that under the condition $\tau > \sigma \tan \varphi$ the soft-plastic and flowing-plastic clay soils undergo viscous-plastic deformations, we come to the following conclusion.

When developing the anti-landslide measures, independent of the consistency of the soils forming the slope, the $\tau_f = \sigma \tan \varphi$ component should be taken as the slip resistance index in the formula for calculating the slip resistance taking into account the condition $C_c + \Sigma_w = 0$.

1. Introduction

The landslide movements in the mountain slopes occur mainly within the layering (bedding) boundaries, through the lythogen and tectonic cracks and existing old landslide slip surfaces that serve as potential weak zones for that kind of earth ground move [1, 2, 5, 6, 7]. Though potential landslide slopes are much dangerous for investigation, sampling of that intervals in the drill holes aren't always possible, because these zones are water saturated and weak. Depending on geology construction, hydrogeological conditions and ground consistency of the landslide area determination of the coefficients of landslide slip resistance must be carried out by following this scheme:

- Standard method [4]. Determined coefficients ca be applied for estimation of unmoved blocks of ground or blocks of landslide stability, where the stability coefficient does determines by the method of Tertsagi's circle-cylinder sliding surface.
- The coefficient of the landslide slip resistance determined by the double cut method experiment without of additional wetting can be applied for earth move through the

Table 1. Indices of soil hydrophysical properties.

Drill hole N	Interval	Density g/cm^3			Moisture	Void ratio	Plasticity			Liquidity index	Degree of saturation
		Bulk	Particle	Dry			Liquid limit	Plastic limit	Plasticity index		
		ρ	ρ_s	ρ_d	W	e	W_L	W_P	I_P	I_L	S_r
Rigid, semi rigid, tight-plastic consistency, $I_L < 0$, $0 < I_L < 0.5$ (area I)											
Bh-5	20.2–20.4	2.09	2.72	1.805	0.158	0.507	0.298	0.208	0.090	< 0	0.850
Rigid, semi rigid, strength plastic consistency, $I_L < 0$, $0 < I_L < 0.5$ (area II)											
Bh-5	18.0–18.2	1.99	2.70	1.63	0.221	0.656	0.302	0.230	0.072	< 0	0.909
Rigid, semi rigid, tight-plastic consistency, $I_L < 0$, $0 < I_L < 0.5$ (area III)											
Bh-1	16.1–16.3	1.99	2.72	1.66	0.198	0.639	0.327	0.215	0.112	< 0	0.843
Bh-9	17.0–17.3	1.89	2.71	1.55	0.214	0.748	0.328	0.154	0.174	0, 345	0.775
Rigid, semi rigid, tight-plastic consistency $0.50 < I_L < 1$ (area II)											
Bh-10	16.4–16.6	1.53	2.26	1.059	0.445	1.134	0.457	0.345	0.112	0.893	0.887
Bh-13	22.1–22.3	1.91	2.64	1.60	0.191	0.650	0.256	0.142	0.114	0.675	0.850
Bh-2	36.0–36.2	1.90	2.74	1.39	0.368	0.986	0.395	0.201	0.194	0.861	1.00

landslide slip surface or tectonic crack. Such kind of landslide move may become under impact of the slope cut and seismic forces.

- The coefficients of the landslide slip determined by the method of double cut of sliding resistance through the wetting surfaces must be applied for sliding surfaces in case when the slopes are water saturated or wetted making possible to get the ground moved

2. Experimental part

Experiments on ground's sliding resistance determination test in this work were performed on samples taken from the landslide slope near the International School in Dilijan, Tavush Marz, RA. Sample test schemes were selected to model the nature of the ground strength loss and deformation in the slope mass. The naturally occurring clay soils were in different physical states, with consistency values ranging from rigid: $I_L < 0$ to flowing plastic: $0.75 < I_L < 1$. Slip resistance tests were performed with a GGP-30 [4] type slide cutting device at the "Hydro Project" institute. When selecting the test mode, the consistency of the clay soils was taken into account: solid, semi-solid, tight-plastic clay samples were tested in a consolidated slow mode, and soft plastic, flowing plastic samples in a non-consolidated fast mode. The indices of soil hydrophysical properties are provided in table 1.

3. Discussion of the obtained results

Diagrams of three different test schemes, carried out by the standard, wetted and without wetting surface and double cutting slip resistance methods are shown in figures 1 and 2. Table 2 provides the different consistencies clayish grounds slip resistance coefficients obtained under different conditions, that is, the double-cut non-wetted, with wetted surface cutting scheme and indices determined by the standard scheme of the sliding resistance test: internal coefficient's ($\tan\varphi_{\text{double}}/\tan\varphi_{\text{standard}}$; $\tan\varphi_{\text{double wet surface}}/\tan\varphi_{\text{standard}}$) and relationship index ratios ($C_{\text{double}}/C_{\text{standard}}$; $C_{\text{double wet surface}}/C_{\text{standard}}$).

Analyzing the results of the experimental studies in table 2, figures 1 and 2, we conclude as follows:

- In the case of solid consistency, the $C_{\text{double}}/C_{\text{standard}}$ ratio varies from 0.170 to 0.267.

Journal of Physics: Conference Series 2231 (2022) 012017 doi:10.1088/1742-6596/2231/1/012017

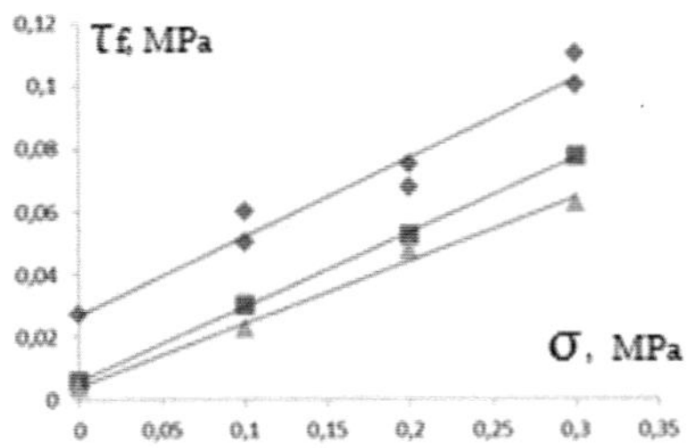

a. (Bh-5; 20,2-20,4)

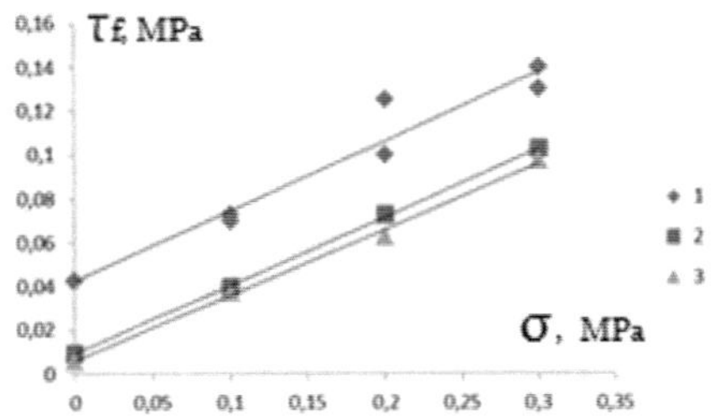

b. (Bh-5; 18,0,-18,2)

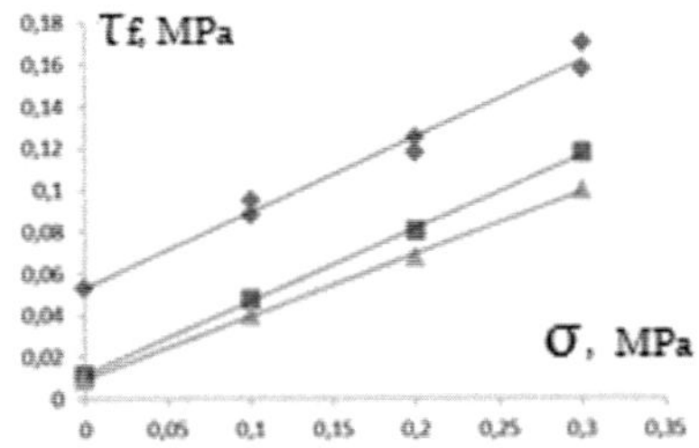

c. (Bh-1; 16,1-16,3)

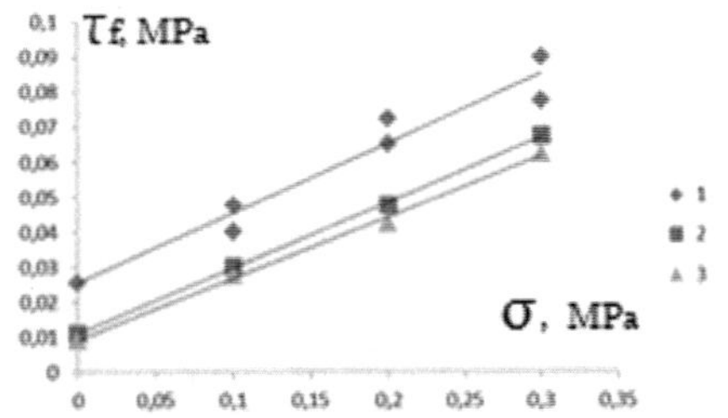

d(Bh-9; 17,0-17,3)

Figure 1. Area-3, sandy clay with a semi-solid consistency. The legend: 1. Standard method, 2. Double cutting method without surface wetting, 3. Dual cutting method with surface wetting.

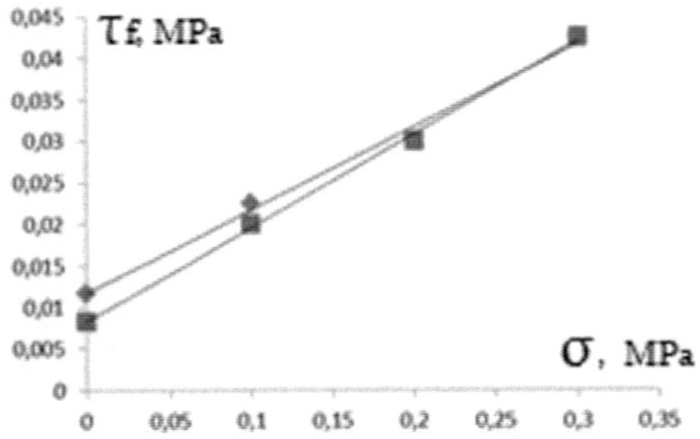

a(Bh-13; 22,1-22,3)

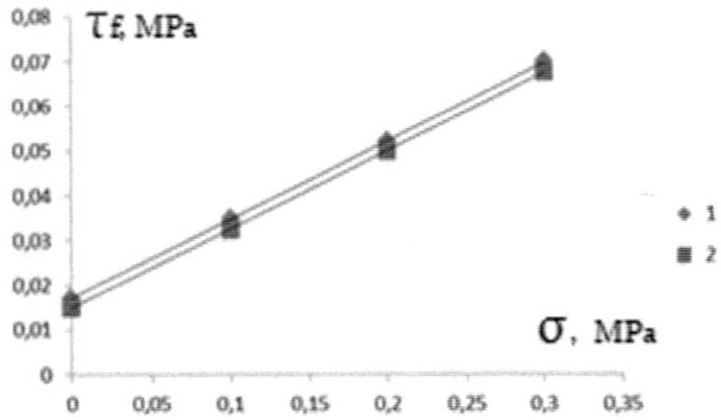

b(Bh-10; 16,4-16,6)

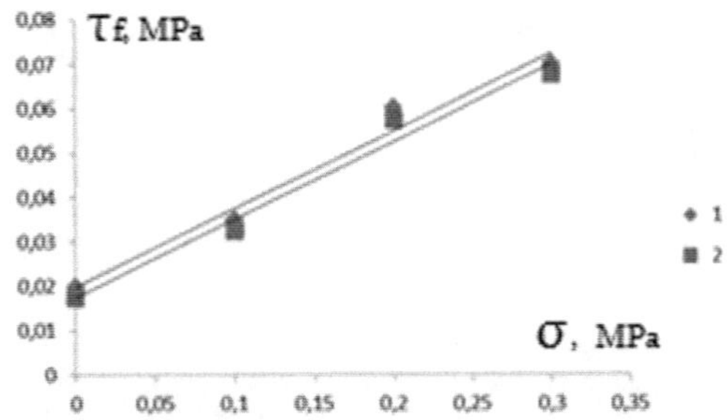

c(Bh-2; 36,0-36,2)

Figure 2. Sandy clay of soft-plastic and flow-plastic consistency. The legend: 1. Standard method, 2. Double cutting method.

Table 2. Consistencies clayish grounds slip resistance coefficients obtained under different conditions.

Drill hole N	Interval	Consistency index	Sliding resistance indice (standard test)		Sliding resistance indice (double, standard, test)		Sliding resistance indice (double sliding test, wetted surface)		$\frac{\tan\varphi_{\text{double}}}{\tan\varphi_{\text{standatd}}}$	$\frac{C_{\text{double}}}{C_{\text{standatd}}}$	$\frac{\tan\varphi_{\text{double wetted surface}}}{\tan\varphi_{\text{standatd}}}$	$\frac{C_{\text{double wetted surface}}}{C_{\text{standatd}}}$
		I_L	$\tan\varphi$	C, MPa	$\tan\varphi$	C, MPa	$\tan\varphi$	C, MPa				
Rigid, semi rigid, tight-plastic consistency, $I_L < 0$, $0 < I_L < 0.5$ (area I)												
Bh-5	20.2–20.4	< 0	0.413	0.041	0.400	0.007	0.350	0.004	0.968	0.171	0.847	0.102
Rigid, semi rigid, strength plastic consistency, $I_L < 0$, $0 < I_L < 0.5$ (area II)												
Bh-5	18.0–18.2	< 0	0.362	0.030	0.362	0.008	0.313	0.007	1.0	0.267	0.865	0.223
Rigid, semi rigid, tight-plastic consistency, $I_L < 0$, $0 < I_L < 0.5$ (area III)												
Bh-1	16.1–16.3	< 0	0.362	0.053	0.350	0.012	0.300	0.009	0.967	0.226	0.829	0.174
Bh-9	17.0–17.3	0.345	0.200	0.025	0.188	0.011	0.175	0.009	0.940	0.440	0.875	0.368
Rigid, semi rigid, tight-plastic consistency $0.50 < I_L < 1$ (area I)												
Rigid, semi rigid, tight-plastic consistency $0.50 < I_L < 1$ (area II)												
Bh-13	22.1–22.3	0.675	0.250	0.012	0.238	0.009	—	—	0.952	0.750	—	—
Bh-10	16.4–16.6	0.893	0.187	0.014	0.163	0.012	—	—	0.872	0.857	—	—
Bh-2	36.0–36.2	0.861	0.175	0.020	0.175	0.017	—	—	1.0	0.850	—	—

- The reduction of the internal friction coefficient is incomparably small, the ratio $\tan\varphi_{\text{double}}/\tan\varphi_{\text{standard}}$ varies from 0.807 to 1.0.
- In the case of semi-rigid consistency, the $C_{\text{double}}/C_{\text{standard}}$ ratio varies from 0.171 to 0.238 and $\tan\varphi_{\text{double}}/\tan\varphi_{\text{standard}}$ varies from 0.822 to 0.952; in the case of tight-plastic consistency the $C_{\text{double}}/C_{\text{standard}}$ ratio varies from 0.292 to 0.440 and $\tan\varphi_{\text{double}}/\tan\varphi_{\text{standard}}$ varies from 0.846 to 1.0.
- In the case of a flow-plastic consistency ($I_L > 0.75$), the $C_{\text{double}}/C_{\text{standard}}$ ratio ranges from 0.600 to 0.941 and $\tan\varphi_{\text{double}}/\tan\varphi_{\text{standard}}$ from 0.872 to 1.0 and correlation and internal contact coefficients change slightly.

In order to clarify the reasons for the changes in the slip resistance indices tested by the three different schemes above, we present the Sh. Kulon–N. Maslov slip resistance formula, which is expressed as follows:

$$\tau_f = \sigma \tan\varphi + C_c + \Sigma_w, \quad (1)$$

where C_c is a solid cementation bond; Σ_w is a water-colloidal binding.

The difference between the slip resistance values (bonds-C) obtained by double-cutting test with standard non-wetted surface is due to the elimination of solid cementitious bond in case of double cutting — $C_c = 0$, in this case only the recoverable water colloid bond — $\Sigma w \neq 0$ is valid. We do need to note that there is no significant difference between the values of the internal friction coefficients ($\tan\varphi$) obtained by the double-cut test and the value obtained in the standard test. Analysis of the results of experimental studies show that the total cementation determined by the standard method, which includes the two components of the bond ($C = C_c + \Sigma_w$, C_c is cementation, Σ_w is aqueous colloid) decreases with increasing consistency index (I_L). For soft-plastic, flowing plastic states, the ratio of standard and double-cut test bonds to the $C_{\text{double surface}}/C_{\text{standard}}$ increases to 1. The decrease in slip resistance is mainly due to the decrease of consistency index (I_L) mainly of rigid cementitious nature. Thus, based on the obtained experimental data, it is proposed to reduce the total bond in the formula for estimating the stability of landslide slopes made of solid, semi-rigid, tight-plastic clay soils by the value of solid cementitious bond (C_c). According to the obtained data, in case of double cutting of clay soils with solid consistency and with natural moisture, the total bond reduction is 73.3–83.0%, in case of semi-rigid it is 76.2–82.9%, and for tight-plastic: 56–70.8%. In case of wetting of the sliding cutting surface the decrease in consistency in rigid grounds is 77.7–91.1%, in case of semi-hard soils: 80–87.8%, and in the case of tight-plastic soils: 80.8–63.2%. From this it is clear that the bonding nature of the hard, semi-rigid plastic and tight-plastic clay soils mainly is cementitious, so in the formula for estimating landslide stability, the slip bond strength value should be taken as the colloidal bonding component, taking in an account the condition, that cementation bonding Cc value equals to zero. In case of soft plastic and flowing plastic clay soils, the binding reduction at $0.75 > I_L > 0.5$ consistency is 5.9–40%, and the decrease is insignificant in case of $1.0 > I_L > 0.75$ consistency, it is 11.8–14.3%. Although the double-cut test is slightly reduced in soft-plastic, flow-plastic clay soils when testing the slope stability, the slope stability assessment should take into account the fact that in the case of mobilization of the aqueous colloidal component, they are capable of adhesive-plastic landslide deformations and the landslide events on the slopes have a plastic or landslide flow mechanism. According to N. N. Maslov's theory, the stabilized landslide (flow) is due to the mobilization of water-colloidal binding nature (Σ_w), and for its development it is necessary this condition:

$$\sigma \tan\varphi + C_c + \Sigma_w > \tau > \sigma \tan\varphi + C_c. \quad (2)$$

Conclusions

At the same time, taking into account that in the landslide slopes there are no rigid structural connections in the area of the slip surface: $C_c = 0$, in the slope stability assessment formula the component $\tau_f = \sigma \tan\varphi$ should be taken as the slip resistance index ensuring the slope stability. From here we come to a very important conclusion, the essence of which lies in the following: for older landslide areas (where slip surfaces already exist), the $\tau_f = \sigma \tan\varphi$ component should be taken as the slip resistance index in the formula for calculating the slip resistance when developing anti-slip measures independent of the consistency of the soils forming the slope, assuming the condition $C_c + \Sigma_w = 0$.

References

[1] Bondarik G K, Pendin V V, and Yarg L A 2007 *Engineering Geodynamics* (Moscow: Izdat. KDU) [in Russian]
[2] Emel'yanova E P 1972 *Main Regularities of the Landslide Process* (Moscow: Nedra) [in Russian]
[3] Maslov N N 1984 *Physical and Technical Theory of Clayey Soils Creep in the Construction Practice* (Moscow: Stroyizdat [in Russian]
[4] Meschyan S R 2008 *Experimental Fundamentals of Rheology of Clay Soils* (Yerevan: Gitutyun NAN RA) [in Russian]
[5] Schuster R L and Raymond J K 1978 *Landslides. Analysis and Control.* Special report (National Research Council (U.S.). Transportation Research Board) (Washington : National Academy of Sciences)
[6] Trofimov V T, Korolev V A, Voznesensky E A, et al. 2005 *Soil Science* (Moscow: Izdat. MGU) [in Russian]
[7] Hairoyan S G 2005 Impact of shrinkable fracturing of shear strength in clay soils In *Madrid, 2005. 6–14 June, European Association of geosciense engineers* G005

Investigation of mechanical properties and fracture surface of cylindrical specimens of Al 6101 alloy under static tension

D K Magomedova[1*], A A Churakova[2], and D V Gunderov[2]
[1]Saint Petersburg State University, St. Petersburg, Russia
[2]Ufa State Aviation Technical University, Ufa, Russia
E-mail: *magmedva.dasha@mail.ru

Abstract. The data on the study of mechanical properties of cylindrical specimens of Al alloy 6101 in different states under static tension at room temperature are presented. The specimens were tested after natural aging, artificial aging, and ECAP. The surface of fractures of these specimens was also investigated using an electron microscope, and a chemical analysis of elements on the surface of fractures was carried out. Basing on the performed study, it can be concluded that alloy 6101 shows the highest strength (ultimate tensile strength 350 MPa) after ECAP treatment in UFG state. It was confirmed due to the chemical analysis that destruction pores in this material originate on particles of secondary phases based on impurity iron.

1. Introduction

Nowadays, aluminum alloys are widely used in various fields, such as light industry, electrical and aircraft engineering. These materials are highly valued for the combination of such characteristics as high strength-to-weight ratio, high conductivity, corrosion resistance, technological ductility, thermal stability, etc. [1–9]. Further enhancement of properties of aluminum alloys is an important area of modern materials science.

One of the most widely used aluminum-based alloys is the thermostrengthened alloy 6101 of Al-Mg-Si system. [1, 3, 9]. This alloy has a significantly higher strength in comparison to pure Al and increased operational and technological characteristics for aircraft industry and production of wires for power lines.

It is known that the methods of severe plastic deformation (SPD) can improve the properties of metallic materials, including through grain refinement to a nanostructural state [2, 10]. One of the SPD techniques is the method of equal channel angular pressing (ECAP) [10]. There are many studies that confirm the increase in the strength and other properties of Al alloys, including 6101, under ECAP treatment [3, 5, 9].

The aim of this research is to study the destruction features of alloy 6101 of Al-Mg-Si system in various states – after natural aging, artificial aging and ECAP. The destruction nature of alloys of this system, processed by ECAP method, has previously been studied only in a few works, for example [6, 7, 11].

Table 1. The chemical composition of Al 6101 alloy (wt.%).

Grade	Tensile strength	Si	Fe	Cu	Mg	Zn	B	Al	Ti+V+Cr+Mn
6101	Max	0.60	0.30	0.01	0.60	0.02	0.01	—	0.015

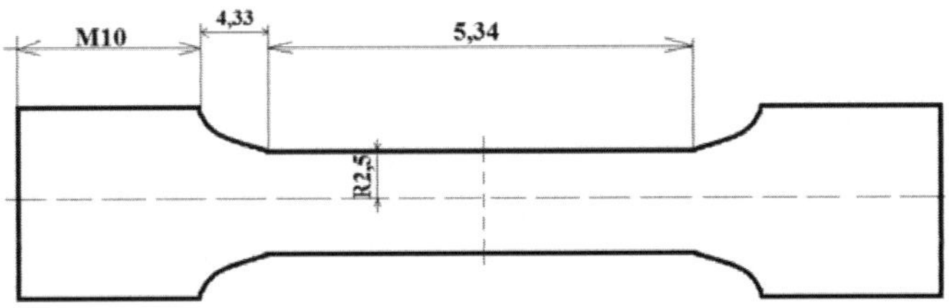

Figure 1. General form of a specimen.

2. Research methods

Cylindrical specimens made of 6101 alloy rods by mechanical treatment according to recommendations given in [4] were used as the object of the study. Table 1 shows the chemical composition of this alloy.

Figure 1 shows the geometry of the studied specimens.

This alloy falls into the thermostrengthened category, therefore, aging procedures were applied to it in order to increase the strength characteristics [1, 8].

The three types of processing have been considered to identify the dependence of strength characteristics on material microstructure:

a) CG-NA: the specimens have been subjected to annealing at a temperature of 550°C for 2 hours with subsequent quenching in water at room temperature, then the material has been naturally hardened for 6 days. This alloy state is designated as CG-NA, where CG is coarse-grained.

b) CG-AA: the specimens have been subjected to annealing at a temperature of 550°C for 2 hours with subsequent quenching in water at room temperature, then the material was aged for 12 hours by annealing at a temperature of 170°C. This alloy state is designated CG-AA.

c) UFG-ECAP: the specimens have been subjected to annealing at 550°C for 2 hours, quenching in water at room temperature, 4 cycles of ECAP at room temperature via Bc route, the channel conjugation angle is 120°C. According to the data of [3, 9, 11], it is possible to form an ultrafine-grained (UFG) structure with a grain size in the range of $400 - 600\,\text{nm}$ as a result of SPD treatment under similar conditions in Al-Mg-Si alloys. Let's designate this state as UFG-ECAP.

Mechanical tests on static tensile with a fixed tensile rate equal to $1.4 \times 10^{-4}\text{s}^{-1}$ have been carried out on the Shimadzu AG-50kNX machine at room temperature; the change in the length of the specimens has been recorded using a video extensometer. Geometrical dimensions/geometry of specimens for mechanical tests are shown in figure 1. The fracture surfaces have been examined using the Zeiss Supra 40VP scanning electron microscope (SEM) in the secondary electron imaging (SEI) and backscattered electron (BSE) modes.

3. Results

Figure 2 shows engineering tensile diagrams at the "stress-strain" coordinates $\sigma - \varepsilon$, obtained as a result of static tensile tests of specimens of the selected geometry (figure 1). The tensile

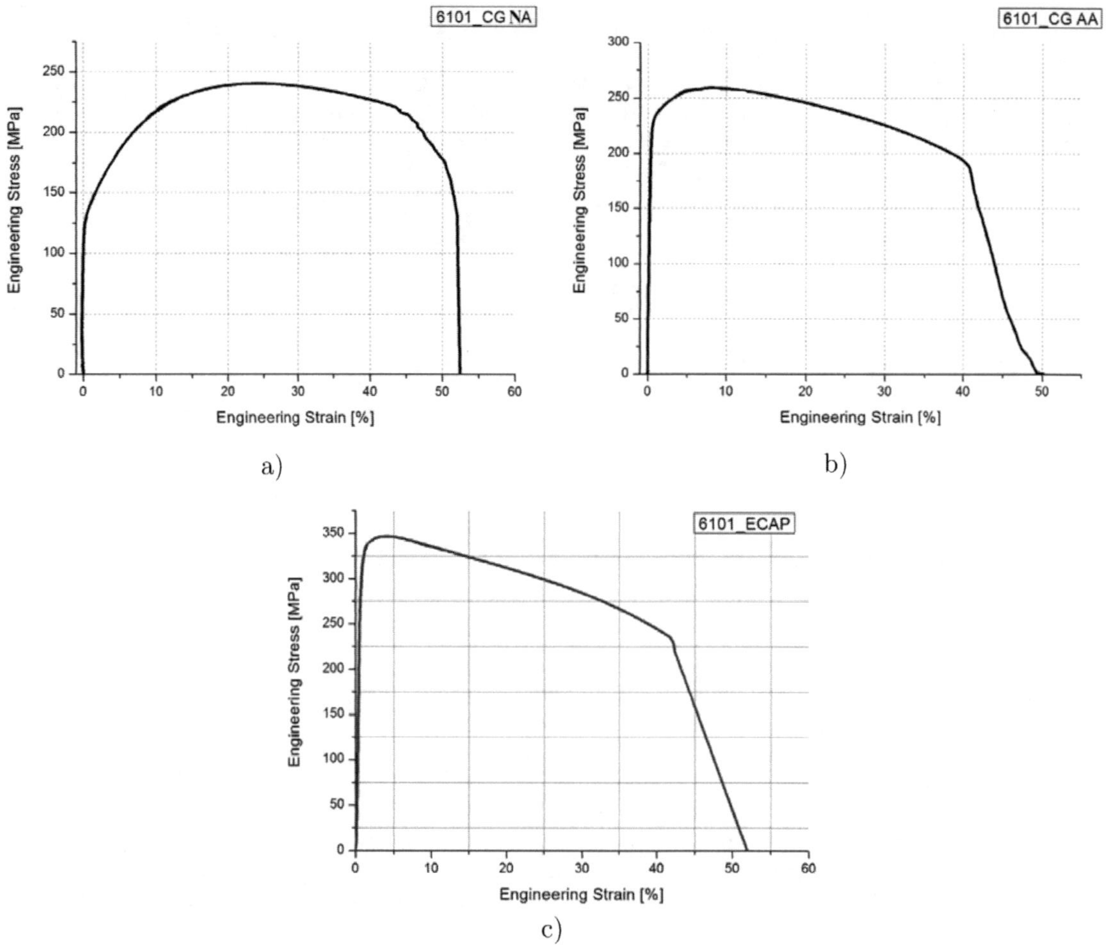

Figure 2. σ-ε diagrams of alloy specimens in the state: a) CG-NA, b) CG-AA, c) UFG-ECAP.

tests have been carried out until their fracture.

As can be seen from the presented diagrams, CG-AA specimens (ultimate tensile strength (σ_{UTS}) is 260 MPa) demonstrate higher strength in comparison to CG-NA state (σ_{UTS} = 230 MPa). The maximum strength is demonstrated by the alloy in UFG-ECAP state (σ_{UTS} = 350 MPa). It is also worth mentioning that the magnitude of the relative elongation in all studied states is practically the same, as can be clearly seen in figure 2.

Investigation of the surface of fractures by SEM method showed that it has a dimpled character in all the studied states (figure 3) and bear the marks of predominantly ductile fracture. Fractures have a homogeneous structure in general.

The results of the chemical microanalysis carried out from the surface of fractures are shown in figures 3, 4 and in tables 2–4.

The main hardening mechanism for these alloys is the formation of the Guinier–Preston (GP) zones and nanosized particles of aluminides based on alloying elements Mg and Si in aluminum during natural and artificial aging, respectively [1, 8, 9]. The formation of GP with dimensions of tens of Angstrom and particles in the form of needles with a thickness of several nanometers and length up to 100 nm as a result of NA and AA leads to precipitation hardening of the alloy. It is known that particles of secondary phases formed during artificial aging cause a more significant

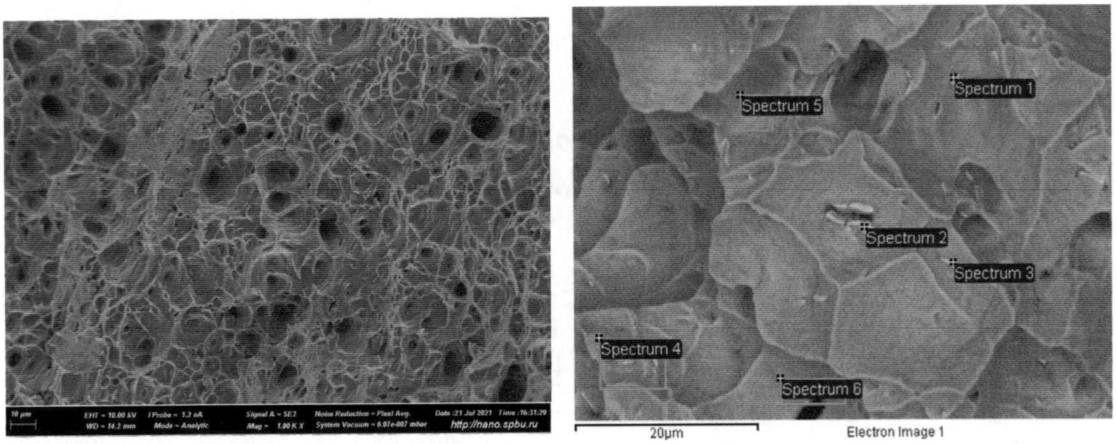

a) b)

Figure 3. Images for CG-NA: a) the general view of the surface of fractures b) the area of the surface of fractures, corresponding to the performed chemical analysis.

Table 2. Spectral analysis for the main elements of the fracture surface corresponding to figure 3*b*.

Spectrum No.	Mg	Al	Si	Fe
1	4.69	74.64	1.01	2.08
2	–	64.23	13.12	22.66
3	1.04	91.47	1.94	3.70
4	0.94	96.11	0.97	
5	1.28	97.96	0.75	
6	1.04	98.96		
Mean	1.5	87.23	2.97	4.74

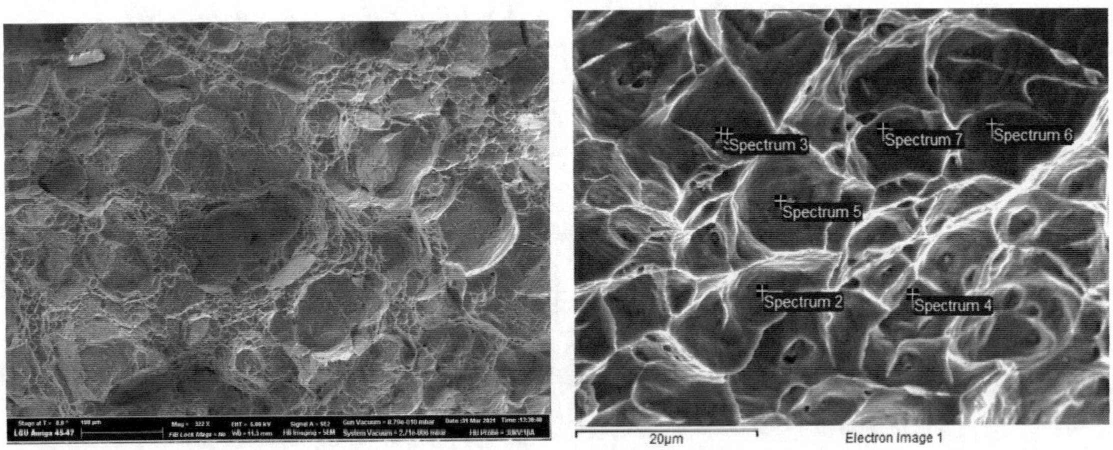

a) b)

Figure 4. Images for CG-AA: a) the general view of the surface of fractures b) the area of the surface of fractures, corresponding to the performed chemical analysis.

Table 3. Spectral analysis for the main elements of the fracture surface corresponding to figure 4*b*.

Spectrum No.	Mg	Al	Si	Fe
1	0.46	77.97	4.84	16.73
2	0.62	91.75	2.23	5.40
3	0.39	64.85	7.79	26.96
4	0.42	77.05	5.25	17.28
5	0.56	82.28	4.34	12.82
6	0.57	84.97	3.28	11.18
7	0.56	85.60	3.14	10.71
Mean	0.51	80.64	4.41	14.44

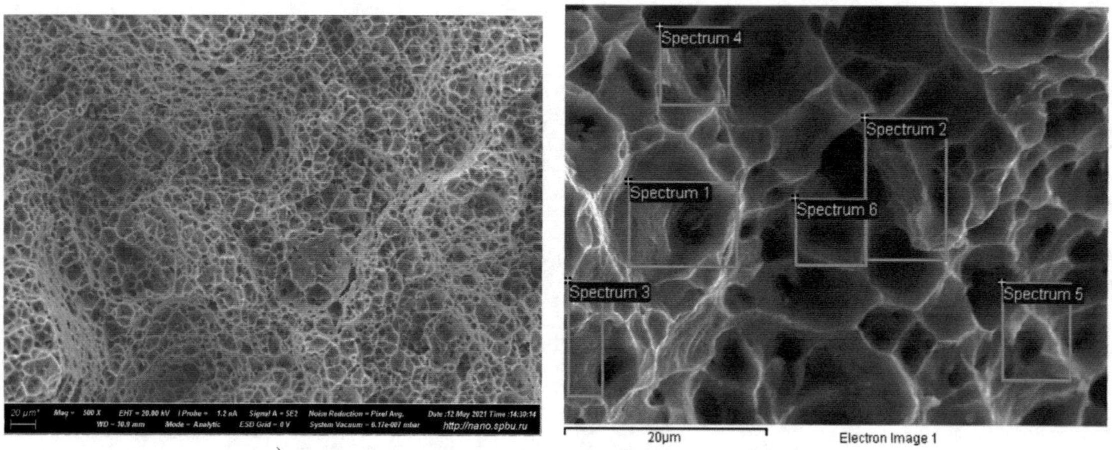

a) b)

Figure 5. Images for UFG-ECAP: a) the general view of the surface of fractures b) the area of the surface of fractures, corresponding to the performed chemical analysis.

Table 4. Spectral analysis for the main elements of the fracture surface corresponding to figure 5*b*.

Spectrum No.	Mg	Al	Si	Fe
1	1.15	97.91	0.61	0.33
2	0.91	98.00	0.74	0.35
3	1.31	97.84		0.85
4	1.15	96.82	1.40	0.63
5		100.00		
6	1.03	98.43		0.54
Mean	0.93	98.17	0.46	0.45

hardening in comparison with GP zones in aluminum alloys after natural aging [8]. It's not unexpected that higher strength is achieved in the alloy after NA. ECAP, on the other hand, leads to the achievement of the highest strength since, in addition to precipitation hardening in UFG state, a significant contribution to the strength of the alloy is made by grain boundary hardening due to grain refinement and dislocation hardening [2, 9].

As mentioned above, the fracture has a dimple character in all the studied states. The dimples observed on the fracture surface in UFG-ECAP state have the smallest diameter. This indicates that a large number of microdefect nucleation centers have been involved at once in UFG-ECAP state during fracture.

Analysis of the chemical composition of the fractured specimens shows that Fe content on the fracture surface of the specimens in CG-NA and CG-AA states is increased relative to its average content in the alloy (tables 1 and 2). Fe does not dissolve in aluminum matrix [8]. As can be seen from table 1, its amount in the alloy does not exceed 0.3 wt%, however, the large precipitates of a micron-sized phase are formed on the basis of Fe during the melt crystallization. It is these intermetallic Al_3Fe particles (a few microns in size) [8] that are the centers for the nucleation of discontinuities in the form of nanopores and subsequent fracture during mechanical tests of Al alloys in all the studied states. Accordingly, these intermetallic particles in a high content are located on the fracture surface deep in the dimples, and the spectrum of fractures shows an increased Fe content, which is many times higher than the average Fe content in the alloy. At the same time, while analyzing the chemical composition of the fracture surface of the specimens in UFG-ECAP state (table 4), it can be seen that Fe content on the fracture surface in UFG-ECAP state is noticeably lower than that of CG-NA and CG-AA states. Presumably, this is explained by the fact that the nature of fracture changes somewhat after ECAP, because the regions of nucleation of nanosized pores and subsequent fracture can be other structural elements, for example, regions with a higher density of dislocations and/or boundaries of ultrafine grains formed during ECAP. In addition, on the basis of the data from [9, 10, 12], ECAP could lead to a redistribution of intermetallic Al_3Fe particles in aluminum as well as to their refinement, which could also affect the nature of fracture and, consequently, the results of the chemical microanalysis of the fracture surface of the specimens. Based on the obtained results, it can be assumed that Al_3Fe particles in UFG-ECAP state contribute much less to the fracture propagation of the studied material under static tension.

4. Conclusion

On the basis of this study, it can be concluded that alloy 6101 demonstrates the highest strength (ultimate tensile strength is 350 MPa) after ECAP treatment (in UFG-ECAP state).

The fracture surface of the specimens after static tension at room temperature is formed by dimples and is predominantly viscous in all the studied states.

Analysis of the chemical composition of the surface of the fractured specimens made it possible to establish an increased Fe content in comparison with Fe content in the alloy. Fe content at the fractures of the specimens in UFG-ECAP state is lower than at the fractures of the specimens in the states of natural and artificial aging (CG-NA and CG-AA). It is assumed that the reason for the established difference in Fe content is that ECAP treatment of the alloy could lead to the redistribution and refinement of intermetallic Al_3Fe particles in aluminum, which are the centers of nucleation of pores/dimples during the fracture.

It is conceivable that iron-containing particles in UFG-ECAP state contribute to a lesser extent to the fracture propagation of alloy 6101 under static tension.

Acknowledgments

The present work was supported by the Russian Science Foundation (grant No. 17-19-01311).

The authors thank the Interdisciplinary Resource Centre for Nanotechnology of SPbU and the Resource Centre for Applied Aerodynamics of SPbU.

References

[1] Karabay S 2006 Modification of AA-6201 alloy for manufacturing of high conductivity and extra high conductivity wires with property of high tensile stress after artificial aging heat treatment for all-aluminium alloy conductors *Materials & Design* **27** (10) 821–32 doi: 10.1016/j.matdes.2005.06.005

[2] Sabirov I, Murashkin M Yu, and Valiev R Z 2013 Nanostructured aluminium alloys produced by severe plastic deformation: New horizons in development *Materials Science and Engineering: A* **560** 1–24 doi: 10.1016/j.msea.2012.09.020

[3] Murashkin M Yu, Sabirov I, Sauvage X, and Valiev R Z 2016 Nanostructured Al and Cu alloys with superior strength and electrical conductivity *Journal of Materials Science* **51** (1) 33–49 doi: 10.1007/s10853-015-9354-9

[4] Argon A S, Im J, and Needleman A 1975 Distribution of plastic strain and negative pressure in necked steel and copper bars *Metallurgical Transactions A* **6** (4) 815 doi: 10.1007/BF02672305

[5] Murashkin M Yu, Sabirov I, Kazykhanov V U, et al. 2013 Enhanced mechanical properties and electrical conductivity in ultra-fine grained Al alloy processed via ECAP-PC *Journal of Materials Science* **48** (13) 4501–9 doi: 10.1007/s10853-013-7279-8

[6] Magomedova D K and Murashkin M Yu 2018 Influence of grain size and second phase particles on the process of void initiation *Journal of Physics: Conference Series* **991** 012055 doi: 10.1088/1742-6596/991/1/012055

[7] Magomedova D K, Murashkin M Yu, and Efimov M A 2018 Technique development for conducting mechanical tests to study the pore formation process in case of material fracture *AIP Conference Proceedings* **1959** 070021 doi: 10.1063/1.5034696

[8] Mondolfo L F 1979 *Aluminum Alloys: Structure and Properties* (Boston: Butterworths)

[9] Medvedev A, Arutyunyan A, Lomakin I, et al 2018 Fatigue properties of ultra-fine grained Al-Mg-Si wires with enhanced mechanical strength and electrical conductivity *Metals* **8** (12) 1034 doi: 10.3390/met8121034

[10] Valiev R Z and Alexandrov I V 2000 *Nanostructured Materials Produced by Severe Plastic Deformation* (Moscow: Logos Pub.)

[11] Vinogradov A Y and Khasimoto S 2004 Fatigue in ultrafine-grained materials processed by equal-channel angular pressing *Russ. Metall* **2004** (1) 42–51

[12] Medvedev A, Murashkin M, Enikeev N, et al. 2021 Influence of morphology of intermetallic particles on the microstructure and properties evolution in severely deformed Al-Fe alloys *Metals* **11** (5) 815 doi: 10.3390/met11050815

On a system of independent arguments for constitutive tensor functions on the growing surface in micropolar continuum

E V Murashkin

Ishlinsky Institute for Problems in Mechanics of the Russian Academy of Sciences, Moscow, Russia

E-mail: evmurashkin@gmail.com

Abstract. The present study is devoted to the boundary value problems statements for the growing materials with microstructural features. The general form of tensor relations on the propagating growing surface is derived as a consequence of the conservation laws of momentum and angular momentum. The necessary system of independent arguments of constitutive differential constraints on the growing surface in micropolar continuum is determined. A complete set of joint rational invariants of the system of tensors and vectors that determine the thermodynamics of the production process of a woven 3D material is given and discussed. An invariant-complete formulation of the constitutive relations on the growing surface is obtained.

Introduction

Additive manufacturing technologies are currently an interdisciplinary field of science that attracts the attention of an extremely large number of researchers [1]. Literary search shows that there are fundamentally two types of growth (volumetric and surface) and, as a result, two approaches to the mathematical description of growth processes. Volumetric growth processes [2, 3] include the formation of a solid component during a chemical reaction, the growth of biological tissues, bones, and the natural formation of fruits [4].

Additive manufacturing processes are associated with the synthesis of products by sequentially attaching material to the growing surface of an arbitrary shape. Modern woven composites are also produced by additive manufacturing procedures [5–9]. A derivation of constitutive equations in mechanics of growing bodies is an extremely important and rather complex problem [19–21]. Furthermore, 3D materials used in additive manufacturing, as well as final products, have microstructural (micropolar) features and mechanical properties that are fine described by asymmetric (micropolar) continuum. Therefore, to develop mathematical models of such technological processes for processing 3D woven materials and manufacturing 3D products, it is necessary to use the mechanics of growing solid in combination with the concepts of oriented manifolds and pseudotensors. Literature search shows that relative tensors in the theories of continuum mechanics is not widespread, despite deep mathematical research (algebra, invariant theory and differentiation of relative tensors) [10–18].

Journal of Physics: Conference Series 2231 (2022) 012019 doi:10.1088/1742-6596/2231/1/012019

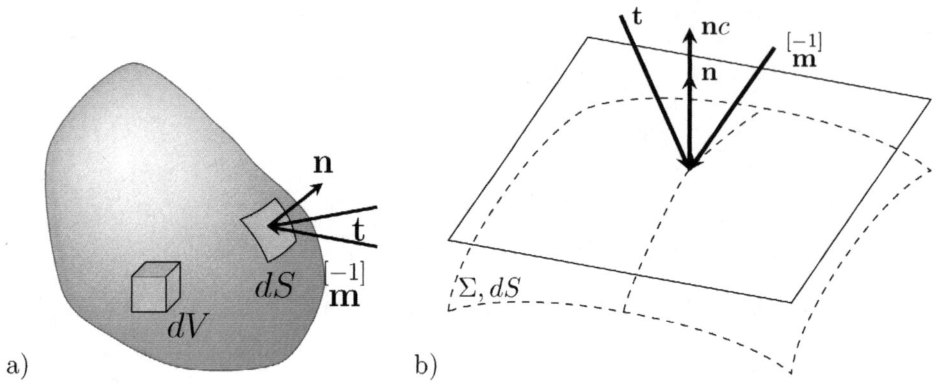

Figure 1. Growing solid and vectors acting on growing surface.

1. Boundary conditions on the propagating growing surface in materials with characteristic directions

Throughout the article, the description in Euler coordinates is used. Terminology and notations mainly correspond to the studies [22–25]. A propagating growing surface Σ in three-dimensional Euclidean space can be given by the implicit equation

$$t = \underset{*}{\tau}(x^i). \tag{1}$$

The traction vectors of forces $\mathbf{t}$ and couples $\overset{[-1]}{\mathbf{m}}$ act on a plane tangent element on the growing surface dS with a unit normal vector n_i (see figure 1), expressed through the components of the actual force $\boldsymbol{\sigma}$ and couples $\overset{[-1]}{\boldsymbol{\mu}}$ stress tensors by following equations

$$\mathbf{t} = \mathbf{n}\cdot\boldsymbol{\sigma}, \quad \overset{[-1]}{\mathbf{m}} = \mathbf{n}\cdot\overset{[-1]}{\boldsymbol{\mu}}. \tag{2}$$

The unit normal vector n_i to a growing surface Σ directed in the way of its propagation is calculated through the spatial gradient in virtue of (1) by

$$n_i = c\,\partial_i\underset{*}{\tau}, \quad c = |\boldsymbol{\nabla}\underset{*}{\tau}|^{-1} \quad (t = \underset{*}{\tau}), \tag{3}$$

where c is the linear propagation velocity of the growing surface in the normal direction n_k.

Earlier in the papers (see, for example, [22–24]) the boundary conditions on the propagating growing surface have been derived in detail for symmetric and asymmetric theories. Here the final formulae are presented:

(i) for the actual components of the force stress tensor σ^{ij}

$$\sigma^{ij} = \int\limits_{\underset{*}{\tau}+0}^{t} [\partial_.\sigma^{ij}(x^s,t')]dt' + \mathcal{S}^{ji} + \underset{*}{\sigma^{ij}}(x^s), \tag{4}$$

$$\mathcal{S}^{ij} = \int\limits_{\underset{*}{\tau}-0}^{\underset{*}{\tau}+0} [\partial_.\sigma^{ij}(x^s,t')]dt', \tag{5}$$

$$c[\nabla_j\underset{*}{\sigma^{ji}}(x^s) + \nabla_j\mathcal{S}^{ji} + \underset{*}{X^i}(x^s)] - n_j\partial_.\sigma^{ji}(x^s,t) = 0 \quad (t = \underset{*}{\tau}+0); \tag{6}$$

(ii) for the components of the couple stress tensor

$$\overset{[-1]}{\mu}{}^{i\cdot}_{\cdot k} = \int\limits_{\underset{*}{\tau}+0}^{t} [\partial.\overset{[-1]}{\mu}{}^{i\cdot}_{\cdot k}(x^s,t')]dt' + \overset{[-1]}{\mathcal{M}}{}^{i\cdot}_{\cdot k} + \underset{*}{\overset{[-1]}{\mu}}{}^{i\cdot}_{\cdot k}(x^s), \tag{7}$$

$$\overset{[-1]}{\mathcal{M}}{}^{i\cdot}_{\cdot k} = \int\limits_{\underset{*}{\tau}-0}^{\underset{*}{\tau}+0} [\partial.\overset{[-1]}{\mu}{}^{i\cdot}_{\cdot k}(x^s,t')]dt', \tag{8}$$

$$c[\nabla_i \underset{*}{\overset{[-1]}{\mu}}{}^{i\cdot}_{\cdot k}(x^s) + \nabla_i \overset{[-1]}{\mathcal{M}}{}^{i\cdot}_{\cdot k} - 2\underset{*}{\overset{[-1]}{\tau}}_k + \underset{*}{\overset{[-1]}{Y}}_k] - n_i\partial.\overset{[-1]}{\mu}{}^{i\cdot}_{\cdot k}(x^s,t) = 0 \quad (t = \underset{*}{\tau}+0). \tag{9}$$

In the above equations (4)–(9) the following notation are accepted: $\mathcal{S}^{ji}$ — the integral associated with the stress jump, $\underset{*}{\sigma}{}^{ij}(x^s) = \sigma^{ij}(x^s,t)|_{t=\underset{*}{\tau}(x^s)-0}$ — stress tensor components, calculated at the moment $t = \underset{*}{\tau}(x^s) - 0$ right before the element is deposited in the main solid, $\underset{*}{X}{}^i(x^s) = X^i(x^s,t)\big|_{t=\underset{*}{\tau}(x^s)+0}$. Time $t = \underset{*}{\tau}(x^s) + 0$ corresponds to the moment immediately after the element is deposited to the propagating growing surface, $\overset{[-1]}{\mathcal{M}}{}^{i\cdot}_{\cdot k}$ — the integral associated with the jump of couple stresses, $\underset{*}{\overset{[-1]}{\mu}}{}^{i\cdot}_{\cdot k}(x^s) = \overset{[-1]}{\mu}{}^{i\cdot}_{\cdot k}(x^s,t)|_{t=\underset{*}{\tau}(x^s)-0}$ components of the couple stress tensor calculated at time $t = \underset{*}{\tau}(x^s) - 0$.

2. Constitutive tensor functions for the actual force and couple stresses and the stresses in the growing element, taking into account the selected direction

In the production processes of woven materials, characteristic directions can be selected on the growing surface. Let us denote the vector of a direction $\boldsymbol{\imath}$ (see figure 2). In this case, the constitutive tensor functions expressing the force $\underset{*}{\sigma}{}^{ij}$ and couple stresses $\underset{*}{\overset{[-1]}{\mu}}{}^{i\cdot}_{\cdot k}$ in terms of actual stresses and couples on the propagating growing surface can be taken in the form

$$\underset{*}{\sigma}{}^{ij} = \mathfrak{F}^{ij}(\sigma^{ij}, \overset{[-1]}{\mu}{}^{i\cdot}_{\cdot j}, n_j, \imath_j, \ldots), \quad \underset{*}{\overset{[-1]}{\mu}}{}^{i\cdot}_{\cdot j} = \overset{[-1]}{\mathfrak{Z}}{}^{i\cdot}_{\cdot j}(\sigma^{ij}, \overset{[-1]}{\mu}{}^{i\cdot}_{\cdot j}, n_j, \imath_j, \ldots), \tag{10}$$

and in the opposite case

$$\underset{*}{\sigma}{}^{ij} = \mathfrak{K}^{ij}(\underset{*}{\sigma}{}^{ij}, \underset{*}{\overset{[-1]}{\mu}}{}^{i\cdot}_{\cdot j}, n_j, \imath_j, \ldots), \quad \underset{*}{\overset{[-1]}{\mu}}{}^{i\cdot}_{\cdot j} = \overset{[-1]}{\mathfrak{M}}{}^{i\cdot}_{j}(\sigma^{ij}, \overset{[-1]}{\mu}{}^{i\cdot}_{\cdot j}, n_j, \imath_j, \ldots), \tag{11}$$

Equations (10) and (11) may be elucidate by experimental determination of the functions $\mathfrak{F}_{ij}$, $\overset{[-1]}{\mathfrak{Z}}{}^{i\cdot}_{\cdot k}$, $\mathfrak{K}_{ij}$ and $\overset{[-1]}{\mathfrak{M}}{}^{i\cdot}_{k}$. Schematically, the necessary parameters determination in the growing process is shown in figure 3. In this case, the "black box" means possible changes in the parameters of the stress-strain state of the material in the time interval from the moment of growing element creation to the moment of its deposition to the main solid, i.e. in the time interval $\underset{*}{\tau} - 0 \leq t \leq \underset{*}{\tau} + 0$. The "black box" can be associated with various physical phenomena. In particular, constitutive tensor functions $\mathfrak{F}_{ij}$, $\overset{[-1]}{\mathfrak{Z}}{}^{i\cdot}_{\cdot k}$, $\mathfrak{K}_{ij}$ and $\overset{[-1]}{\mathfrak{M}}{}^{i\cdot}_{k}$ in the production of woven materials may depend on the moving selected directions associated with the propagating growing surface. The physical meaning of additional selected direction can be associated with

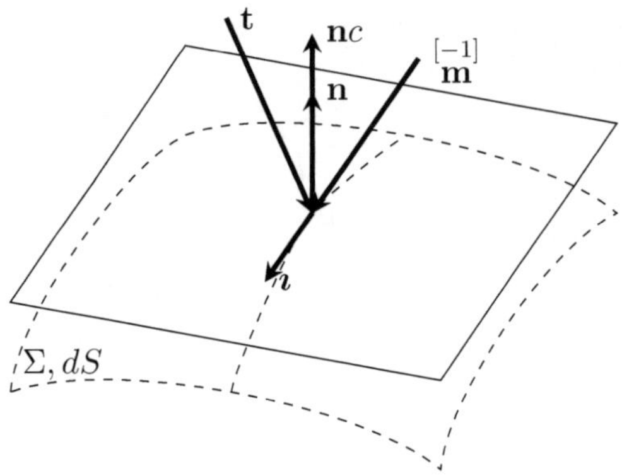

Figure 2. Geometric visualization of vectors acting on a plane tangent element T to the propagating growing surface Σ.

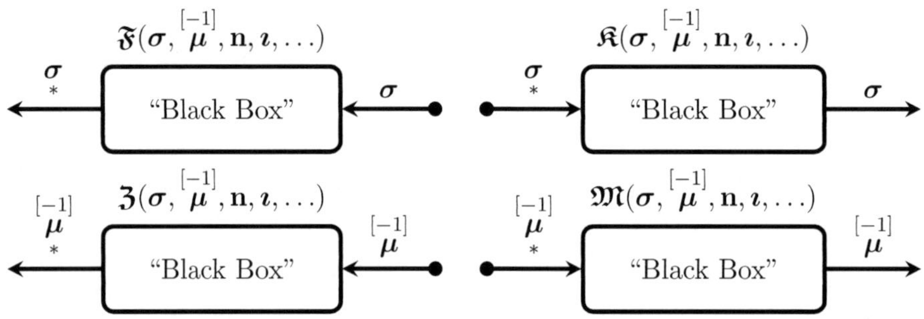

Figure 3. "Black Box" Scheme for Mechanics of Growing Solids.

the characteristic direction of laying fibers in woven composite materials, the directions of reinforcement with stiffer fibers, the direction of winding threads in a bobbin, etc. An important restriction on constitutive tensor functions $\mathfrak{F}_{ij}$, $\overset{[-1]}{\mathfrak{Z}}{}^{i\cdot}_{\cdot k}$, $\mathfrak{K}_{ij}$ and $\overset{[-1]}{\mathfrak{M}}{}^{i\cdot}_{k}$ is the insensitivity of their arguments under rotations of the moving coordinate system around the unit normal vector n_j to the growing surface. In this case, it is necessary to choose a system of joint invariants of tensors σ^{ij}, $\overset{[-1]}{\mu}{}^{i\cdot}_{\cdot j}$ and vectors $n_j, \imath_j$, satisfying the condition of rotational invariance with respect to the vector n_j.

3. Procedures for constructing systems of joint algebraic relative invariants of second rank tensors and vectors

As we can see from the discussion in the previous sections, in order to further specify the constitutive tensor functions on the propagating growing surface, it is necessary to determine a system of joint invariants of pseudotensors σ^{ij}, $\overset{[-1]}{\mu}{}^{i\cdot}_{\cdot j}$ and vectors $n_j, \imath_j$. In what follows, it is reasonable to take into account that the unit normal vector n_i and the vector $\imath_j$ tangent to the growing surface determine its local geometry. Let us introduce a special orthonormal coordinate system with basis vectors: a unit normal vector n_i, a tangent vector $\imath_j$ and a vector in the tangent plane to the growing surface, which can be expressed through the vectors n_i and $\imath_j$ according to the rule

$$\boldsymbol{\tau} = \mathbf{n} \times \boldsymbol{\imath}. \tag{12}$$

In this case, the system of three mutually orthogonal vectors $\mathbf{n}$, $\boldsymbol{\imath}$ and $\boldsymbol{\tau}$ is a right-handed one.

The algorithm for constructing systems of joint rational invariants is in-depth considered in the monograph [13]. Following the methodology noted in [13], for the case considered here the system of algebraic rational invariants of the force stress tensor σ^{ij} and the vectors n_k and $\imath_i$ can be assumed as follows

$$\begin{gathered}\mathbf{t}\cdot\mathbf{n}=\mathbf{n}\cdot\boldsymbol{\sigma}\cdot\mathbf{n},\quad \underset{2}{\mathbf{t}}\cdot\mathbf{n}=\mathbf{n}\cdot\boldsymbol{\sigma}^2\cdot\mathbf{n},\quad \mathbf{t}\cdot\mathbf{t},\quad \mathbf{t}_\perp\cdot\mathbf{t}_\perp,\\ \mathbf{t}\cdot\underset{2}{\mathbf{t}},\quad \mathbf{t}_\perp\cdot\underset{2}{\mathbf{t}}_\perp,\quad \underset{2}{\mathbf{t}}\cdot\underset{2}{\mathbf{t}},\quad \underset{2}{\mathbf{t}}_\perp\cdot\underset{2}{\mathbf{t}}_\perp.\end{gathered}\tag{13}$$

then for the pseudo-affinor of couple stresses $\overset{[-1]}{\mu}{}^{i\cdot}_{\cdot k}$ and vectors n_k, $\imath_i$ is furnished by

$$\begin{gathered}\overset{[-1]}{\mathbf{m}}\cdot\mathbf{n}=\mathbf{n}\cdot\overset{[-1]}{\boldsymbol{\mu}}\cdot\mathbf{n},\quad \underset{2}{\overset{[-2]}{\mathbf{m}}}\cdot\mathbf{n}=\mathbf{n}\cdot\overset{[-1]}{\boldsymbol{\mu}}\cdot\overset{[-1]}{\boldsymbol{\mu}}\cdot\mathbf{n},\quad \overset{[-1]}{\mathbf{m}}\cdot\overset{[-1]}{\mathbf{m}},\quad \overset{[-1]}{\mathbf{m}}_\perp\cdot\overset{[-1]}{\mathbf{m}}_\perp,\\ \overset{[-1]}{\mathbf{m}}\cdot\underset{2}{\overset{[-2]}{\mathbf{m}}},\quad \overset{[-1]}{\mathbf{m}}_\perp\cdot\underset{2}{\overset{[-2]}{\mathbf{m}}}_\perp,\quad \underset{2}{\mathbf{m}}\cdot\underset{2}{\overset{[-2]}{\mathbf{m}}},\quad \underset{2}{\overset{[-2]}{\mathbf{m}}}_\perp\cdot\underset{2}{\overset{[-2]}{\mathbf{m}}}_\perp.\end{gathered}\tag{14}$$

Joint invariants of tensorσ^{ij} and pseudo-affinor $\overset{[-1]}{\mu}{}^{i\cdot}_{\cdot k}$ reads by

$$\begin{gathered}\mathbf{t}\cdot\overset{[-1]}{\mathbf{m}},\quad \mathbf{t}_\perp\cdot\overset{[-1]}{\mathbf{m}}_\perp\quad \mathbf{t}\cdot\underset{2}{\overset{[-2]}{\mathbf{m}}},\quad \mathbf{t}_\perp\cdot\underset{2}{\overset{[-2]}{\mathbf{m}}}_\perp,\\ \overset{[-1]}{\mathbf{m}}\cdot\underset{2}{\mathbf{t}},\quad \overset{[-1]}{\mathbf{m}}_\perp\cdot\underset{2}{\mathbf{t}}_\perp,\quad \underset{2}{\mathbf{t}}\cdot\underset{2}{\overset{[-2]}{\mathbf{m}}},\quad \underset{2}{\mathbf{t}}_\perp\cdot\underset{2}{\overset{[-2]}{\mathbf{m}}}_\perp.\end{gathered}\tag{15}$$

The system of invariants (13)–(15) consists vectors $\mathbf{t}_\perp$, $\overset{[-1]}{\mathbf{m}}_\perp$, $\underset{2}{\mathbf{t}}_\perp$ and $\underset{2}{\overset{[-2]}{\mathbf{m}}}_\perp$ which are projections of vectors $\mathbf{t}$, $\overset{[-1]}{\mathbf{m}}$, $\underset{2}{\mathbf{t}}$ and $\underset{2}{\overset{[-2]}{\mathbf{m}}}$, respectively, in the tangent plane to the growing surface.

Obviously, the system of joint algebraic rational relative invariants represented by the equations (13)–(15) is complete. At the same time, the joint invariants included in it are not independent and can be partially excluded from consideration due to rational syzygies. Moreover, high-order joint algebraic rational invariants, including cubes and biquadras of the force and the couple stress tensor, should be excluded by virtue of the Hamilton–Cayley theorem [13].

Thus, the irreducible complete system of joint algebraic rational relative invariants of the force stress tensor σ^{ij}, the couple stress pseudo-affinor $\overset{[-1]}{\mu}{}^{i\cdot}_{cdotk}$ and vectors n_k and $\imath_i$ can be taken as

$$\mathbf{t}\cdot\mathbf{n},\quad \mathbf{t}\cdot\mathbf{t},\quad \overset{[-1]}{\mathbf{m}}\cdot\mathbf{n},\quad \overset{[-1]}{\mathbf{m}}\cdot\overset{[-1]}{\mathbf{m}},\quad \mathbf{t}\cdot\overset{[-1]}{\mathbf{m}},\quad \mathbf{t}_\perp\cdot\overset{[-1]}{\mathbf{m}}_\perp.\tag{16}$$

We note again that the system of invariants (16) is insensitive to rotations of a local orthogonal coordinate system with basis vectors n_i, $\imath_i$, τ_i around the vector n_i.

4. Constitutive tensor functions in a specific orthogonal coordinate system

Consider the invariant-geometric interpretation in a special orthogonal coordinate system with unit vectors n_i, $\imath_i$, τ_i. Let force $\mathbf{t}$ and couples $\overset{[-1]}{\mathbf{m}}$ traction vectors be known on the propagating growing surface. In this case, constitutive tensor functions (10) on the growing surface in terms of the complete system of joint invariants (16) for the restriction to the two-dimensional plane tangent element T of the $ten\overset{*}{s}or\tau$, takes the form

$$\underset{*}{\sigma}{}^{ij}=\overset{[-1]}{\mathfrak{F}}{}^{ij}(\mathbf{t}\cdot\mathbf{n},\mathbf{t}\cdot\mathbf{t},\overset{[-1]}{\mathbf{m}}\cdot\mathbf{n},\overset{[-1]}{\mathbf{m}}\cdot\overset{[-1]}{\mathbf{m}},\mathbf{t}\cdot\overset{[-1]}{\mathbf{m}},\mathbf{t}_\perp\cdot\overset{[-1]}{\mathbf{m}}_\perp).\tag{17}$$

$$\underset{*}{\overset{[-1]}{\mu}}{}^{i\cdot}_{\cdot j}=\overset{[-1]}{\mathfrak{Z}}{}^{i\cdot}_{\cdot j}(\mathbf{t}\cdot\mathbf{n},\mathbf{t}\cdot\mathbf{t},\overset{[-1]}{\mathbf{m}}\cdot\mathbf{n},\overset{[-1]}{\mathbf{m}}\cdot\overset{[-1]}{\mathbf{m}},\mathbf{t}\cdot\overset{[-1]}{\mathbf{m}},\mathbf{t}_\perp\cdot\overset{[-1]}{\mathbf{m}}_\perp).\tag{18}$$

For the projections $\mathbf{t}_\perp$ and $\overset{[-1]}{\mathbf{m}}_\perp$ in the tangent plane T to the growing surface is valid the Pythagorean theorem

$$|\mathbf{t}_\perp|^2 = |\mathbf{t}_\perp \cdot \boldsymbol{\imath}|^2 + |\mathbf{t}_\perp \cdot \boldsymbol{\tau}|^2, \qquad (19)$$
$$|\overset{[-1]}{\mathbf{m}}_\perp|^2 = |\overset{[-1]}{\mathbf{m}}_\perp \cdot \boldsymbol{\imath}|^2 + |\overset{[-1]}{\mathbf{m}}_\perp \cdot \boldsymbol{\tau}|^2.$$

The squares of the joint invariants and the lengths of the vectors (16) and the lengths of the projections (19) are easily calculated using the actual values of $\mathbf{t}$ and $\overset{[-1]}{\mathbf{m}}$ on the growing surface, i.e. through the actual components of force and couple stresses, according to the formulae

$$|\mathbf{t} \cdot \mathbf{n}| = |\sigma_{33}|, \quad |\mathbf{t}_\perp \cdot \boldsymbol{\imath}|^2 = \sigma_{31}^2, \quad |\mathbf{t}_\perp \cdot \boldsymbol{\tau}|^2 = \sigma_{32}^2,$$
$$|\overset{[-1]}{\mathbf{m}} \cdot \mathbf{n}| = |\mu_{33}|, \quad |\overset{[-1]}{\mathbf{m}}_\perp \cdot \boldsymbol{\imath}|^2 = \mu_{31}^2, \quad |\overset{[-1]}{\mathbf{m}}_\perp \cdot \boldsymbol{\tau}|^2 = \mu_{32}^2, \qquad (20)$$
$$|\mathbf{t} \cdot \mathbf{t}| = |\sigma_{3s}\sigma_{s3}| = |\sigma_{31}\sigma_{13} + \sigma_{32}\sigma_{23} + \sigma_{33}^2|,$$
$$|\overset{[-1]}{\mathbf{m}} \cdot \overset{[-1]}{\mathbf{m}}| = |\mu_{3s}\mu_{s3}| = |\mu_{31}\mu_{13} + \mu_{32}\mu_{23} + \mu_{33}^2|.$$

Thus, constitutive tensor functions (17) and (18) on the growth surface given expressions (20) and adopting the following notation for invariants,

$$I = |\sigma_{33}|, \quad II = \sigma_{31}^2 + \sigma_{32}^2, \quad III = |\sigma_{31}\sigma_{13} + \sigma_{32}\sigma_{23} + \sigma_{33}^2|,$$
$$IV = |\mu_{33}|, \quad V = \mu_{31}^2 + \mu_{32}^2, \quad VI = |\mu_{31}\mu_{13} + \mu_{32}\mu_{23} + \mu_{33}^2|, \qquad (21)$$
$$VII = |\sigma_{33}\mu_{33}|.$$

can be rewritten in the form

$$\underset{*}{\sigma}^{ij} = \overset{[-1]}{\mathfrak{F}}{}^{ij}(I, II, III, IV, V, VI, VII). \qquad (22)$$

$$\underset{*}{\overset{[-1]}{\mu}}{}^{i\cdot}_{\cdot j} = \overset{[-1]}{\mathfrak{Z}}{}^{i\cdot}_{\cdot j}(I, II, III, IV, V, VI, VII). \qquad (23)$$

Boundary conditions in the form of differential constraints on the growing surfaces (6), (9) have the necessary properties of sensitivity to the geometry of the growth surface and the characteristic directions of material laying out in the processes of winding threads or the production of woven composites.

5. Conclusion

The paper deals with the problems of derivation constitutive tensor functions for 3D woven materials is discussed. The general form of the tensor relation on the growing surface is given, taking into account the additional preferred direction. The system of independent arguments of the constitutive tensor function on the growing surface in this case is determined. A complete set of joint rational invariants of the force and couple stress tensor and characteristic directors is determined. An invariant-complete formulation of the constitutive relations on the growing surface is given. Statements of boundary value problems simulating the processes of synthesis of woven 3D materials are proposed. The resulting differential constraints are specified for orthogonal coordinate systems sensitive to the geometry of the growing surface.

Acknowledgments

This work was carried out within the framework of a state assignment (state registration No. AAAA-A20-120011690132-4) and with the support of the Russian Foundation for Basic Research (project No. 20-01-00666).

References

[1] Berman B 2012 3-D printing: The new industrial revolution *Business Horizons* **55** (2) 155–62 doi: 10.1016/j.bushor.2011.11.003

[2] Epstein M and Maugin G A 2000 Thermomechanics of volumetric growth in uniform bodies, *International Journal of Plasticity* **16** (7) 951–78 doi: 10.1016/S0749-6419(99)00081-9

[3] Maugin G A 2009 On inhomogeneity, growth, ageing and the dynamics of materials *Journal of Mechanics of Materials and Structures*, **4** (4) 731-41 doi: 10.2140/jomms.2009.4.731

[4] Goriely A 2017 *The Mathematics and Mechanics of Biological Growth* (New York: Springer) doi: 10.1007/978-0-387-87710-5

[5] Dickson A N, Ross K A, and Dowling D P 2018 Additive manufacturing of woven carbon fibre polymer composites *Composite Structures* **206** 637–43 doi: 10.1016/j.compstruct.2018.08.091

[6] Dickson A N and Dowling D P 2019 Enhancing the bearing strength of woven carbon fibre thermoplastic composites through additive manufacturing *Composite Structures* **212** 381–8 doi: 10.1016/j.compstruct.2019.01.050

[7] Liu Y and Chou T-W 2020 Additive manufacturing of multidirectional preforms and composites: from three-dimensional to four-dimensional *Materials Today Advances* **5** 100045 doi: 10.1016/j.mtadv.2019.100045

[8] Goh G D, Agarwala S, Goha G L, et al. 2017 Additive manufacturing in unmanned aerial vehicles (UAVs): Challenges and potential *Aerospace Science and Technology* **63** 140–51 doi: 10.1016/j.ast.2016.12.019

[9] Sitotaw D B, Ahrendt D, Kyosev Y, and Kabish A K 2020 Additive manufacturing and textilesstate-of-the-art *Applied Sciences* **10** (15) 5033 doi: 10.3390/app10155033

[10] Veblen O and Thomas T Y 1924 Extensions of Relative Tensors *Transactions of the American Mathematical Society* **26** (3) 373–7 doi: 10.2307/1989146

[11] Veblen O 1927 *Invariants of Quadratic Differential Forms* (Cambridge: Cambridge University Press)

[12] Levi-Civita T 1927 *The Absolute Differential Calculus (Calculus of Tensors)* (London & Glasgow: Blackie & Son Limited)

[13] Gurevich G B 1964 *Foundations of the Theory of Algebraic Invariants* (Gröningen, P. Noordhoff)

[14] Einstein A 1979 *General Relativity; An Einstein Centenary Survey* (Cambridge: Cambridge University Press)

[15] Schouten J A 1951 *Tensor Analysis for Physicist* (Oxford: Clarendon Press)

[16] Sokolnikoff I S 1964 *Tensor Analysis: Theory and Applications to Geometry and Mechanics of Continua* (New York: John Wiley & Sons Inc)

[17] Synge J L and Schild A 1978 *Tensor calculus* Vol. 5 (New York: Courier Corporation)

[18] Das A J 2007 *Tensors: The Mathematics of Relativity Theory and Continuum Mechanics* (Springer Science & Business Media) doi: 10.1007/978-0-387-69469-6

[19] Southwell R V 1936 *An Introduction to the Theory of Elasticity. For Engineers and Physicists* (London: Oxford Univ. Press)

[20] Kovalev V A and Radayev Yu N 2014 On a form of the first variation of the action integral over a varied domain *Izvestiya Saratovskogo Universiteta. Novaya Seriya. Seriya Matematika. Mekhanika. Informatika* **14** (2) 199–209 doi: 10.18500/1816-9791-2014-14-2-199-209

[21] Arutyunyan N Kh and Naumov V E 1984 The boundary value problem of the theory of viscoelastic plasticity of a growing body subject to aging *Journal of Applied Mathematics and Mechanics* **48** (1) 1–10 doi: 10.1016/0021-8928(84)90099-6

[22] Murashkin E V and Radayev Yu N 2019 On a differential constraint in asymmetric theories of the mechanics of growing solids *Mechanics of Solids* **54** (8) 1157–64 doi: 10.3103/S0025654419080053

[23] Murashkin E V and Radayev Yu N 2019 On a differential constraint in the continuum theory of growing solids *Journal of Samara State Technical University, Ser. Physical and Mathematical Sciences* **23** (4) 646–56 doi: 10.14498/vsgtu1696

[24] Murashkin E V and Radayev Yu N 2019 On a class of constitutive equations on propagating growing surface *Vestnik Chuvashskogo Gosudarstvennogo Pedagogicheskogo Universiteta. Seriya: Mekhanika Predel'nogo Sostoyaniya* No. 3(41) 11–29 doi: 10.26293/chgpu.2019.40.2.012

[25] Murashkin E V 2021 On the boundary conditions formulation in the problems of synthesis of woven 3d materials *Vestnik Chuvashskogo Gosudarstvennogo Pedagogicheskogo Universiteta. Seriya: Mekhanika Predel'nogo Sostoyaniya* No. 1(47) 114–21 doi: 10.37972/chgpu.2021.1.47.010

Quadrature formulas for Cauchy-type integrals with the Cauchy kernel to a integer power and a Jacobi weight function with complex exponents

H A Amirjanyan[1*], **A V Sahakyan**[1**], **and A K Kukudzhanov**[2***]

[1]Institute of Mechanics of the National Academy of Sciences of Armenia, Yerevan, Republic of Armenia

[2]Moscow Institute of Physics and Technology (National Research University), Dolgoprudny, Russia

E-mail: [*]amirjanyan@gmail.com, [**]avsahakyan@gmail.com, [***]alexeykukudzhanov@gmail.com

Abstract. The paper presents quadrature formulas for hypersingular integrals of various orders. It is assumed that the density of these integrals is represented as a product of a function that satisfies the Hölder condition and a weight function of Jacobi orthogonal polynomials. In this case, the exponents of the weight function can be complex numbers, the real part of which is greater than -1. Numerical analysis of the dependence of the root-mean-square deviation of the quadrature formula of order 8 on the value of the hypersingular integral calculated using standard software packages is carried out for various complex values of the weight function exponents. For hypersingular integrals up to the fourth order inclusive, a numerical analysis of the convergence of quadrature formulas is carried out for certain complex values of the exponents of the weight function.

1. Introduction

The concept of a hypersingular integral was introduced by Hadamard back in 1903 [1], but its application to the solution of applied problems can be attributed only to the middle of the 20th century. The first work where hypersingular integrals found practical application was probably the work of Nekrasov [2]. Further, the theory of hypersingular integral equations has been widely developed in various fields: the theory of elasticity [3–5], fracture mechanics [6–8], the theory of wave diffraction [9, 10], the theory of fractal antennas [11, 12], aerodynamics [13, 14] and many others. The solution of these equations, mainly, is constructed by approximate methods, which are studied in many works, in particular [15–23]. It should be noted that most of the works of an applied nature refer to the most common special case, when the behavior of the density of a singular or hypersingular integral at the ends of the interval of integration is described by a root function. There are significantly fewer works where the density of an integral contains the weight function of Jacobi polynomials with arbitrary admissible real exponents. In [7, 24–25], quadrature formulas are given for the Cauchy-type integral and the hypersingular integral when the exponents of the Jacobi weight function are complex.

In this paper, we present interpolation quadrature formulas for hypersingular integrals with the Cauchy kernel to a power greater than two. As before, it is assumed that the densities of these integrals contain as a factor the Jacobi weight function with arbitrary, including complex, admissible exponents. A numerical analysis of the convergence of the presented formulas is carried out.

2. Quadrature formulas

Let us consider the integral

$$J_p(z) = \int_{-1}^{1} \frac{\Phi(x)}{(x-z)^p} dx \quad (z \in C,\ z \neq \pm 1), \tag{1}$$

where p is a natural number.

According to [16, 26], the integral (1), at $z \in (-1, 1)$, is understood in the sense of the Cauchy–Hadamard principal value

$$J_p(z) = \lim_{\varepsilon \to 0} \left[\int_{-1}^{z-\varepsilon} \frac{\Phi(x)}{(x-z)^p} dx + \int_{z+\varepsilon}^{1} \frac{\Phi(x)}{(x-z)^p} dx + \frac{\xi(\varepsilon)}{\varepsilon^{p-1}} \right],$$

where $\xi(\varepsilon)$ is some function chosen so that the specified limit exists.

With respect to the density of the integral $\Phi(x)$, we assume that it can be represented in the form

$$\Phi(x) = \varphi(x)(1-x)^{\alpha}(1+x)^{\beta} \quad (\mathrm{Re}\alpha, \mathrm{Re}\beta > -1), \tag{2}$$

where $\varphi(x)$ satisfies the Hölder condition on the closed interval $[-1, 1]$ and has continuous derivatives up to the order p.

We replace the function $\varphi(x)$ by the polynomial

$$\varphi_n(x) = \sum_{j=1}^{n} \frac{X_j P_n^{(\alpha,\beta)}(x)}{(x-\xi_j) P_n'^{(\alpha,\beta)}(\xi_j)}, \tag{3}$$

where ξ_j $(j = \overline{1, n})$ are the roots of the Jacobi polynomial $P_n^{(\alpha,\beta)}(x)$.

In the case when the exponents α and β are real, it is obvious that (3) is the Lagrange interpolation polynomial and $X_j = \varphi(\xi_j)$. What is to be understood by X_j when the exponents α and β, therefore, the roots ξ_j $(j = \overline{1, n})$ are the complex number?

If the function $\varphi(x)$ in (2) is the sought-for solution of the integral equation, then the unknown coefficients X_j will be found from the corresponding system of linear algebraic equations.

Now let the function $\varphi(x)$ be given and defined on the segment $[-1, 1]$. The complex roots of the Jacobi polynomial $P_n^{(\alpha,\beta)}(x)$ are located on an ellipse-like curve, framing a segment $[-1, 1]$ either above or below. A change in the real part of the exponents α and β leads to a displacement of the roots along the indicated curve, an increase in the imaginary parts leads to an expansion of this curve and an increase of the order n of the Jacobi polynomial leads to the contraction of these roots to the segment $[-1, 1]$. If $\varphi(x) = p_m(x)$ $(m < n)$ is the polynomial of the order less than n, it is obvious that $X_j = p_m(\xi_j)$. If the function $\varphi(x)$ analytically continues into the complex plane, then, again, $X_j = \varphi(\xi_j)$ can be accepted, but with the possibility of an error, the magnitude of which will depend on the imaginary parts of the exponents.

In the last case, as well in the case when function $\varphi(x)$ is defined only on the segment, it is expedient to change $\varphi(x)$ by the following Lagrange interpolation polynomial with Chebychev's nodes

$$\varphi(x) \approx t_n(x) = \frac{1}{n} \sum_{i=1}^{n} \frac{\varphi(t_i) T_n(x)}{(x-t_i) U_{n-1}(t_i)}, \quad t_i = \cos\left(\frac{2i-1}{2n}\pi\right) \tag{4}$$

and, then in (3) put $X_j = t_n(\xi_j)$. In this case the exactness of the quadrature formula will not depend from exponents α and β.

In [24, 25, 27] the quadrature formula for a singular integral, i.e. integral (1) at $p = 1$ was presented in the form

$$J_1(z) \approx \sum_{j=1}^{n} \frac{w_j X_j}{\xi_j - z} \left[1 - \frac{R_n^{(\alpha,\beta)}(z)}{R_n^{(\alpha,\beta)}(\xi_j)} \right] \quad (z \in C,\ z \neq \pm 1,\ \mathrm{Re}(\alpha, \beta) > -1), \tag{5}$$

where X_j and ξ_j were descripted above,

$$w_j = \frac{2}{n+\alpha+\beta+1} \frac{R_n^{(\alpha,\beta)}(\xi_j)}{P_{n-1}^{(\alpha+1,\beta+1)}(\xi_j)},$$

$$R_n^{(\alpha,\beta)}(\zeta) = -\left(\frac{2}{\zeta - 1} \right)^{n+1} 2^{\alpha+\beta} B(n+\alpha+1, n+\beta+1) F \left[n+1, n+\alpha+1; 2n+\alpha+\beta+2; \frac{2}{1-\zeta} \right].$$

Here $B(a, b) = \Gamma(a)\Gamma(b)/\Gamma(a+b)$ is the known beta-function and $F[a, b; c; z]$ is hypergeometric series [28], $R_n^{(\alpha,\beta)}(\zeta)$ is closely related to Jacobi function of the second kind $Q_n(\zeta)$

$$Q_n(\zeta) = \frac{1}{2(\zeta - 1)^\alpha (\zeta + 1)^\beta} \int_{-1}^{1} \frac{P_n^{(\alpha,\beta)}(s)(1-s)^\alpha (1+s)^\beta}{s - \zeta} ds = \frac{R_n^{(\alpha,\beta)}(\zeta)}{2(\zeta - 1)^\alpha (\zeta + 1)^\beta}$$

and is defined in the entire complex plane cut along the segment $[-1, 1]$, and at the points of this segment is determined as the half-sum of the limits when approaching them from the upper and lower half-planes [29].

According to [30, 26], for the integrals $J_p(z)$, $p \in N$, $p > 1$ we have:

$$J_p(z) = \frac{1}{(p-1)!} \frac{d^{p-1}}{dz^{p-1}} J_1(z) \tag{6}$$

Therefore, differentiating the quadrature formula (5) $(p - 1)$ times, we obtain the quadrature formula for $J_p(z)$.

In [7], directly, without taking into account (6), the quadrature formula for $J_2(z)$ was obtained in the form

$$J_2(z) = -\sum_{j=1}^{n} \frac{A_j X_j}{z - \xi_j} \left[2(n+1) R_{n+1}^{(\alpha-1,\beta-1)}(z) + \frac{R_n^{(\alpha,\beta)}(z) - R_n^{(\alpha,\beta)}(\xi_j)}{z - \xi_j} \right] \tag{7}$$
$$(z \in C,\ z \neq \pm 1,\ \mathrm{Re}(\alpha, \beta) > -1),$$

where $A_j = 2[(n+\alpha+\beta+1) P_{n-1}^{(\alpha+1,\beta+1)}(\xi_j)]^{-1}$, and the rest of the designations, taking into account the above about X_j, coincide with the previous ones.

Writing out the quadrature formula (5) in the notation (7)

$$J_1(z) \approx \sum_{j=1}^{n} \frac{A_j X_j}{z - \xi_j} [R_n^{(\alpha,\beta)}(z) - R_n^{(\alpha,\beta)}(\xi_j)]$$

and taking into account the following differentiation formula

$$\frac{d}{dz} R_n^{(\alpha,\beta)}(z) = -2(n+1) R_{n+1}^{(\alpha-1,\beta-1)}(z),$$

it is easy to verify that the quadrature formula (7) is indeed the derivative of (5).

Using formula (6), for the integral $J_p(z)$ we can write out the following quadrature formula:

$$J_p(z)=(-1)^{p-1}\sum_{j=1}^{n}\frac{A_jX_j}{z-\xi_j}\left[\frac{R_n^{(\alpha,\beta)}(z)-R_n^{(\alpha,\beta)}(\xi_j)}{(z-\xi_j)^{p-1}}+\sum_{m=1}^{p-1}\frac{2^m R_{n+m}^{(\alpha-m,\beta-m)}(z)}{B(m+1,n+1)(n+m+1)(z-\xi_j)^{p-m-1}}\right]. \quad (8)$$

If it is necessary to calculate the integral $J_p(z)$ at the roots ξ_j, quadrature formula (8) should be used in the following form:

$$J_p(\xi_m) = -\frac{2^p\Gamma(n+p+1)}{\Gamma(p+1)\Gamma(n+1)}R_{n+p}^{(\alpha-p,\beta-p)}(\xi_m)A_mX_m$$
$$+(-1)^{p-1}\sum_{\substack{j=1\\ j\neq m}}^{n}\frac{A_jX_j}{\xi_m-\xi_j}\left[\frac{R_n^{(\alpha,\beta)}(\xi_m)-R_n^{(\alpha,\beta)}(\xi_j)}{(\xi_m-\xi_j)^{p-1}}+\sum_{m=1}^{p-1}\frac{2^m R_{n+m}^{(\alpha-m,\beta-m)}(\xi_m)}{B(m+1,n+1)(n+m+1)(\xi_m-\xi_j)^{p-m-1}}\right]. \quad (9)$$

Thus, we have obtained a quadrature formula for calculating the hypersingular integral $J_p(z)$ $(p \in N)$.

3. Numerical analysis of the convergence of the quadrature formula

Obviously, quadrature formula (9) is satisfied exactly if the function $\varphi(x)$ is a polynomial of order less than n. Consequently, the proximity of the value obtained by the quadrature formula to the true value of the integral is determined exclusively by how well the function $\varphi(x)$ is interpolated by the polynomial. The latter dictates the choice of the approximation order n. Numerical analysis will be carried out using the example of a fairly well interpolated function

$$\varphi(x) = e^x \sin x. \quad (10)$$

The selected function is an entire function, and on its example it is possible to clarify the situation with the choice of coefficients X_j when exponents α and β are complex. Table 1 shows the values of the standard deviation of the selected function from the polynomial (3) at $X_j = \varphi(\xi_j)$ and at $X_j = t_n(\xi_j)$ (4) for various values of the exponents α and β.

We calculate the indicated deviations using the formulas

$$\varepsilon_J = \sqrt{\int_{-1}^{1}\left[e^x\sin x - \sum_{j=1}^{n}\frac{\varphi(\xi_j)P_n^{(\alpha,\beta)}(x)}{(x-\xi_j)P_n^{\prime(\alpha,\beta)}(\xi_j)}\right]^2 dx},$$
$$\varepsilon_C = \sqrt{\int_{-1}^{1}\left[e^x\sin x - \sum_{j=1}^{n}\frac{t_n(\xi_j)P_n^{(\alpha,\beta)}(x)}{(x-\xi_j)P_n^{\prime(\alpha,\beta)}(\xi_j)}\right]^2 dx}. \quad (11)$$

To calculate the integral $J_p(z)$ using software packages, in particular, "Wolfram Mathematica" (WM), it should be represented in the following, regularized, form:

$$J_p(z) = \int_{-1}^{1}\frac{\Phi(x)-\sum_{k=0}^{p-2}\frac{\Phi^{(k)}(z)}{k!}(x-z)^k}{(x-z)^p}dx + \sum_{k=0}^{p-2}\frac{\Phi^{(k)}(z)}{k!}\frac{(1-z)^{k-p+1}-(-1-z)^{k-p+1}}{k-p+1}. \quad (12)$$

The first integral is calculated in the sense of the Cauchy principal value; as a rule, there is such a possibility in software packages.

To estimate the proximity of the values of the integral $J_p(z)$ calculated using WM ($J_p^{\mathrm{WM}}(z)$) and by formula (8) ($J_p^Q(z)$), we calculate the standard deviation by the formula:

$$\Delta_p = \frac{1}{M}\sqrt{\sum_{k=1}^{M}[J_p^{\mathrm{WM}}(x_k) - J_p^Q(x_k)]^2} \quad \left(x_k = -1 + \frac{2k}{M+1}\right). \tag{13}$$

In table 1, we also present the deviations Δ_{2J} and Δ_{2C} for the integral $J_2(z)$, calculated for different X_j. Quadrature formulas are calculated at $n = 8$.

Table 1. Moduli of deviations for the density $\varphi(x)$ and for integral $J_2(z)$ for $n = 8$ and different α and β.

α, β	$\lvert\varepsilon_J\rvert$	$\lvert\varepsilon_C\rvert$	$\lvert\Delta_{2J}\rvert$	$\lvert\Delta_{2C}\rvert$
$\alpha = -0.63 + 8i$ $\beta = 0.25 - 3i$	1.1×10^{-3}	2.6×10^{-7}	$M = 20$ 0.21 $M = 40$ 0.34	$M = 20$ 3.5×10^{-5} $M = 40$ 4.4×10^{-5}
$\alpha = -0.63 + 2i$ $\beta = 0.25 - i$	1.2×10^{-6}	2.6×10^{-7}	$M = 20$ 1.3×10^{-4} $M = 40$ 1.8×10^{-4}	$M = 20$ 1.1×10^{-5} $M = 40$ 1.1×10^{-5}
$\alpha = -0.63 + 2i$ $\beta = 0.25 + 2i$	5.4×10^{-7}	2.6×10^{-7}	$M = 20$ 4.6×10^{-5} $M = 40$ 7.5×10^{-5}	$M = 20$ 1.0×10^{-5} $M = 40$ 1.1×10^{-5}
$\alpha = 0.42 + 2i$ $\beta = 2i$	9.1×10^{-7}	2.6×10^{-7}	$M = 20$ 4.5×10^{-6} $M = 40$ 4.9×10^{-6}	$M = 20$ 1.0×10^{-6} $M = 40$ 5.3×10^{-7}
$\alpha = -0.42$ $\beta = -0.15$	2.5×10^{-7}	2.6×10^{-7}	$M = 20$ 3.5×10^{-6} $M = 40$ 4.0×10^{-6}	$M = 20$ 3.5×10^{-6} $M = 40$ 4.0×10^{-6}

The data in the table confirm the fact that when calculating the integral with density of the form (2), when the exponents are complex numbers, it is advisable to choose $X_j = t_n(\xi_j)$. In addition, we note that the deviations of the values of the integral on different grids are practically the same.

For clarity, we write out the quadrature formulas for $J_3(z)$ and $J_4(z)$ in expanded form:

$$J_3(z) = \sum_{j=1}^{n} \frac{A_j X_j}{z-\xi_j}\left[\frac{R_n^{(\alpha,\beta)}(z) - R_n^{(\alpha,\beta)}(\xi_j)}{(z-\xi_j)^2} + \frac{2(n+1)R_{n+1}^{(\alpha-1,\beta-1)}(z)}{z-\xi_j} + 2(n+1)(n+2)R_{n+2}^{(\alpha-2,\beta-2)}(z)\right],$$

$$J_4(z) = -\sum_{j=1}^{n} \frac{A_j \varphi(\xi_j)}{z-\xi_j}\left[\frac{R_n^{(\alpha,\beta)}(x) - R_n^{(\alpha,\beta)}(\xi_j)}{(z-\xi_j)^3} + 2(n+1)\frac{R_{n+1}^{(\alpha-1,\beta-1)}(z)}{(z-\xi_j)^2}\right. \tag{14}$$

$$\left. + 2(n+1)(n+2)\frac{R_{n+2}^{(\alpha-2,\beta-2)}(x)}{(z-\xi_j)} + \frac{4}{3}(n+1)(n+2)(n+3)R_{n+3}^{(\alpha-3,\beta-3)}(z)\right]$$

We now investigate the convergence of the quadrature formula (8) for certain exponents α and β for the integrals $J_p(z)$ ($p = \overline{1,4}$). Let, for example, $\alpha = -0.37+5.1i$ and $\beta = -0.535+1.4i$. To calculate the standard deviation, we take $X_j = t_n(\xi_j)$. Table 2 shows the moduli of deviations of the function $\varphi(x)$ from the interpolation polynomial and integrals $J_p(z)$ ($p = \overline{1,4}$) calculated by applied programs and by the quadrature formula (8) for various orders of approximation n.

From table 2, we note that the deviation of the Cauchy-type integral $J_1(z)$ from its quadrature value is 2–3 times less than the deviation of the density of the integral from the interpolation polynomial. The integral $J_0 = \int_{-1}^{1}\Phi(x)dx$ has the same deviation. This fact can apparently be explained by the fact that the quadrature formula for the integral $J_0 \approx \sum_{j=1}^{n} w_j\varphi(\xi_j)$ is a

Table 2. Convergence of the quadrature formula for the integrals $J_p(z)$ $\left(p=\overline{1,4}\right)$.

n	$\lvert\varepsilon_C\rvert$	$\lvert\Delta_{1C}\rvert$	$\lvert\Delta_{2C}\rvert$	$\lvert\Delta_{3C}\rvert$	$\lvert\Delta_{4C}\rvert$
6	3.5×10^{-4}	1.7×10^{-4}	4.6×10^{-3}	0.13	2.85
7	2.3×10^{-5}	7.9×10^{-6}	2.7×10^{-4}	9.1×10^{-3}	0.20
8	2.6×10^{-7}	1.4×10^{-7}	6.8×10^{-6}	2.2×10^{-4}	5.2×10^{-3}
9	1.8×10^{-7}	7.1×10^{-8}	2.2×10^{-6}	8.5×10^{-5}	2.1×10^{-3}
10	1.7×10^{-8}	5.2×10^{-9}	1.4×10^{-7}	6.7×10^{-6}	1.9×10^{-4}
11	7.5×10^{-10}	2.8×10^{-10}	3.5×10^{-9}	2.4×10^{-7}	7.9×10^{-6}
12	3.8×10^{-12}	1.4×10^{-12}	7.0×10^{-11}	3.0×10^{-9}	8.7×10^{-8}

formula of the highest algebraic accuracy. The quadrature formula for the integral $J_1(z)$ is also a formula of the highest algebraic accuracy, but only when z is a root of the function $R_n^{(\alpha,\beta)}(z)$. In addition, we note that the same degree of accuracy in calculating the integrals $J_p(z)$ $(p=\overline{1,4})$ can be provided for the orders of approximation $(n+p-1)$. The latter fact is due to formula (6).

It should also be noted that the quadrature formula allows you to calculate the integral for almost any α and β, satisfying the conditions (Reα, Re$\beta > -1$), while software packages already at not very large values Imα or Imβ begin to display messages about possible errors.

Conclusion

Quadrature formulas for calculating Cauchy-type integrals with a kernel of positive integer power containing the weight function of orthogonal Jacobi polynomials with complex exponents are presented. The convergence of the values of the quadrature formula to the true value obtained by numerical integration using software packages is substantiated by numerical analysis. Formulas are also given that make it possible to use the quadrature formula in the case when the external parameter of the Cauchy-type integral coincides with one of the nodes of the quadrature formula. The obtained quadrature formulas can be successfully used to solve hypersingular integral equations of various types.

References

[1] Hadamard J 1949 *Leçons sur la Propagation des Ondes et les Équations de l'Hydrodynamique* (New York: Chelsea)

[2] Nekrasov A I 1947 *Wing Theory in a Nonstationary Flow* (Moscow: Izdat. Akademii Nauk SSSR) [in Russian]

[3] Linkov A M and Mogilevskaya S C 1994 Complex hypersingular integrals and integral equations in plane elasticity. *Acta Mechanica* **105** (1-4) 189–205 doi: 10.1007/BF01183951

[4] Ventzel E S, Dzhan-Temirov K E, Trofimov A M, and Negolsha E V 1992 *Method of Compensating Loads in Problems of the Theory of Thin Plates and Shells* (Kharkov: KhVVKIURV) [in Russian]

[5] Hildenbrand J and Kuhn G 1992 Numerical computation of hypersingular integrals and application to the boundary integral equation for the stress tensor *Engineering Analysis with Boundary Elements* **10** (3) 209–17 doi: 10.1016/0955-7997(92)90004-Q

[6] Ioakimidis N I 1982 Application of finite-part integrals to the singular integral equations of crack problems in plane and three-dimensional elasticity *Acta Mechanica* **45** (1-2) 31–47 doi: 10.1007/BF01295569

[7] Sahakyan A V 2020 Solution a problem for edge crack with a hypersingular governing equation by the mechanical quadrature method *Mechanics — Proceedings of National Academy of Sciences of Armenia* **73** (2) 44–57 doi: 10.33018/73.2.4

[8] Ang W-T 2013 *Hypersingular Integral Equations in Fracture Analysis* (Oxford: Woodhead Publishing House)

[9] Lu J-F and Hanyga A 2004 Scattering of SH wave by a crack terminating at the interface of a bimaterial *Computational Mechanics* **34** (1) 74–85 doi: 10.1007/s00466-004-0555-3

[10] Nazarchuk Z T 1989 *A Numerical Study of Wave Diffraction on Cylindrical Structures* (Kiev: Naukova Dumka) [in Russian]

[11] Boykov I and Aikashev P 2019 To the numerical method for synthesis of fractal antennas In *2019 International Seminar on Electron Devices Design and Production (SED)* (IEEE) pp. 1–6 doi: 10.1109/SED.2019.8798475
[12] Lifanov I K and Nenashev A S 2005 Hypersingular integral equations and the theory of wire antennas *Differential Equations* **41** (1) 126–45 doi: 10.1007/s10625-005-0142-1
[13] Ashley H and Landahl M 1965 *Aerodynamics of Wings and Bodies* (New York: Dover Publications)
[14] Lifanov I K, Poltavskii L N, and Vainikko G M 2003 *Hypersingular Integral Equations and Their Applications* (CRC Press)
[15] Boykov I V 2005 *Approximate Methods for Calculating Singular and Hypersingular Integrals. Part 1. Singular Integrals* (Penza: Izdat. Penzenskogo Gosudarstvennogo Universiteta) [in Russian]
[16] Boykov I V 2009 *Approximate Methods for Calculating Singular and Hypersingular Integrals. Part 2. Hypersingular Integrals* (Penza: Izdat. Penzenskogo Gosudarstvennogo Universiteta) [in Russian]
[17] Linz P 1985 On the approximate computation of certain strongly singular integrals *Computing* **35** (3-4) 345–53 doi: 10.1007/BF02240199
[18] Kaya A C and Erdogan F 1987 On the solution of integral equations with strongly singular kernels *Quarterly of Applied Mathematics* **45** 105–22 doi: 10.1090/qam/885173
[19] Diligenti M and Monegato G 1993 Finite-part integrals: Their occurence and computation In Altomare F and Mastroianni G (Editors) *Proceedings of the Second International Conference in Functional Analysis and Approximation Theory: Supplemento ai Rendiconti del Circolo Matematico di Palermo.* Ser. 2, 33 pp. 39–61
[20] Boykov I V, Dobrynina N F, and Domnin L N 1996 *Approximate Methods for Calculating Hadamard Integrals and Solving Hypersingular Integral Equations* (Penza: Izdat. Penzenskogo Gosudarstvennogo Tekhnicheskogo Universiteta) [in Russian]
[21] Criscuolo G 1997 A new algorithm for Cauchy principal value and Hadamard finite-part integrals *Journal of Computational and Applied Mathematics* **78** (2) 255–75 doi: 10.1016/S0377-0427(96)00142-2
[22] Plieva L Yu 2016 Interpolation-type quadrature formulas for hypersingular integrals in the interval of integration *Numerical Analysis and Applications* **9** (4) 326–34 doi: 10.1134/S1995423916040066
[23] Boikov I V and Syomov M A 2016 On a method of calculation of hypersingular integrals *Russian Mathematics* **60** (3) 1–13 doi: 10.3103/S1066369X16030014
[24] Saakyan A V 2012 Quadrature formulas of maximum algebraic accuracy for auchy type integrals with complex exponents of the Jacobi weight function *Mechanics of Solids* **47** (6) 695–9 doi: 10.3103/S002565441206012X
[25] Sahakyan A V and Amirjanyan H A 2018 Method of mechanical quadratures for solving singular integral equations of various types *Journal of Physics: Conference Series* **991** 012070 doi: 10.1088/1742-6596/991/1/012070
[26] Chikin L A 1953 Special cases of the Riemann boundary value problem and singular integral equations *Uchenye Zapiski Kazanskogo Gosudarstvennogo Universiteta imeni V.I. Ul'yanova-Lenina* **113** (10) 57–105
[27] Sahakyan A V 2011 Method of discrete singularities for solution of singular integral and integro-differential equations *Proceedings of A. Razmadze Mathematical Institute* **156** 101–11
[28] Batemen H and Erdelyi A 1953 *Higher Transcendental Functions.* Vol. 1 (McGraw-Hill Book Company)
[29] Batemen H and Erdelyi A 1953 *Higher Transcendental Functions.* Vol. 2 (McGraw-Hill Book Company)
[30] Muskhelishvili N I 1968 Singular integral equations (Moscow: Nauka) [in Russian]

Stress analysis of composites with interfacial bridged cracks and damaged zones

M Perelmuter

Ishlinsky Institute for Problems in Mechanics of the Russian Academy of Sciences, Moscow, Russia

E-mail: perelm@ipmnet.ru

Abstract. Stress analysis of composite structures with bridged cracks and damaged zones is performed with the direct boundary integral equations method in axisymmetric formulation. It is assumed that: a) the crack faces interact in some zones behind of the cracks edges (bridged zones); b) these zones width can be comparable to the whole crack size; c) adjacent parts of the structure can be connected by mechanical ligaments (including interfaces ahead of cracks fronts) and the relative displacements of these parts can occur under the loading (these zones consider as damaged zones). The boundary element software to solve axisymmetric problems of elasticity with interfacial bridged cracks and damaged zones was developed. The results for different properties of composite materials and bonds are presented. Applications the bridged crack approach to modeling of cracks self-healing is remarked.

1. Introduction

Cracks growth modelling in composite materials and in adhesive joints is based on models of a crack process zones (cohesive or bridged) and it includes as the main stage the computation of displacements and stresses in the crack process zones and in the crack fronts vicinity. In this paper the bridged crack model [1] is used for cracks process zone modelling. It should be emphasized that in the bridged crack model, unlike to cohesive models, the total stress intensity factor (due to the action of the external loading and the bridging tractions) is not zero. Bonding in the crack bridged zone reduces the stress intensity factors. This effect depends on the bridged zone length and bonds properties, but within the bridged model tractions in the bridged are assumed to exist together with a stress singularity at the crack front. Damaged zones in materials (zones of weak bonds) also are considered here as parts of bridged zones of cracks. When a crack grows size of the bridged zone is decreased due to bonds breaking. Inverse process of a bridged zone formation can be considered to model cracks self-healing [2].

The problem of displacements and stresses analysis for bridged cracks in homogeneous infinite media was considered in [3, 4] (planar problem) and in [5] (problem with axial symmetry). For bridged interfacial cracks between two semi-infinite plates the problems were solved in [6, 7] by the singular integral-differential equation method (planar problem). In the last several decades a number of papers were devoted to the application of the boundary element method (BEM) to computation of the stress intensity factors (SIF) for cracks on bi-material interfaces in finite size structures, see, for example, [8—12]. In all these papers interaction between crack faces was not assumed nor considered. Only few papers have been directed to analysis of bridged cracks in finite size structures, two-dimensional problems were considered in [13—15], axisymmetric problems were considered for bridged penny-shaped cracks in homogeneous [16] and transversely isotropic [17] elastic materials. The main objective of this paper is the stress analysis of composite structures with bridged interfacial cracks by the direct boundary integral equations method in axisymmetric formulation.

2. Bridged interfacial crack

Let us consider a plane crack of radius R on the interface of two dissimilar elastic materials, see figure 1. The annular layer of width d, adjoining to the crack front, is the bridged zone of this crack. In this zone the crack faces interact with each other; this interaction restrains the crack opening. To describe the interaction between the crack faces, it is assumed that there are bonds between the crack faces at the interfacial crack bridging zone. Under the external loading action in these bonds the bridged tractions are appeared. These tractions have (in particular, in the simplest case of an uniaxial tension) the axial t_z and radial t_r components, figure 1. The crack faces in the bridged zone are loaded by the axial and radial stresses, which are numerically equal to these tractions components. The relation between the bond tractions and the crack opening (bonds deformation law) depends on the physical nature of bonds and their properties. It will be assumed here that bonds deformation law is predefined. The general nonlinear form of this spring-like bonds deformation law can be written similar to [6, 7]

$$t_{r,z} = k_{r,z}(r,\sigma)(u_{r,z}^{(1)} - u_{r,z}^{(2)}), \quad k_{r,z}(\sigma) = \phi_{1,2}(r,\sigma)\tfrac{E_b}{H}, \quad \sigma = \sqrt{t_r^2 + t_z^2}, \tag{1}$$

where the functions $k_{r,z}$ are the effective bond stiffness in the radial and axial directions, $\phi_{1,2}$ are dimensionless functions used for the description of non-uniform behavior of stiffness over the bridged zone, H is the linear scale being proportional to the thickness of the bonding zone, E_b is the effective elasticity modulus of the bond, σ is the modulus of the bond tractions vector. Note that the second formula in (1) presents the stiffness of linear-elastic bonds if the functions $\phi_{1,2}$ depend only on the coordinate. In particular, the stiffness has a uniform distribution over the crack bridged zone if the functions $\phi_{1,2}$ are constants.

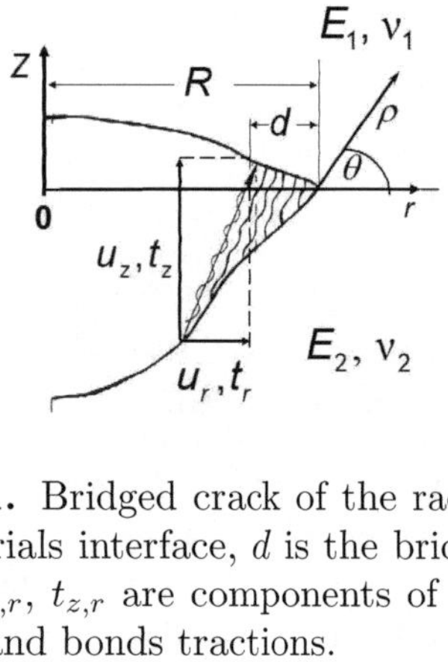

Figure 1. Bridged crack of the radius R at the materials interface, d is the bridged zone width, $u_{z,r}$, $t_{z,r}$ are components of the crack opening and bonds tractions.

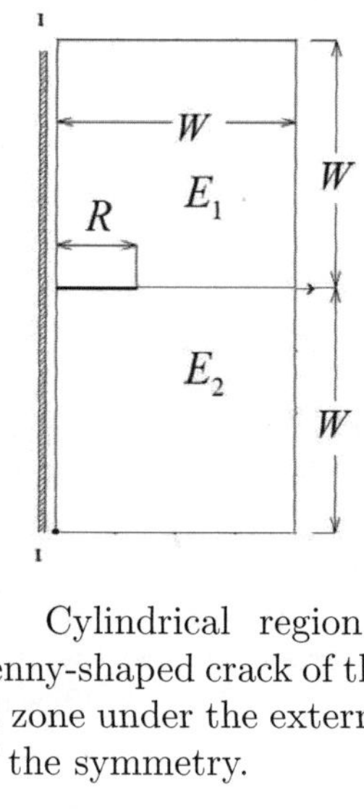

Figure 2. Cylindrical region with the interfacial penny-shaped crack of the radius R with bridged zone under the external tension, I–I is axis of the symmetry.

3. Boundary element formulation for bridged interfacial crack

The modelling of interfacial cracks is based on the multi-domain boundary element formulation. In this case the direct boundary integral equations (BIE) for elasticity problems are used for each homogeneous subregion of the structure and cracks are located between subregions [18]. The supplementary boundary conditions at interfacial subregions boundaries and at cracks faces (including bridged zones) are introduced and used to eliminate additional variables on subregions jointed boundaries. This approach can be used for any finite size structures with arbitrary loading.

For axisymmetric elasticity problems without body forces the direct BIE for any subregion of the structure is given by [19]

$$c_{ij}(p)u_i(p) = \int_\Gamma [G_{ij}(q,p)r(q)t_i(q) - F_{ij}(q,p)r(q)u_i(q)]dS(q), \quad i,j = r,z, \tag{2}$$

where Γ is the line which is rotated about the axis symmetry Oz, and it generates the axisymmetric surface S ($dS = 2\pi d\Gamma$), the kernel functions $G_{ij}(q,p)$, $F_{ij}(q,p)$ are displacements and tractions in the direction j at field point q, due to unit ring load in the direction i, applied at source point p. The expressions of the kernel function $G_{ij}(q,p)$ can be derived by integration of the fundamental solution for three-dimensional case over an axisymmetric ring for radial and axial loads and the expressions for kernel function $F_{ij}(q,p)$ are calculated by differentiation respect to r and z according to stress-displacement relations in cylindrical polar coordinates [19, 20]. If the point p lies on z-axis then must to consider limiting relations of the axisymmetric kernels [20]. The coefficients $c_{ij}(p)$ in equation (2) depend on local geometry of the boundary at a point p, for a smooth boundary these coefficients are defined through the Kronecker delta $c_{ij}(p)=0.5\delta_{ij}$.

The displacements continuity and the tractions equilibrium supplementary conditions are imposed at boundaries between subregions with ideal contact

$$u_i^k(p) = u_i^f(p), \quad t_i^k(p) = -t_i^f(p), \quad i = r,z, \tag{3}$$

where k and f are joint subregions numbers, $u_i(p)$ and $t_i(p)$ are displacements and tractions components at the boundary point p.

Along bridged zones of interfacial cracks or along damaged zones between subregions are imposed the relationships between bond tractions and displacements difference of the upper and lower crack faces or displacements difference of jointed subregions boundaries (the bonds deformation law, see (1))

$$t_i^k(p,\sigma) = \kappa_i(p,\sigma)\Delta u_i(p), \quad t_i^f(p,\sigma) = -t_i^k(p,\sigma), \quad i = r,z, \tag{4}$$

where $t_i^k(p,\sigma)$ and $t_i^f(p,\sigma)$ are the components of traction vector, $\Delta u_i(p) = u_i^k(p) - u_i^f(p)$ are the difference of crack face displacements, belonging to adjacent subregions and $\kappa_i(p,\sigma)$ are bonds stiffness depended in the general case on the distance from the crack front and bonds tension.

For numerical solution of the BIE (2) the boundaries of the structure subregions are subdivided into quadratic isoparametric elements. The special crack tip boundary elements [15] are used for the interfacial crack displacements and stresses asymptotic modelling. The collocation scheme is used to discretize BIE and collocation points coincide with the nodal points of boundary elements. The kernels in equation (2) are singular: the kernels $G_{ij}(q,p)$ are weakly singular and the kernels $T_{ij}(q,p)$ have the strong singularity. In the collocation scheme these singularities are transferred to the discrete equations. For traction kernels $G_{ij}(q,p)$ containing weak singularity is used transformation of local coordinate system and the integration is performed by using Gaussian quadrature with the logarithmic weights. The displacement kernels $T_{ij}(q,p)$ with strong singularity which exist only in the sense of Cauchy principal value, and free term $c_{ij}(p)$ are computed by application of the rigid-body movement technique, which applicable in axisymmetric problem only for integrals connected to z-axis direction. The rigid movement does not exist in radial direction. Therefore these integrals are computed by using finite-part integration technics [21].

After discretization of the BIE the matrix equation for each subregion of the structure has the standard form

$$[\mathbf{T}]\{\mathbf{u}\} = [\mathbf{G}]\{\mathbf{t}\}, \tag{5}$$

where $[\mathbf{G}]$ and $[\mathbf{T}]$ are fully populated, nonsymmetric matrices composed of integrals of the BIE fundamental solutions, $\{\mathbf{u}\}$ and $\{\mathbf{t}\}$ are vectors of nodal displacements and tractions over a subregion boundary. Formation of vectors and matrices in (5) is performed taking into account displacements continuity between boundary elements and possible discontinuity of tractions between neighboring boundary elements. Therefore, the displacement vector $\{\mathbf{u}\}$ dimension (number of equations for a subregion) is mM, here M is number of nodes on boundary of each subregion. The corresponding dimension of the matrix $[\mathbf{T}]$ is $mM \times mM$. The traction vector $\{\mathbf{t}\}$ dimension is determined as $m\alpha N$, where N is the number of elements on a subregion boundary, α is the number of nodes in each element. The corresponding dimension of matrix $[\mathbf{G}]$ is $mM \times m\alpha N$. Here for quadratic boundary elements: $\alpha = 3$ and $M = 2N$.

4. SIF computation for bridged interfacial cracks

The SIF for interfacial cracks can be defined on the basis of asymptotical relations for the crack opening and the stresses in the polar coordinates system ρ, θ with the origin point at the crack tip (see figure 1). Using the asymptotic crack tip displacements solution [22] we can write the radial $u_{r,j}$ and axial $u_{z,j}$ components of the crack faces displacements for $\theta_j = (3-2j)\pi$ ($j = 1, 2$ denotes parameters for the upper and lower subregions)

$$u_{r,j} = \Phi_j\sqrt{\frac{\rho}{2\pi}}(K_\mathrm{I} f_1 + K_\mathrm{II} f_2), \quad u_{z,j} = \Phi_j\sqrt{\frac{\rho}{2\pi}}(K_\mathrm{I} f_2 + K_\mathrm{II} f_1), \tag{6}$$

where $K_{\mathrm{I,II}}$ are SIF for interface crack, ρ is a distance to the crack tip, parameter Φ_j is defined as

$$\Phi_j = \frac{2\sin\frac{\theta_j}{2}\left(1-\nu_j^2\right)}{E_j\left(0.25+\beta^2\right)cosh(\pi\beta)}, \quad \beta = \frac{\ln\,\alpha}{2\pi}, \quad \alpha = \frac{\mu_1+\mu_2\kappa_1}{\mu_2+\mu_1\kappa_2}. \tag{7}$$

Here E_j, ν_j are materials Young's moduli and Poisson's ratios, the functions $f_{1,2}$ are

$$f_1 = 0.5\sin\left(\beta\ln\frac{\rho}{D}\right) - \beta\cos\left(\beta\ln\frac{\rho}{D}\right), \quad f_2 = 0.5\cos\left(\beta\ln\frac{\rho}{D}\right) + \beta\sin\left(\beta\ln\frac{\rho}{D}\right). \tag{8}$$

In the last relation D is an arbitrary length to normalize the distance ρ. The factor $\cosh(\pi\beta)$ included in the displacements relations (6) and (7) to use the conventional definitions of the SIF given in [23].

From relation (6)–(8) the radial $\Delta u_r = u_{1,r} - u_{2,r}$ and axial crack opening $\Delta u_z = u_{1,z} - u_{2,z}$ can be written as

$$\Delta u_r = A\sqrt{\frac{\rho}{2\pi}}(K_\mathrm{I} f_1 + K_\mathrm{II} f_2), \quad \Delta u_z = A\sqrt{\frac{\rho}{2\pi}}(K_\mathrm{I} f_2 + K_\mathrm{II} f_1), \tag{9}$$

where

$$A = \frac{2}{\cosh(\pi\beta)(0.25+\beta^2)}\left(\frac{1-\nu_1^2}{E_1} + \frac{1-\nu_2^2}{E_2}\right). \tag{10}$$

SIF $K_{\mathrm{I,II}}$ can be found from relations (9), but these parameters depend on arbitrary length parameter D. To avoid this dependence we can obtain from (9) SIF module as

$$K(\rho) = \sqrt{K_\mathrm{I}^2 + K_\mathrm{II}^2} = \frac{\cosh(\pi\beta)\sqrt{1+4\beta^2}}{(1-\nu_1^2)/E_1 + (1-\nu_2^2)/E_2}\sqrt{\frac{2\pi(\Delta u_r^2 + \Delta u_z^2)}{\rho}}. \tag{11}$$

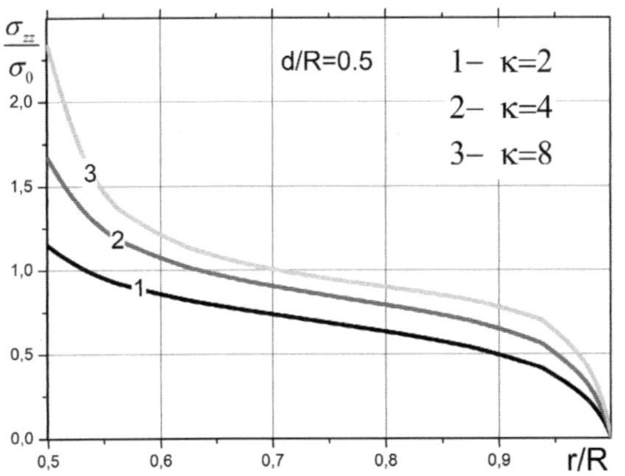

Figure 3. Normal bonds stresses (σ_{zz}) over the bridged zone r/R, variation of bonds stiffness κ, $E_2/E_1 = 5$.

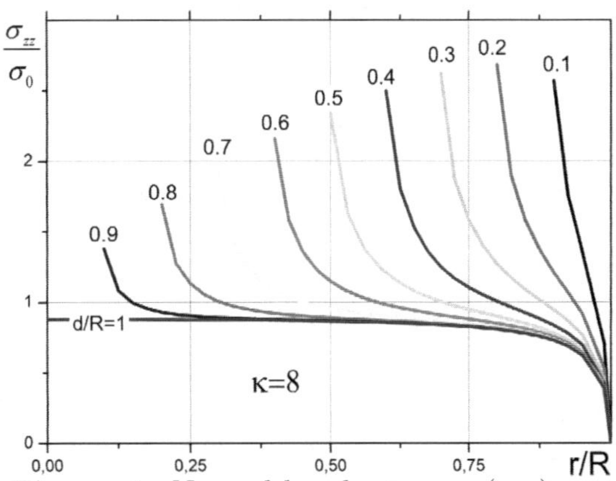

Figure 4. Normal bonds stresses (σ_{zz}) over the bridged zones r/R, variation of bridged zones width d/R, $E_2/E_1 = 5$.

The SIF module for the quarter-point boundary element can be computed from displacements on the element adjoining to the crack tip by letting $\rho = 0.25L$ in relation (11) (L is the element length adjoining to the crack tip) and therefore in this relation Δu_r and Δu_z are the components of the crack opening at the point $\rho = 0.25L$.

The SIF module can be computed also by the linear extrapolation over element length to the crack tip the value of the SIF module at the quarter-point node of the element and at the edge of the element ($\rho = L$)

$$K|_{\rho=0} = \frac{4K(L/4) - K(L)}{3}, \tag{12}$$

where $K(L)$ is defined by (11) for $\rho = L$.

5. Numerical results

The algorithms of numerical solution BIE system (2) for structures with bridged interfacial cracks and the SIF computation were implemented as the computer code for axisymmetric problems of elasticity. The SIF module is calculated according to relation (12).

To verify the solution algorithm the problem for circular interface crack (radius R) between two half-space under internal pressure applied to the crack faces was considered. The computation was performed for the finite cylindrical region divided into two subregions ($W/R = 10$), W is the external radius of the region. The numerical results have an error in compare to analytical solution [24] lower than 3%.

The bond deformation law used in the computation results presented below for the bridged crack was specified by the relationship (1) with assumption that $\phi_{1,2}(r, \sigma) = 1$. Therefore, the bonds stiffness in the radial and axial directions $\kappa_{r,z}$ and the relative bond stiffness κ are defined as follows

$$\kappa_{r,z}(p, \sigma) = \frac{E_\mathrm{b}}{H}, \quad \kappa = \kappa_{r,z}(p, \sigma)\left(\frac{E_2}{R}\right)^{-1} = \frac{E_\mathrm{b} R}{E_2 H}. \tag{13}$$

Dependencies of the crack bridged stresses and the SIF modulus versus relative bonds stiffness were computed for finite cylindrical region divided into two subregions ($W/R = 5$) under external tension σ_0 applied at $z = \pm W$. A circular interface crack is placed at the materials interface. The calculations were performed with assumption that the radial component of the bonds stiffness is zero (κ_r). The following material parameters $E_2/E_1 = 5$, $\nu_1 = \nu_2 = 0.3$ were used for computation.

The normal bonds stresses (tractions) distributions over the crack bridged zone of width $d/R = 0.5$ for various relative bonds stiffness are shown in figure 3. It is observed the significant increasing of bonds stresses if the stiffness of bonds became bigger. The maximum stress value for linearly elastic bonds is observed at the edge of the crack bridged zone. The obtained results make it possible to evaluate that stresses at the edge of the crack bridged zone ($d = 0.5R$) are proportional to the square root of the relative bonds stiffness. For example, for $\kappa_1 = 4$ and $\sigma_{zz}/\sigma_0 = 1.668$, and for $\kappa_2 = 8$ and $\sigma_{zz}/\sigma_0 = 2.334$, we obtain $2.334/1.668 \approx 1.4 \approx \sqrt{\kappa_2/\kappa_1} = \sqrt{2}$.

The distributions of the normal bond stresses (tractions) with $\kappa = 8$ and for different relative sizes of the bridged zones width are given in figure 4. As the relative bridged zone width is increasing then the stresses distribution tends to be a uniform with small nonuniform region near the crack tip. The bonds stress attain a maximum at the trailing edge of the crack bridged zone. The value of this maximum is varied with variation of the width of crack bridged zone and for k=8 the absolute maximum of the bonds stresses is attained at $d/R \approx 0.2$.

Results of the SIF module calculation versus the relative bonds stiffness for different width of the bond zone are shown in figure 5. The SIF module reduction can be used as a criterion of crack healing efficiency. From the results obtained it follows that the efficiency of SIF reduction at $d/R > 0.5$ (0.75 of the crack surface is filled with bonds) varies slightly. Thus, sufficient efficiency of crack healing is achieved already at $d/R > 0.5$. This conclusion is also supported by the results presented in figure 6. The SIF module dependence on width of the crack filling zone with bonds stabilizes at $d/R \geq 0.5$ and this stabilization zone expands with stiffness increasing.

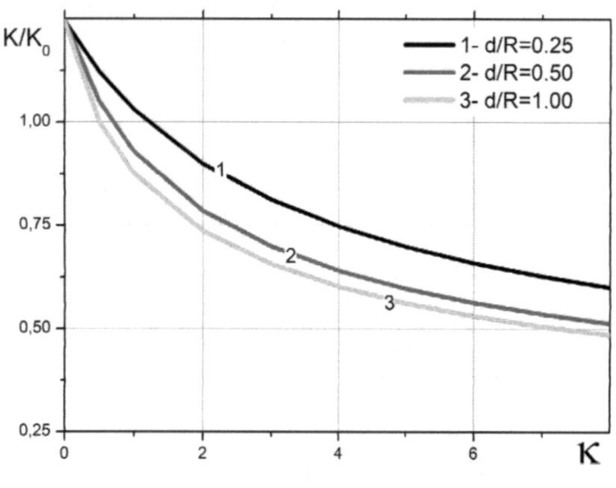

Figure 5. Dependence of the SIF module on relative bonds stiffness κ, variation of the crack bridged zone width d/R, $K_0 = 2\sigma_0\sqrt{R/\pi}$.

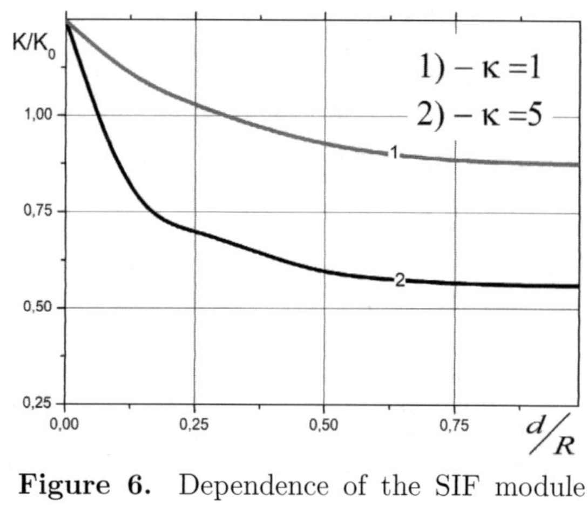

Figure 6. Dependence of the SIF module on crack bridged zone relative width d/R, $K_0 = 2\sigma_0\sqrt{R/\pi}$.

Closing

The computation of the crack opening, bridged stresses and the SIF module analysis is the first step in bridged cracks growth modeling. Analysis of bridged cracks growth can be implemented in the frame of the nonlocal criterion of bridged cracks growth [6] which can be easily incorporated into the developed boundary element code. The bridged crack model can be used also to model cracks self-healing assuming that the self-healing process is the result of the crack bridged zone formation [2].

Acknowledgments

Work was supported by the Russian Science Foundation (project No. 19-19-00616).

References

[1] Cox B N and Marshall D B 1994 Concepts for bridged cracks in fracture and fatigue *Acta Metallurgica et Materialia* **42** (2) 341—63 doi: 10.1016/0956-7151(94)90492-8

[2] Perelmuter M N 2020 Modeling of crack self-healing kinetics *Physical Mesomechanics* **23** (4) 301–8 doi: 10.1134/S1029959920040037

[3] Weitsman Y 1986 Nonlinear analysis of crazes *Journal of Applied Mechanics* **53** (1) 97–103 doi: 10.1115/1.3171745

[4] Rose L R F 1987 Crack reinforcement by distributed springs *Journal of the Mechanics and Physics of Solids* **35** (4) 383–405 doi: 10.1016/0022-5096(87)90044-5

[5] Movchan N V and Willis J R 1998 Penny-shaped crack bridged by fibres *Quarterly of Applied Mathematics* **56** (2) 327—40

[6] Goldstein R and Perelmuter M 1999 Modeling of bonding at an interface crack *International Journal of Fracture* **99** (1-2) 53–79 doi: 10.1023/A:1018382321949

[7] Perel'muter M N 2011 An interface crack with non-linear bonds in a bridged zone *Journal of Applied Mathematics and Mechanics* **75** (1) 106–18 doi: 10.1016/j.jappmathmech.2011.04.016

[8] Yuuki R and Cho S B 1989 Efficient boundary element analysis of stress intensity factors for interface cracks in dissimilar materials *Engineering Fracture Mechanics* **34** (1) 179–88 doi: 10.1016/0013-7944(89)90251-8

[9] Raveendra S T and Banerjee P K 1991 Computation of stress intensity factors for interfacial cracks *Engineering Fracture Mechanics* **40** (1) 89–103 doi: 10.1016/0013-7944(91)90130-S

[10] Tan C L and Gao Y L 1991 Axisymmetric boundary integral equation analysis of interface cracks between dissimilar materials *Computational Mechanics* **7** (5-6) 381–96 doi: 10.1007/BF00350167

[11] Hadjesfandiari A R and Dargush G F 2011 Analysis of bi-material interface cracks using complex weighting functions and non-standard quadrature *International Journal of Solids and Structures* **48** (10) 1499–512 doi: 10.1016/j.ijsolstr.2011.01.036

[12] Gu Y and Zhang C 2020 Novel special crack-tip elements for interface crack analysis by an efficient boundary element method *Engineering Fracture Mechanics* **239** 107302 doi: 10.1016/j.engfracmech.2020.107302

[13] Liu Y-F, Masuda C, and Yuuki R 1998 An efficient BEM to calculate weight functions and its application to bridging analysis in an orthotropic medium *Computational Mechanics* **22** (5) 418–24 doi: 10.1007/s004660050374

[14] Selvadurai A P S 2010 Crack-bridging in a unidirectionally fibre-reinforced plate *Journal of Engineering Mathematics* **68** (1) 5–14 doi: 10.1007/s10665-010-9365-0

[15] Perelmuter M 2013 Boundary element analysis of structures with bridged interfacial cracks *Computational Mechanics* **51** (4) 523–34 doi: 10.1007/s00466-012-0817-4

[16] Atkinson C 1970 An iterative scheme for solving problems relating to cracks opening under a displacement-dependent internal stress *International Journal of Fracture Mechanics* **6** (2) 193–8 doi: 10.1007/BF00189827

[17] Selvadurai A P S 2015 Bridged defects in continuously and discretely reinforced solids *Journal of Engineering Mathematics* **95** 359–80 doi: 10.1007/s10665-014-9779-1

[18] Blandford G E, Ingraffea A R, and Liggett J A 1981 Two-dimensional stress intensity factor computations using the boundary element method *International Journal for Numerical Methods in Engineering* **17** (3) 387—404 doi: 10.1002/nme.1620170308

[19] Banerjee P and Butterfield R 1981 *Boundary Element Methods in Engineering Science* (McGraw-Hill)

[20] Becker A A 1992 *The Boundary Element Methods in Engineering. A Complete Course* (McGraw-Hill)

[21] Guiggiani M and Casalini P 1987 Direct computation of Cauchy principal value integrals in advanced boundary elements *International Journal for Numerical Methods in Engineering* **24** (9) 1711—20 doi: 10.1002/nme.1620240908

[22] Chen E P 1985 Finite element analysis of a bimaterial interface crack *Theoretical and Applied Fracture Mechanics* **3** (3) 257—62 doi: 10.1016/0167-8442(85)90035-7

[23] Rice J R 1988 Finite element analysis of a bimaterial interface crack *Journal of Applied Mechanics* **55** 98–103 doi: 10.1016/0167-8442(85)90035-7

[24] Kassir M K and Bregman A M 1972 The stress-intensity factor for a penny-shaped crack between two dissimilar materials *Journal of Applied Mechanics* **39** (1) 308–10 doi: 10.1115/1.3422648

Modeling of self-healing of microcracks in the process of longitudinal electroplastic rolling

K V Kukudzhanov

Ishlinsky Institute for Problems in Mechanics of the Russian Academy of Sciences, Moscow, Russia

E-mail: kconstantin@mail.ru

Abstract. Multiscale model to describe the process of self-healing of microcracks in thin metal sheets during longitudinal electroplastic rolling is developed. The simultaneous treatment of a mechanical load and a pulsed high-energy electromagnetic field on the sheet is assumed. Finite element modeling has shown that under certain conditions microcracks and micropores can self-heal partially in the working zone. Impact of external pulsed high-energy electromagnetic field leads to welding of the shores of microcracks by molten metal in their tops, while the length and volume of cracks decrease.

1. Introduction

There are various methods of self-healing of metals, alloys, ceramics and composite materials [1–15]. Meanwhile, the method of self-healing of microdefects in conductive materials under impacted to them by short pulses of a high-energy electromagnetic field (hereinafter referred to as HEEMF) [1–10] is of considerable practical interest. This is due to the possibility of using this method to heal structural defects (microcracks and micropores) in any materials conducting an electric current, unlike other less universal and more energy- and labor-intensive methods of self-healing of microdefects of metals, alloys, ceramics and composite materials [11–16]. The method under consideration uses the effect of concentration of the electromagnetic field on inhomogeneities in the conducting media. This method of healing microdefects is widely used in modern high-tech materials processing, for example, in the electroplastic processing of metals, alloys and conductive composites. However, due to the complexity of multiphysical modeling of ongoing processes, this method has not been studied theoretically.

It has been experimentally established that the considered self-healing of microdefects can be achieved under impact to the working zone by an external intense short-term electromagnetic field during a time 10^{-5}–10^{-3} s, so that the value of the specific dissipated electromagnetic energy in the conductor is in the range of 10^8–10^{10} J/m^3. We will call such a field a pulsed high-energy electromagnetic field (HEEMF).

Rolling is the main approach in the processing of metals by pressure. The vast majority of semi-finished products and products of the metallurgy are produced in the form of rolled products. Moreover, about 90% of all rolled products are produced by means of longitudinal rolling.

Longitudinal electroplastic rolling is schematically shown in figure 1. The figure shows the upper half of the rolled sheet, where H_0 is the initial half-thickness of the sheet, H_1 is the half-

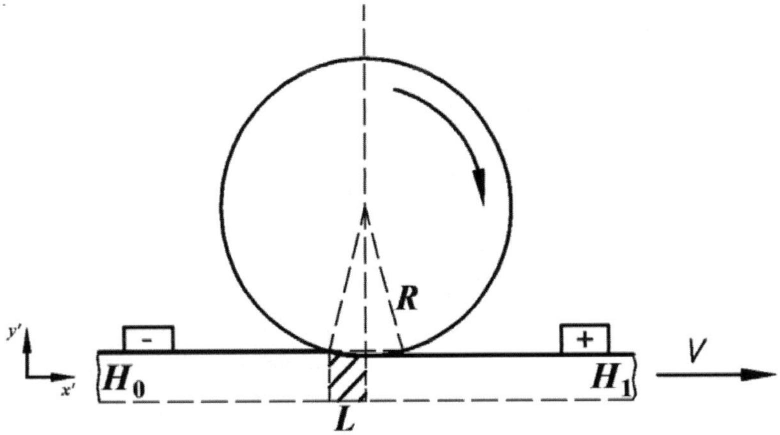

Figure 1. Diagram of electroplastic rolling with impact of the high-energy electromagnetic field.

thickness of the sheet after rolling. The working (deformation) rolling zone is located between the upper and lower rolls. The upper area of the working zone is shaded in figure 1. The working zone has size L along the x axis. A method of supplying current to the working zone using four sliding contacts is considered. The contacts are placed before and after the working zone. This method is optimal in terms of, firstly, ease of implementation in manufacturing and, secondly, the safety of the working rolls of the rolling mill from premature electrocorrosive destruction. The short pulses of an electromagnetic field (current) is applied to the sliding contacts.

Electro-plastic rolling (or pulse rolling) with impact by pulsed HEEMF is in demand in the manufacture of thin sheets of hard-to-form conducting materials, when traditional rolling methods lead to the nucleation and growth of multiple microcracks in the sheet. Microcracks nucleation and growth occur when the sheet is captured by working rolls in the zone of entry into the working zone. The working zone itself is begun just behind the sheet capture zone and is located directly under/between working rolls. It has been experimentally established that microcracks can coalense, forming mesocracks coming out on the surface of a sheet (on both its planes). This makes the sheet unusable. While the impact of pulsed HEEMF on the rolled sheet leads to partial self-healing of these cracks [10].

This article is aimed at developing a multiscale model and investigating the physical mechanism of self-healing of microcracks in the process of longitudinal electroplastic rolling of thin metal sheets.

Modeling of self-healing of microcracks during electroplastic rolling has not been attempted before.

2. Multiscale model

The multiscale (multilevel) model of longitudinal rolling of thin sheets under the influence of mechanical load and pulsed HEEMF was developed to describe the process of self-healing of microcracks in the material during electroplastic rolling.

The nonlinear rolling problem in itself presents a certain complexity in view of the intense plastic flow of metal in the working zone. The development model of a rolling with the impact of pulsed HEEMF in order to study the mechanism of self-healing of microdefects in the working zone complicates the problem additionally.

Meanwhile, the developing of the model can be simplified if we take into account the

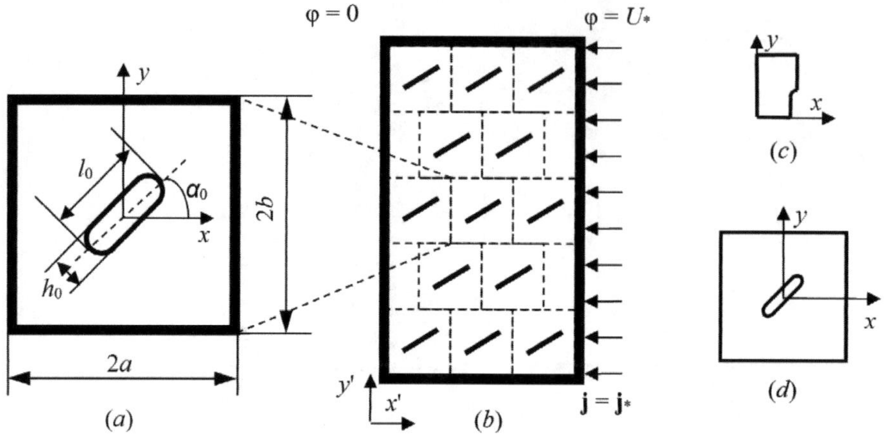

Figure 2. Representative cell with a micro-crack (a); the ordered structure of micro-cracks (b), here the arrows show the direction of the electric current density vector at the sheet; domains of integration used in the calculations (c, d).

differences in the temporal and spatial scales of the simulated processes. Note that the duration of HEEMF pulses is on the order of 100 μs, while the time spent by the material in the working zone is at least 10 ms (for high speeds of movement of the rolled sheet). At the same time, the sizes of microcracks are 0.1–10 μm, whereas the sizes of the sheet in the working zone are 1–10 mm. Therefore, the process of macro-forming of the sheet and the process of micro-forming of a representative element under impact of HEEMF pulse can be considered in the working zone separately, as described below.

In view of the above, the following models of micro- and macro-scale levels and the hypothesis of the connection of these levels are applied.

Microanalysis of the mechanism of self-healing of microcracks in the working zone is carried out on the basis of a microscale model. This model was developed in [15–18] and modified in this article taking into account the conditions realized during electroplastic rolling. A pre-damaged conductive material with microdefects (microcracks or micropores) with sizes $l_0 \times h_0$ is considered. All microdefects have uniform shape, size and spatial orientation (figure 2). We consider a plane strain and the size of the defect along the z-axis is assumed to be infinite. We assume that the defects in the material are arranged at the nodes of the rectangular lattice (figure 2b). In this case, separation of a representative cell is easy possible (figure 2a). The sheet sizes are much larger than microdefect size.

The solution is obtained in the domains of integration used in the calculations shown in figures 2c and 2d and containing either one quarter or one whole representative cell [14–16].

The basic assumptions, equations of the microscale model, the initial, boundary, contact conditions and conditions on the interphases (solid–melt, melt–gas) boundaries are described in detail in [17–20]. The complete system of equations of the considered micromodel has the form:

$$\int_V \nabla\delta\varphi\sigma^{\mathrm{E}}(T)\nabla\varphi\, dV = \int_S \delta\varphi j\, dS, \quad \mathbf{j} = \sigma^{\mathrm{E}}(T)\mathbf{E} = -\sigma^{\mathrm{E}}(T)\frac{\partial\varphi}{\partial\mathbf{x}}, \tag{1}$$

$$\int_V \boldsymbol{\sigma} : \delta\boldsymbol{\varepsilon}\, dV = \int_S \mathbf{t}\cdot\delta\mathbf{u}\, dS + \int_V \mathbf{f}\cdot\delta\mathbf{u}\, dV, \quad \dot{\boldsymbol{\varepsilon}} = \dot{\boldsymbol{\varepsilon}}^{\mathrm{el}} + \dot{\boldsymbol{\varepsilon}}^{\mathrm{pl}} + \dot{\boldsymbol{\varepsilon}}^{\mathrm{th}}, \tag{2}$$

$$\dot{\boldsymbol{\sigma}} = \lambda(T)\dot{\boldsymbol{\varepsilon}}^{\mathrm{el}} : \mathbf{I} + 2\mu(T)\dot{\boldsymbol{\varepsilon}}^{\mathrm{el}}, \quad \dot{\boldsymbol{\varepsilon}}^{\mathrm{pl}} = \dot{\Lambda}\frac{\partial\Phi}{\partial\boldsymbol{\sigma}} = \dot{\Lambda}\mathbf{s}, \quad \bar{\sigma} = \sigma_Y(T), \quad \bar{\sigma} = \sqrt{\tfrac{3}{2}\mathbf{s}:\mathbf{s}}, \quad d\boldsymbol{\varepsilon}^{\mathrm{th}} = \alpha(T)\mathbf{I}\, dT, \tag{3}$$

$$\rho(T)c(T)\dot{T} = r^{\mathrm{E}} + r^{\mathrm{pl}} + r^{\mathrm{melt}} + r^{\mathrm{evap}}, \quad r^{\mathrm{E}} = \eta^{\mathrm{E}}\mathbf{j}\cdot\mathbf{E} = \eta^{\mathrm{E}}\nabla\varphi\cdot\sigma^{\mathrm{E}}\cdot\nabla\varphi, \quad r^{\mathrm{pl}} = \eta^{\mathrm{pl}}\boldsymbol{\sigma} : \dot{\boldsymbol{\varepsilon}}^{\mathrm{pl}}, \tag{4}$$

$$T = T_{\mathrm{melt}}, \quad t_{\mathrm{sol}} \le t \le t_{\mathrm{liq}}, \quad \int_{t_{\mathrm{sol}}}^{t_{\mathrm{liq}}} (r^{\mathrm{E}} + r^{\mathrm{pl}})\, dt = \rho\Lambda_{\mathrm{melt}}, \tag{5}$$

$$T = T_{\mathrm{evap}}, \quad t_{\mathrm{eliq}} \le t \le t_{\mathrm{evap}}, \quad \int_{t_{\mathrm{eliq}}}^{t_{\mathrm{evap}}} (r^{\mathrm{E}} + r^{\mathrm{pl}})\, dt = \rho\Lambda_{\mathrm{evap}}. \tag{6}$$

The notation in this equations is given in [19, 20].

In the considered multiscale model, the relationship between the same type of variables of different scale levels is carried out by setting (determining) the appropriate boundary conditions. To connect the macro- and micro-scale levels and obtain a solution to the problem at the micro-scale level, it is necessary to determine the average values of the macroscopic fields of stresses $\sigma'(x')$ and electrical potential $\varphi'(x')$ at the boundaries of the representative element in the working zone (a touch above the variables denotes that they belong to the macroscale level).

Otherwise, the boundary and contact conditions for the electrical potential φ, displacements $\mathbf{u}$ and traction forces at the boundaries of the integration domain and phase boundaries used in this model, as well as the initial conditions, coincide with those considered in [19, 20].

Consider the region of the integration shown in Figure 2c and the part of the boundary Γ_1: $y = \Phi(x)$, that does not pass along the crack surface and the axes of symmetry of the representative element. At the Γ_1 the normal component (cross-sectional averages) of the stress tensor $\sigma'_n(x')$ and the electric potential $\varphi'(x')$ were set and the absence of a tangent component of the stress tensor was also set:

$$\sigma_n|_{\Gamma_1} = \sigma'_n(x'), \quad \sigma_\tau|_{\Gamma_1} = 0, \tag{7}$$

$$\varphi(x, y, t)|_{\Gamma_1} = \varphi'(x'). \tag{8}$$

Therefore, at the macroscale level, it is advisable to use a one-dimensional rolling model to obtain cross-sectional averages of macro stresses in the working zone. To study the stress state of thin sheets, the one-dimensional rolling model, first proposed by Karman, has become widespread. In this model, it is assumed that the values of the velocity components in the rolling direction $v'_x = v'_x(x')$ and the normal stress tensor $\sigma'_{xx} = \sigma'_{xx}(x')$, $\sigma'_{yy} = \sigma'_{yy}(x')$ depend only on the coordinate along the rolling direction and do not depend on the coordinate perpendicular to the rolling direction (figure 1) or that any flat section at the entrance to the working zone remains flat during the entire deformation time. To obtain average macro stresses in the working zone, a modified Karman's model is used, described and investigated in [21]. The model with good accuracy allows us to obtain the required values of the average cross-sectional stresses for thin sheets (when $L/H > 1.5$, where $H = (H_1 + H_2)/2$ is the average half-thickness of the rolled sheet (figure 1)). The complete system of equations of the model and boundary conditions in the working zone has the form [21]:

$$\frac{d\sigma'_{xx}}{dx'} = -(\sigma'_{xx} - \sigma'_{yy})\frac{\dot{H}(x')}{H(x)} - \tau'\frac{1+\dot{H}^2(x')}{H(x')}, \tag{9}$$

$$\frac{d(\sigma'_{xx} - \sigma'_{yy})}{dx'} = \begin{cases} -4\mu\frac{\dot{H}(x')}{H(x')}, & |\sigma'_{xx} - \sigma'_{yy}| < 2\tau'_s, \\ 0, & |\sigma'_{xx} - \sigma'_{yy}| = 2\tau'_s, \end{cases} \tag{10}$$

$$v'_x = \frac{H_0}{H(x')}v_0, \quad v'_y|_{y'=H(x')} = \dot{H}(x')v'_x, \quad \dot{H}(x') = \frac{dH}{dx'}, \tag{11}$$

$$\tau' = \begin{cases} \gamma(v'_r - v'_t) = t', & |t'| < \phi\tau'_s, \\ \pm\phi\tau'_s, & |t'| \geq \phi\tau'_s, \end{cases} \tag{12}$$

$$\sigma'_{xx}(0) = \sigma'_{yy}(0) = 0, \quad \sigma'_{yy}(L) = 0. \tag{13}$$

where v_0 is velocity of sheet in enter to the working zone, μ is shear modulus, τ'_s is shear yield stress, ϕ, γ are coefficients, characterizing sliding on the surface of the roller, v'_t is tangential velocity of the material on the surface of the roller, v'_r is velocity of rotation of the roller.

The laws of conservation of charge (1) is used to determine the macroscopic field of electrical potential $\varphi'(x', y')$ in the sheet. The cross-sectional averages of the potential $\varphi'(x')$ are obtained by averaging over the corresponding cross-section.

3. Results of numerical simulation

Based on the proposed multiscale model, the problem of short-term field action on typical internal microcracks in the sheet during electroplastic rolling is solved.

Equations (1)–(8) of the micromechanical problem coupled with boundary, contact, and initial conditions were solved by the finite element method. The computations were performed for the plane strain using linear four-node isoparametric and three-node finite elements. The restructuring of the finite element grid is performed on the basis of the mixed Euler-Lagrangian method. The time derivative is approximated by the inverse difference.

Equations (9)–(13) were solved by the Runge-Kutta method. Equations for electrical potential and temperature of the macromechanical problem coupled with boundary and initial conditions were solved by the finite element method.

Simulation was done for aluminium sheet, the physical and mechanical properties of which and their dependence on temperature were taken in accordance with [22–24]. The sizes of the sheet are $H_1 = 1.0\,\mathrm{mm}$, $H_2 = 0.9\,\mathrm{mm}$, and the size of the working zone is $L = 2.85\,\mathrm{mm}$. The distance between the sliding contacts is 9.4 mm. A pulse in the form of a meander of duration $t = 50\,\mu\mathrm{s}$ with a potential difference of 4.40 mV is applied to the sliding contacts.

The microcracks with two parallel edges and rounded tips (figure 2) were considered ($l_0 = 10\,\mu\mathrm{m}$, $h_0 = 1\,\mu\mathrm{m}$). The representative cell sizes (a and b) varied in the range of 40–60 μm. The angle is equal to $\alpha_0 = 90°$.

Changes in time of electromagnetic and temperature fields, stress-strain state, phase boundaries (melting and evaporation regions) in the vicinity of microcracks in the working zone are investigated.

With the help of modeling, it was found that under certain conditions microcracks inside the sheet in the working zone of electroplastic rolling can partially self-heal. The mechanism of self-healing is as follows: under treatment of the external HEEMF pulse, a very strong concentration of the electric field and, accordingly, the electric current density field occurs at the tips of microcracks. This causes extreme heating and melting of the metal at the tips of microcracks (figure 3). This, in turn, initiates the process of self-healing, which proceeds by

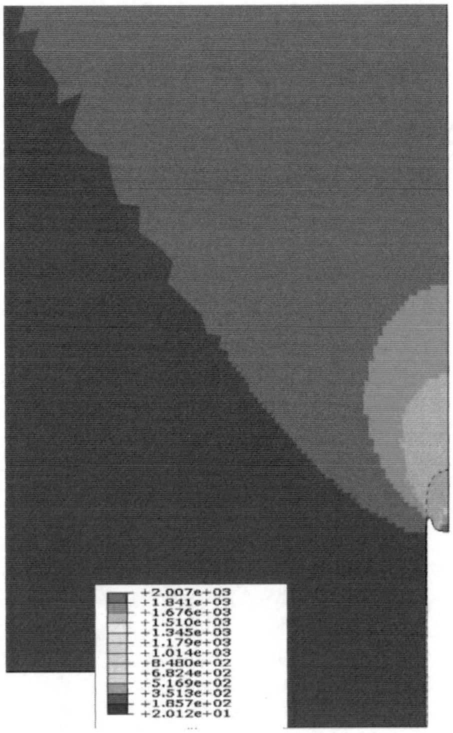

Figure 3. Microcrack's geometry (dashed line is initial at $t = 0$ and solid is at $t = 50\,\mu s$) and contours of temperature (°C) in the vicinity of the microcrack at $t = 50\,\mu s$ (just after pulse of HEEMF ending).

simultaneously reducing the length of the microcrack, ejecting a jet of molten metal from the tip into the crack and closing its shores in the vicinity of the crack tip. Thus, the shores of the microcrack at its top are welded with molten metal. At the same time, the length and volume of microcracks decrease (figure 3).

Repeated impacts of pulses of HEEMF can lead to further self-healing of the microcracks during rolling.

The simulation results were compared with the available experimental data for the electroplastic rolling [10]. The above-described partial self-healing of microcracks inside the sheet led to the fact that their coalense into meso-cracks coming out on the surface of the sheet did not occur. A similar effect of partial disappearance of surface meso-cracks in the sheet under pulsed electric current treatment was observed in existing experiments [10].

It confirms the capabilities of the considered multiscale model to qualitatively describe the healing of microdefects in non-ferrous metal sheets during rolling with the treatment by pulses HEEMF.

Acknowledgments

This investigation is supported by the Russian Science Foundation (project No. 22-19-00616).

References

[1] Beklemishev N N, Koryagin N I, and Shapiro G C 1984 Influence of locally inhomogeneous pulsed electromagnetic field on ductility and strength of conductive materials *Izvestiya AN SSSR. Metally* **4** 184–7

[2] Beklemishev N N, Kukudzhanov V N, Porokhov V A, et al. 1989 *Plasticity and Strength of Metallic Materials with the Pulse Action of a High-Energy Electromagnetic Field Taken into Account* Preprint No. 372 (Moscow: IPM AN SSSR) [in Russian]
[3] Klyushnikov V D and Ovchinnikov I V 1988 Plane problem of effect of an instantaneous point heat source *Mechanics of Solids* **23** (4) 113–7
[4] Ovchinnikov I V 1993 Determination of the reserve of plasticity under the action of a current *Strength of Materials* **25** (6) 437–41 doi: 10.1007/BF00773032
[5] Kukudzhanov V N and Kolomiets-Romanenko A V 2010 Study of the influence of electric current dynamical action on mechanical properties of materials with ordered structure of defects *Mechanics of Solids* **45** (3) 465–75 doi: 10.3103/S0025654410030167
[6] Song H, Wang Z-j and Gao T-j 2007 Effect of high density electropulsing treatment on formability of TC4 titanium alloy sheet *Transactions of Nonferrous Metals Society of China* **17** (1) 87–92 doi: 10.1016/S1003-6326(07)60053-3
[7] Troitskii O A, Baranov Yu V, Avraamov Y S, and Shlyapin A D 2004 *Physical Fundamentals and Technologies of Processing Advanced Materials (Theory, Technology, Structure, and Properties).* Vol. 1 (Moscow-Izhevsk: Institut Komputernykh Issledovaniy) [in Russian]
[8] Conrad H 1989 *A Study into the Mechanism(s) for the Electroplastic Effect in Metals and Its Application to Metalworking, Processing and Fatigue* (Final Report ARO Proposal Number 23090-MS, ARO Funding Document DAAL03-86-K-0015, U.S. Army Research Office, North Carolina State University, Raleigh, NC 27695)
[9] Zheng X G, Shi Y N, and Lu K 2013 Electro-healing cracks in nickel *Materials Science and Engineering: A* **561** 52–9 doi: 10.1016/j.msea.2012.10.080
[10] Ye X, Li X, Song G, and Tang G 2014 Effect of recovering damage and improving microstructure in the titanium alloy strip under high-energy electropulses *Journal of Alloys and Compounds* **616** 173–83 doi: 10.1016/j.jallcom.2014.07.143
[11] Hager M D, Greil P, Leyens C, et al. 2010 Self-healing materials *Advanced Materials* **22** (47) 5424–30 doi: 10.1002/adma.201003036
[12] Ferguson J B, Schultz B F, and Rohatgi P K 2014 Self-healing metals and metal matrix composites *JOM* **66** 866–71 doi: 10.1007/s11837-014-0912-4
[13] Toohey K S, Sottos N R, Lewis J A, et al. 2007 Self-healing materials with microvascular networks *Nature Materials* **6** 581–5 doi: 10.1038/nmat1934
[14] Li G and Meng H (Editors) 2015 *Recent Advances in Smart Self-healing Polymers and Composites* (New York: Woodhead Publishing)
[15] Snoeck D, Malm F, Cnudde V, et al. 2018 Validation of self-healing properties of construction materials through nondestructive and minimal invasive testing *Advanced Materials Interfaces* **5** (17) 1–28 doi: 10.1002/admi.201800179
[16] Dansereau V, Weiss J, Saramito P, and Lattes P 2016 A Maxwell elasto-brittle rheology for sea ice modeling *The Cryosphere* **10** (3) 1339–59 doi: 10.5194/tc-10-1339-2016
[17] Kukudzhanov K V 2015 Modeling the treatment of high-energy pulsed electromagnetic field of the micro-cracks in a polycrystalline metal *PNRPU Mechanics Bulletin* No. 4 138–58 doi: 10.15593/perm.mech/2015.4.09.
[18] Kukudzhanov K V and Levitin A L 2015 Modeling the healing of microcracks in metal stimulated by a pulsed high-energy electromagnetic field. Part I *International Journal of Nanomechanics Science and Technology* **6** 3 233–50 doi: 10.1615/NanomechanicsSciTechnolIntJ.v6.i3.60
[19] Kukudzhanov K V and Levitin A L 2016 Modeling the healing of microcracks in metal stimulated by a pulsed high-energy electromagnetic field. Part II *International Journal of Nanomechanics Science and Technology* **7** 2 123–48 doi: 10.1615/NanomechanicsSciTechnolIntJ.v7.i2.30
[20] Kukudzhanov K V and Levitin A L 2017 Healing of damaged metal by pulsed high-energy electromagnetic field *Procedia IUTAM* **23** 84–100 doi: 10.1016/j.piutam.2017.06.008
[21] Kovalev S I, Koryagin N I, and Shirko I V 1982 *Stresses and Strains During Flat Rolling* (Moscow: Metallurgiya) [in Russian]
[22] Gavrilin I V 2000 *Melting and Crystallization of Metals and Alloys* (Vladimir: VGU) [in Russian]
[23] Pikunov M V 2005 *Metallurgy of the Melts. A Course of Lectures* (Moscow: MISA) [in Russian]
[24] Pikunov M V 1997 *Metal Smelting. Alloy Crystallization. The Solidification of Castings* (Moscow: MISiS) [in Russian]

Finding coefficients in the problem of a cylindrical shell with a circular cutout

S V Kashtanova[1*] and A V Rzhonsnitskiy[2]

[1]Institute for Problems of Mechanical Engineering of RAS, St. Petersburg, Russia
[2]Saint-Petersburg State Institute of Technology, St. Petersburg, Russia

E-mail: *kastasya@yandex.ru

Abstract. In the present paper a cylindrical shell with a circular cutout is considered. There have not yet been explicit formulas for the field of stresses and after reviewing some ideas of the derivation of the solution of this problem, the authors can present how this can be done. A new approach is based on the separation of variables in the basis function by Fourier decomposition and on the proof of reducibility of the system. This way has a strict mathematical foundation, and for the boundary condition in the form of axial tension the limit transition leads for the Kirsch problem, as expected. In this work a finding of coefficients for small parameter is illustrated and the limit transition is shown.

Introduction

In this paper we are considering how to obtain coefficients for the basis solutions, from which explicit formulas for the stresses in the cylindrical shell with a circular hole under axial tension can be derived (figure 1). The same procedure can be done for other boundary conditions such as internal pressure or torsion, but the case of an axial tension is interesting in that it allows to check limit transition that leads us to Kirsch problem. The problem under consideration has a long history and begins from Lurie [1] who set the problem in 1946, but the decomposition into small parameter β (see below), on which the governing equation depends, restricts the model in range of applicability. More than that, the linear infinite system for finding coefficients has a linear dependency that classics could not find and because of it had difficulties with solving it. The authors of the present paper thoroughly studied classical works of Naghdi, Eringen, Lekkerkerker, Murthy, Van Dyke etc [2–4], found that the problem was abandoned [5] and are ready to present an algorithm for finding stresses based on their own previous works [6–8]. The authors proposed an idea by which these coefficients are numerically easy to calculate, but this idea also opened up prospects for their analytical search. In work [7], it was possible to transform the system and reduce it to a finite system, this work is of mathematical nature, in which a proof of reducibility of the system is given. In the present paper, as an example, we will consider small parameter β, as it allows not to overload the reader with heavy formulas since for small β we need fewer basis functions [6], as well as it is easier to see the limit transition to the Kirsch problem.

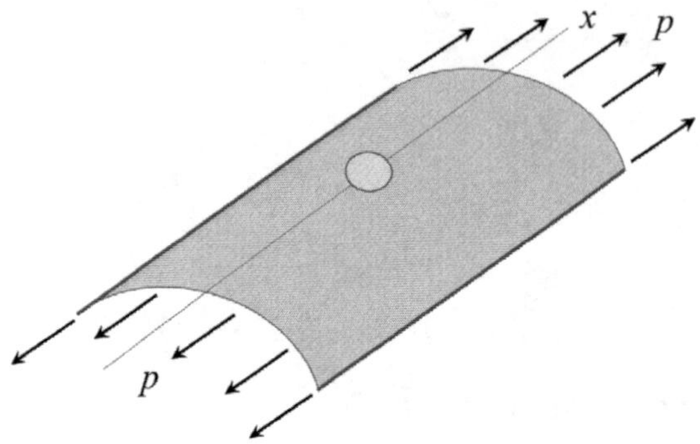

Figure 1. Cylindrical shell with a circular hole under axial tension in x-coordinate.

1. Problem statement with new form of solution

The governing equation of the problem is [1, 6]

$$\Delta\Delta\Phi + 8i\beta^2\frac{\partial^2\Phi}{\partial x^2} = 0. \tag{1}$$

Here next designations used:

$\beta^2 = r_0^2\frac{\sqrt{3(1-\nu^2)}}{4Rh}$ — the ratio between geometric characteristics,
R and h — radius and thickness of cylindrical shell,
r_0 — radius of the hole ($r_0 = 1$ in the present paper),
ν — Poisson's coefficient,
$\Phi = \frac{Eh}{8\beta^2 R}w - iU$,
w — deflection,
U — stress function,
E — Young's modulus.

The stresses T are expressed through the stress function U as

$$T_x = \frac{\partial^2 U}{\partial y^2}, \quad T_{xy} = -\frac{\partial^2 U}{\partial y \partial x}, \quad T_y = \frac{\partial^2 U}{\partial x^2}$$

and median surface stress as $\sigma = T/h$.

We consider boundary conditions in form of axial tension p at infinity along x-coordinate:

- at infinity
$$T_x = p, \quad T_{xy} = 0, \quad T_y = 0, \quad w = 0; \tag{2}$$
- at the boundary of the hole in polar coordinates (r, ϑ) — free edge
$$T_{rr}|_{r=r_0} = 0, \quad T_{r\vartheta}|_{r=r_0} = 0, \quad M_r|_{r=r_0} = 0, \quad Q_r|_{r=r_0} = 0. \tag{3}$$

The solution of (1) is a sum of the unperturbed solution, which depends on boundary conditions applied to a cylindrical shell, and the perturbed solution caused by the presence of the cutout. It can be written as

$$\Phi = \Phi_{\text{unpert.}} + \Phi_{\text{pert}},$$

where $\Phi_{\text{unpert.}} = -ipy^2/2$ that satisfies boundary condition

$$T_x = \frac{\partial^2 U}{\partial y^2} = \frac{\partial^2(-\text{Im}\Phi)}{\partial y^2} = p.$$

In work [3] it was offered the form of solution with unknown coefficients A_n and B_n

$$\Phi = \sum_{n=0}^{\infty}(A_n + iB_n) \cdot \begin{cases} \cos[(1+i)\beta r\cos\vartheta]H_n^{(1)}[(1+i)\beta r]\cos(n\vartheta) & \text{for even } n, \\ \sin[(1+i)\beta r\cos\vartheta]H_n^{(1)}[(1+i)\beta r]\cos(n\vartheta) & \text{for odd } n. \end{cases} \quad (4)$$

But the classics faced with some difficulties in finding coefficients [5], and, in work [6], it was shown that Φ_{pert} can be presented as

$$\Phi_{\text{pert}} = \sum_{n=0}^{\infty}(a_n + ib_n)f_n, \quad (5)$$

$$f_n(r,\vartheta) = \frac{g(r,n,0)}{2} + \sum_{l=1}^{\infty} g(r,n,l)\cos(2l\vartheta), \quad (6)$$

$$g(r,n,l) = (-1)^{[n/2]+l}\frac{H_n^{(1)}[(1+i)\beta r]}{H_n^{(1)}[(1+i)\beta]}\{J_{n+2l}[(1+i)\beta r] + J_{n-2l}[(1+i)\beta r]\}, \quad n,l = 0,1,\ldots,\infty.$$

Here $H_n^{(1)}[(1+i)\beta r]$ is a Hankel function, $J_{n\pm 2l}[(1+i)\beta r]$ is a Bessel function, $[n/2]$ is an integer part of number, $A_n + iB_n = (a_n + ib_n)/H_n^{(1)}[(1+i)\beta]$.

After substituting the solution (5) into the boundary conditions (3) $r_0 = 1$ (in the present work) we get an infinite independent linear system for finding unknown coefficients a_n and b_n [6]:

$$\begin{cases} \operatorname{Im}\sum_{n=0}^{\infty}(a_n + ib_n)g(n,1) = -\frac{p}{4}, & \operatorname{Im}\sum_{n=0}^{\infty}(a_n + ib_n)g'(n,1) = -\frac{p}{2}, \\ \operatorname{Im}\sum_{n=0}^{\infty}(a_n + ib_n)g(n,l) = 0, & \operatorname{Im}\sum_{n=0}^{\infty}(a_n + ib_n)g'(n,l) = 0, \\ \operatorname{Im}\sum_{n=0}^{\infty}(a_n + ib_n)t_3(n,l) = 0, & \operatorname{Im}\sum_{n=0}^{\infty}(a_n + ib_n)t_4(n,l) = 0, \end{cases} \quad (7)$$

where

$$t_3(n,l) = i[-4l^2\nu g(n,l) + \nu g'(n,l) + g''(n,l)],$$
$$t_4(n,l) = i\{12l^2 g(n,l) - [1 + \nu + 4l^2(2-\nu)]g'(n,l) + g'''(n,l)\}.$$

By finding these coefficients a_n and b_n and substituting them in (4), we can find the stress function U and, then, the field of stresses for our problem. In the analytical works [1–3, etc.] of the 40–60s of the twentieth century the approach was that the basis functions and coefficients were expanded in a series by a small parameter β. The difficulties and weaknesses of this approach are described in detail in [5]. A new approach, which leads us to (6), is also proposed and its advantages are described [5, 6].

The next question concerned the proof of the reducibility of an infinite system was made in [7]. And the present paper is devoted to a clear algorithm for finding these stresses. Briefly, the idea that is fully disclosed and justified in [7] is that the search for coefficients is presented in the form of a certain sequence $s_n(\beta)$:

$$a_n + ib_n = (-1)^{n+[n/2]}\varepsilon_n s_n(\beta), \quad n \geq 0, \quad (8)$$

where

$$\varepsilon_n = \begin{cases} 1 & \text{if } n \in \mathbb{N}, \\ \frac{1}{2} & \text{if } n = 0. \end{cases}$$

For negative n the sequence is further defined as $s_{-n}(\beta) = (-1)^n s_n(\beta)$.

In turn, we introduce two more sequences, which completely define each other (note, that sequence $\beta_m(\beta)$ with sub index is not geometric parameter β):

$$\alpha_m(\beta) = \sum_{n\in\mathbb{Z}} s_n(\beta) J_{m-n}[(1+i)\beta], \tag{9}$$

$$\beta_m(\beta) = \sum_{n\in\mathbb{Z}} s_n(\beta) \frac{H_n'[(1+i)\beta]}{H_n[(1+i)\beta]} J_{m-n}[(1+i)\beta], \tag{10}$$

$$\beta_m(\beta) = \sum_{k\in\mathbb{Z}} \left\{ \sum_{n\in\mathbb{Z}} J_{k-n}[(1+i)\beta] \frac{H_n'[(1+i)\beta]}{H_n[(1+i)\beta]} J_{m-n}[(1+i)\beta] \right\} \alpha_k(\beta). \tag{11}$$

In equation (11)

$$\sum_{n\in\mathbb{Z}} J_{k-n}[(1+i)\beta] \frac{H_n'[(1+i)\beta]}{H_n[(1+i)\beta]} J_{m-n}[(1+i)\beta]$$

is the kernel of the Fredholm operator. Let's denote this kernel as $\mathcal{K}(m,k)$, so

$$\beta_m(\beta) = \sum_{k\in\mathbb{Z}} \mathcal{K}(m,k)\alpha_k(\beta). \tag{12}$$

Due to some properties that kernel possesses [7], we assume that for large $m \geq m_0$:

$$\mathcal{K}(m,m) = -\frac{m(1-i)}{2\beta}, \quad \mathcal{K}(m,m\pm 1) = \frac{1}{2}, \quad \mathcal{K}(m,k) = 0 \quad \text{when } k \neq m \text{ and } k \neq m \pm 1. \tag{13}$$

In terms of these sequences, we can rewrite the boundary conditions, since they are relations between these two sequences. The boundary conditions are derived in detail in [7] and the derivation is not given here due to the cumbersomeness, but their form is given here. One of the boundary condition in new terms is $\mathrm{Im}\alpha_2 = p/4$, $\mathrm{Im}\alpha_{2m} = 0$, when $m > 1$ (7), so taking this into account others will be written as (the elements of the first column and matrix $\mathcal{A}$ see in Appendix):

$$("\hat{\alpha}") = \begin{pmatrix} \mathrm{Im}\alpha_0 \\ c_1 \\ d_1 \\ \mathrm{Re}\alpha_2 \\ c_3 \\ d_3 \\ \mathrm{Re}\alpha_4 \\ c_5 \\ d_5 \\ \vdots \\ \mathrm{Re}\alpha_{2m} \\ c_{2m+1} \\ d_{2m+1} \\ \vdots \end{pmatrix} = \mathcal{A} \begin{pmatrix} f_0 \\ g_0 \\ \mathrm{Im}\beta_1 \\ f_2 \\ g_2 \\ \mathrm{Im}\beta_3 \\ f_4 \\ g_4 \\ \mathrm{Im}\beta_5 \\ \vdots \\ f_{2m} \\ g_{2m} \\ \mathrm{Im}\beta_{2m+1} \\ \vdots \end{pmatrix} + p \begin{pmatrix} 1/6 \\ -1/(2\beta) \\ -\beta/[12(1+\nu)] \\ \beta^2(3+\nu)/[12(-1+\nu^2)] \\ -1/\beta \\ -\beta/[12(1+\nu)] \\ -\beta^2/[48(1+\nu)] \\ -1/\beta \\ 0 \\ 0 \\ \vdots \\ -1/\beta \\ 0 \\ 0 \\ \vdots \end{pmatrix}. \tag{14}$$

Here

$$c_{2l+1}(\beta) = \frac{\mathrm{Re}\alpha_{2l+1} + \mathrm{Im}\alpha_{2l+1}}{2}, \quad d_{2l+1}(\beta) = \frac{\mathrm{Re}\alpha_{2l+1} - \mathrm{Im}\alpha_{2l+1}}{2},$$
$$f_{2l}(\beta) = \frac{\mathrm{Re}\beta_{2l} + \mathrm{Im}\beta_{2l}}{2}, \quad g_{2l}(\beta) = \frac{\mathrm{Re}\beta_{2l} - \mathrm{Im}\beta_{2l}}{2}.$$

To shorten the formula entry, we write it in the form

$$("\hat{\alpha}") = \mathcal{A}("\tilde{\beta}") + fr_1. \tag{15}$$

On the other hand, using (12) we can see follow correlation between sequences

$$("\tilde{\beta}") = \mathcal{B}("\alpha") + fr_2. \tag{16}$$

Matrix $\mathcal{B}$ and fr_2 are constructed by the kernel $\mathcal{K}(m,k)$ with already known $\text{Im}\alpha_{2m}$ for $m \geq 1$, so substituting (16) into (15), we can find the sequence $"\alpha"$ as

$$(\mathcal{B} - \mathcal{A}^{-1}Spa)("\alpha") = -fr_2 - \mathcal{A}^{-1}fr_1, \tag{17}$$

where Spa is a sparse matrix which we input as:

$$("\hat{\alpha}") = \begin{pmatrix} \text{Im}\alpha_0 \\ c_1 \\ d_1 \\ \vdots \end{pmatrix} = Spa \cdot \begin{pmatrix} \text{Re}\alpha_0 \\ \text{Im}\alpha_0 \\ c_1 \\ d_1 \\ \vdots \end{pmatrix} = Spa \cdot ("\alpha").$$

So it is the matrix that has units over the main diagonal

$$Spa = \begin{pmatrix} 0 & 1 & 0 & 0 & \cdots \\ 0 & 0 & 1 & 0 & \cdots \\ 0 & 0 & 0 & 1 & \cdots \\ \vdots & \vdots & \vdots & \vdots & \ddots \end{pmatrix}.$$

The column fr_2 consists of an even $\text{Im}\alpha_{2m}$ and will specified below for our case, considered in the next section.

The main idea is that the relations (12) and (14) allow us to find the sequence $\alpha_k(\beta)$, which is displayed in (17). Based on the assumption (13), taking into account the relationships (12) and (14) between the sequences $\alpha_k(\beta)$ and $\beta_m(\beta)$, it follows that all the terms of the sequence $\alpha_k(\beta)$ are equal to zero for $k \geq 2l - 1$, where $l = [m_0/2] + 1$.

So by building the sequence $\beta_m(\beta)$ based on the Kernel $\mathcal{K}(m,k)$, we will determine the sequence $\alpha_k(\beta)$, thereby we will be able to find the sequence $s_n(\beta)$, which gives us the sum $a_n + ib_n$ — the coefficient for basis function in (4) after dividing by $H_n^{(1)}[(1+i)\beta]$.

2. Calculation technique

In the present paper we consider a small parameter β, for which $m_0 = 3$ is sufficient, since (13) is already being executed. We take small β to illustrate the algorithm and not tire the reader with cumbersome formulas. Moreover, we do not take into account higher-order infinitesimals in order to simplify the process. In this case the matrix $\mathcal{B}$ with an accuracy up to infinitesimals of a higher order, based on the kernel $\mathcal{K}(m,k)$, is

$$\mathcal{B}=\begin{pmatrix} -\beta+\dfrac{2\pi}{\beta(\zeta^2+\pi^2)} & \dfrac{2\zeta}{\beta(\zeta^2+\pi^2)} & 1+\dfrac{4\zeta}{\zeta^2+\pi^2} & \dfrac{4\pi}{\zeta^2+\pi^2} & 0 \\ \dfrac{2\zeta}{\beta(\zeta^2+\pi^2)} & \beta-\dfrac{2\pi}{\beta(\zeta^2+\pi^2)} & \dfrac{-4\pi}{\zeta^2+\pi^2} & 1+\dfrac{4\zeta}{\zeta^2+\pi^2} & 0 \\ \dfrac{2\pi}{\zeta^2+\pi^2} & \dfrac{1}{2}+\dfrac{2\zeta}{\zeta^2+\pi^2} & \dfrac{\beta(1-\zeta)}{2} & \dfrac{1}{\beta}+\dfrac{\pi\beta}{2} & 0 \\ 0 & 0 & \dfrac{1}{2} & 0 & \dfrac{\beta}{2} \\ 0 & 0 & 0 & \dfrac{1}{2} & -\dfrac{1}{\beta} \end{pmatrix},$$

$$\zeta=4\ln\beta+4\gamma-2\ln 2,\quad \gamma \text{ is Euler's constant},$$

$$fr_2=\frac{p}{8}\begin{pmatrix} \dfrac{-\pi\beta}{\zeta^2+\pi^2} & \dfrac{-\zeta\beta}{\zeta^2+\pi^2} & 1-\pi\beta^2-\dfrac{1}{2\beta} & -\beta \end{pmatrix}^T.$$

Now, we can rewrite (17) into an equivalent extended matrix and solve:

$$\begin{array}{ccccc|c} \mathrm{Re}\alpha_0 & \mathrm{Im}\alpha_0 & c_1 & d_1 & \mathrm{Re}\alpha_2 & \text{free term} \\ 2\pi-\beta^2(\zeta^2+\pi^2) & 2\zeta & 4\zeta\beta & 4\pi\beta & 0 & \frac{\zeta^2+\pi^2}{2} \\ 2\zeta & -2\pi+\beta^2(\zeta^2+\pi^2) & -4\pi\beta & 4\zeta\beta-\frac{4\beta(\zeta^2+\pi^2)}{-1+\nu} & -2(\zeta^2+\pi^2) & \frac{\zeta^2+\pi^2}{12(1-\nu^2)}\beta^2(3+\nu) \\ 2\pi\beta^2 & \beta^2(2\zeta+\zeta^2+\pi^2) & \frac{\beta^3(1-\zeta)}{2}(\zeta^2+\pi^2) & \frac{\beta(\zeta^2+\pi^2)}{2}(4+\pi\beta^2) & (\zeta^2+\pi^2)(-1+\nu) & \frac{(\zeta^2+\pi^2)}{24(1+\nu)}\beta^2(3+\nu) \\ 0 & 0 & 1 & 0 & \frac{\beta}{2} & \frac{1}{2\beta} \\ 0 & 0 & 0 & \beta & -2 & \frac{\beta^2(8+3\nu-3\nu^2)}{12(1-\nu^2)} \end{array}.$$

The solution for sequence $\alpha_k(\beta)$ small β is

$$\begin{aligned} &\mathrm{Re}\alpha_0=\frac{\pi}{4}+o(\beta),\quad \mathrm{Im}\alpha_0=-1+\frac{\zeta}{4}+o(\beta),\\ &\beta c_1=\frac{1}{2}+o(\beta),\quad \beta d_1=o(\beta),\quad \mathrm{Re}\alpha_2=o(\beta). \end{aligned} \tag{18}$$

Return to sequence $s_n(\beta)$

$$s_n(\beta)=\sum_{k\in\mathbb{Z}}\alpha_k(\beta)J_{k-n}[(1+i)\beta]$$

gives us follow formulas

$$s_0(\beta)=\frac{\pi}{4}+i\frac{\zeta}{4},\quad s_1(\beta)=\frac{1+i}{2\beta},\quad s_2(\beta)=-\frac{i}{4},\quad s_3(\beta)=s_4(\beta)=\ldots=0.$$

3. Results

Using (8) and substituting in (4) we get the potential Φ

$$\begin{aligned} \Phi=&-\frac{ipr^2\sin^2\vartheta}{2}+\left(\frac{\pi}{8}+i\frac{\zeta}{8}\right)\frac{\cos[(1+i)\beta r\cos\vartheta]H_0^{(1)}[(1+i)\beta r]}{H_0^{(1)}[(1+i)\beta]}\\ &-\frac{1+i}{2\beta}\frac{\sin[(1+i)\beta r\cos\vartheta]H_1^{(1)}[(1+i)\beta r]\cos\vartheta}{H_1^{(1)}[(1+i)\beta]}\\ &+\frac{i}{4}\frac{\cos[(1+i)\beta r\cos\vartheta]H_2^{(1)}[(1+i)\beta r]}{H_2^{(1)}[(1+i)\beta]}. \end{aligned}$$

The calculating of the stresses from this potential Φ at the limit transition $\beta \to 0$ leads us to the well-known Kirsch stresses [8]

$$\lim_{\beta \to 0} \mathcal{L}(-\mathrm{Im}[\Phi(\beta)]) = \sigma_{\text{Kirsch}},$$

where $\mathcal{L}$ is the corresponding differentiation operator

$$\mathcal{L}_{11} = \frac{1}{r^2}\frac{\partial^2}{\partial\vartheta^2} + \frac{1}{r}\frac{\partial}{\partial r} \quad \text{for } \sigma_{rr}, \quad \mathcal{L}_{12} = \frac{1}{r^2}\frac{\partial}{\partial\vartheta} - \frac{1}{r}\frac{\partial^2}{\partial r\partial\vartheta} \quad \text{for } \sigma_{r\vartheta}, \quad \mathcal{L}_{22} = \frac{\partial^2}{\partial r^2} \quad \text{for } \sigma_{\vartheta\vartheta}. \quad (19)$$

To illustrate the algorithm for finding coefficients, the authors did not taken into account terms of the order of smallness β^2 for clarity, therefore the authors see no reason to write explicit formulas for the stress field in this form. This can be easily obtained, using the formula $\sigma = \mathcal{L}(-\mathrm{Im}[\Phi(\beta)])$, where $\mathcal{L}$ is defined in (19).

Acknowledgments
The reported study was funded by RFBR, project number 19-31-60008.

Appendix

$$\begin{aligned}
\mathrm{Im}\alpha_0 &= -\frac{1}{30\beta}[-5p\beta - 15d_5(1+\nu) - 15g_0(1-\nu) - 30g_2(1-\nu),\\
&\quad - 30g_4(1-\nu) + 2\beta\mathrm{Im}(30\beta_1 + 10\beta_3 + 3\beta_5)]\\
c_1 &= -\frac{p}{2\beta} + f_0,\\
d_1 &= -\frac{1}{60(1+\nu)}[5p\beta - 30d_5(1+\nu) + 30g_0(1-\nu) - 60g_2(1-\nu)\\
&\quad - 60g_4(1-\nu) + 4\beta\mathrm{Im}(10\beta_3 + 3\beta_5)],\\
\mathrm{Re}\alpha_2 &= \frac{1}{60(1-\nu)(1+\nu)}[-5p\beta^2(3+\nu) + 15\beta d_5(7+8\nu+\nu^2) - 15\beta g_0(1+2\nu-\nu^2)\\
&\quad + 30\beta g_2(3-2\nu-\nu^2) + 30\beta g_4(7-6\nu-\nu^2) - 40\beta^2\mathrm{Im}\beta_3(3+\nu) - 6\beta^2\mathrm{Im}\beta_5(7+\nu)],\\
c_3 &= -\frac{p}{\beta} + f_0 + 2f_2,\\
d_3 &= -\frac{1}{60(1+\nu)}[5p\beta - 60d_5(1+\nu) - 120g_4(1-\nu) + 4\beta\mathrm{Im}(10\beta_3 + 6\beta_5)],\\
\mathrm{Re}\alpha_{2l} &= \frac{\beta}{4l^2(1-\nu)}[(-1+4l-\nu)d_{2l-1}(1+4l+\nu)d_{2l+1} + 2g_{2l}(1-\nu)\\
&\quad + \beta\mathrm{Im}(-2\alpha_{2l} + \alpha_{2l-2} + \alpha_{2l+2} + 2\beta_{2l-1} - 2\beta_{2l+1})],\\
c_{2l+1} &= c_{2l-1} + 2f_{2l}, \quad \text{when } l > 1,\\
d_{2l-1} &= \frac{1}{(4l^2-1)(1+\nu)}\{(4l^2-1)(1+\nu)d_{2l+1} + 2(4l^2-1)(1-\nu)g_{2l} + \beta\mathrm{Im}[2\alpha_{2l}\\
&\quad + (3-6l)\alpha_{2l-2} + 3\alpha_{2l+2} + 6l\alpha_{2l+2} - 2\beta_{2l-1} - 4l\beta_{2l-1} + 2\beta_{2l+1} - 4l\beta_{2l+1}]\} \quad \text{when } l > 2,
\end{aligned}$$

$$\mathcal{A} = \begin{pmatrix}
0 & \frac{1-\nu}{2\beta} & -2 & 0 & \frac{1-\nu}{\beta}\\
1 & 0 & 0 & 0 & 0\\
0 & \frac{-1+\nu}{2(1+\nu)} & 0 & 0 & \frac{1-\nu}{1+\nu}\\
0 & \frac{\beta(-1+\nu)}{4(1+\nu)} & 0 & 0 & \frac{\beta(3+\nu)}{2(1+\nu)}\\
1 & 0 & 0 & 2 & 0
\end{pmatrix}.$$

References

[1] Lur'e A I 1946 Concentration of stresses in the vicinity of an aperture in the surface of a circular cylinder *Prikladnaya Matematika i Mekhanika* **10** (3) 397–406 = Lur'e A I 1960 *Concentration of Stresses in the Vicinity of an Aperture in the Surface of a Circular Cylinder* (New York: Courant Institute of Mathematical Sciences, New York University)

[2] Naghdi A K and Eringen A C 1965 Stress distribution in a circular cylindrical shell with a circular cutout *Ingenieur-Archiv* **34** (3) 161–72 doi: 10.1007/BF00532170

[3] Murthy V V M 1969 Stresses around an elliptic hole in a cylindrical shell *Journal of Applied Mechanics* **36** (1) 39–46 doi: 10.1115/1.3564583

[4] Van Dyke P 1965 Stresses about a circular hole in a cylindrical shell *AIAA Journal* **3** (9) 1733–42 doi: 10.2514/3.3234

[5] Kashtanova S V, Rzhonsnitskiy A V, and Gruzdkov A A 2021 On the issue of analytical derivation of stress state in a cylindrical shell with a circular hole under axial tension *Materials Physics and Mechanics* **47** 186–95 doi: 10.18149/MPM.4722021_3

[6] Kashtanova S V and Rzhonsnitskiy A V 2021 An analytical approach to obtaining the stress field of a cylindrical shell with a circular hole under tension *PNRPU Mechanics Bulletin* No. 2 12–23 doi: 10.15593/perm.mech/2021.2.07

[7] Kashtanova S V and Rzhonsnitskiy A V 2021 Investigation of systems of the stress field problem of a cylindrical shell with a circular cutout under various boundary conditions *Acta Mechanica* [to be published]

[8] Muskhelishvili N I 1963 *Some Basic Problems of the Mathematical Theory of Elasticity* (Groningen: Noordhoff)

Contact problem for a piecewise-homogeneous plane with an interfacial crack under dry friction

V N Hakobyan*, L V Hakobyan, and L L Dashtoyan

Institute of Mechanics of the National Academy of Sciences of Armenia, Yerevan, Republic of Armenia

E-mail: *vhakobyan@sci.am

Abstract. The article considers the contact problem of indentation of an absolutely rigid stamp with dry friction into one of the edges of an interfacial finite crack located in a piecewise homogeneous elastic plane. The governing system of equations for the problem is derived in the form of a system of three singular integral equations with respect to the normal contact pressure under the stamp and dislocation components of the displacements of the crack edge points. The exact solution is constructed. The behavior of contact stresses near the end points of a crack is studied depending on the ratio of the elastic characteristics of heterogeneous materials of half-planes that formed a piecewise-homogeneous plane.

1. Introduction

Many engineering structures contain stress concentrators such as cracks, stamps and inclusions, around which local stress fields with large gradients arise, which can lead to partial or complete destruction of the structure. Therefore, the study of these fields of local stresses and the identification of the features of their change depending on the physical and mechanical and geometric parameters has always been and remains one of the topical areas for contact and mixed boundary value problems in the theory of elasticity and fracture mechanics. In this regard, let us point out the well-known monographs [1–6], where a number of fundamental results in the indicated direction are given. We also cite works [7–10], where exact solutions of a number of problems for a piecewise-homogeneous plane and space with interfacial inclusions are constructed, one side of which is adhered to the matrix, on the other side has moved away from the matrix, creating a crack, or smoothly contacts the matrix. Exact solutions of many problems on the interaction of absolutely rigid inclusions with cracks in an orthotropic elastic plane are given in the monograph [6]. We especially note the work [11, 12] which are more closely related to the present study. In the first, an exact solution of the problem of a plane-deformed state of a homogeneous elastic plane with a finite crack is constructed, into one of the edges of which an absolutely rigid stamp with dry friction is indented. In the second, the plane-deformed stress state of a piecewise-homogeneous elastic plane with an interfacial crack is studied, one of the edges of which is pressed into an absolutely rigid smooth stamp with a flat base. The exact solution of the problem is obtained. In order to expand the field of application of the contact model L. A. Galin [1], this paper studies the plane-deformed state of a piecewise-homogeneous elastic plane with a crack, on one of the edges of which an absolutely rigid stamp with dry friction is pressed.

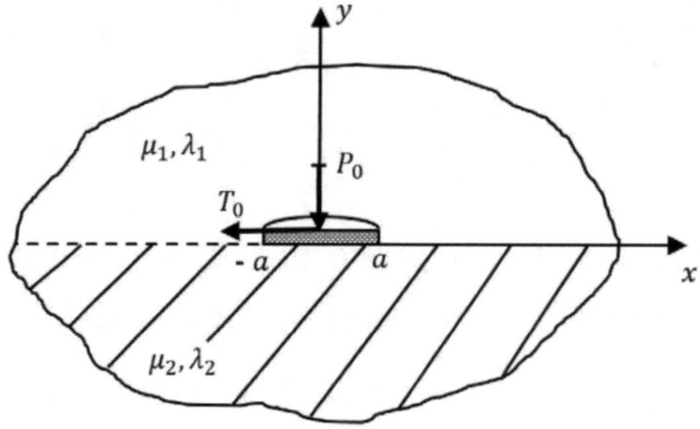

Figure 1. Contact problem.

2. Statement of the problem and derivation of the governing equations

Let a piecewise-homogeneous plane made by joining two heterogeneous half-planes with Lamé coefficients μ_1, λ_1 and μ_2, λ_2, respectively, referred to the Cartesian coordinate system Oxy, on the interval $(-a, a)$ of Ox-axis is weakened by an interfacial crack. Ox-axis is directed along the line of joining of heterogeneous half-planes. The piecewise-homogeneous plane is considered under plane deformation. It is assumed that the piecewise-homogeneous plane is deformed under action of an absolutely rigid stamp intended into the lower edges of the crack under the influence of a normal concentrated load P_0 and shear concentrated load $T_0 = \rho P_0$ applied in the center of the stamp (figure 1). It is also assumed that there is dry friction on Coulomb's law with a coefficient ρ in the contact region.

The problem is to construct an exact solution and to study the behavior of contact stresses acting under the stamp and on the plane of the joint of heterogeneous half-planes in the vicinity of the end points of the crack, depending on the physical and mechanical and geometric parameters of the problem. The problem can be mathematically formulated as the following boundary value problem:

$$\begin{cases} \sigma_y^{(1)}(x,0) = \sigma_y^{(2)}(x,0), \\ \tau_{xy}^{(1)}(x,0) = \tau_{xy}^{(2)}(x,0), \\ u^{(1)}(x,0) = u^{(2)}(x,0), \\ v^{(1)}(x,0) = v^{(2)}(x,0), \end{cases} \quad |x| > a, \tag{1a}$$

$$\begin{cases} \sigma_y^{(1)}(x,0) = \tau_{xy}^{(1)}(x,0) = 0, \\ \tau_{xy}^{(2)}(x,0) = \rho\sigma_y^{(2)}(x,0), \\ v^{(2)}(x,0) = \delta, \end{cases} \quad |x| < a, \tag{1b}$$

where δ is the rigid displacement of the stamp along the direction Oy-axis, and $\tau_{xy}^{(j)}(x,y)$ $(j = 1,2)$ are components of the stress tensor acting in the upper and lower half-planes, respectively, $u^{(j)}(x,y)$ and $v^{(j)}(x,y)$ $(j = 1,2)$ are horizontal and normal components of the displacement vector in the corresponding half-planes.

To construct an exact solution of the boundary value problem (1), we will use discontinuous solutions of the equations of the theory of elasticity for a piecewise homogeneous plane obtained in [7]. Keeping the notation introduced in [7] for the stresses acting on the crack edges and displacements of the points of the crack edges, the following expressions are obtained:

$$\begin{aligned}
\sigma_y^{(1)}(x,0) &= \frac{l_0}{\Delta}\sigma(x) + \frac{l_1}{\Delta}u'(x) + \frac{l_2}{\pi\Delta}\int_{-a}^{a}\frac{\tau(s)}{s-x}ds - \frac{l_3}{\pi\Delta}\int_{-a}^{a}\frac{v'(s)}{s-x}ds,\\
\tau_{xy}^{(1)}(x,0) &= \frac{l_0}{\Delta}\tau(x) - \frac{l_1}{\Delta}v'(x) - \frac{l_2}{\pi\Delta}\int_{-a}^{a}\frac{\sigma(s)}{s-x}ds - \frac{l_3}{\pi\Delta}\int_{-a}^{a}\frac{u'(s)}{s-x}ds,\\
v_2'(x,0) &= -\frac{l_0}{\Delta\vartheta_2^{(2)}}v'(x) - \frac{d_0}{\Delta}\tau(x) - \frac{d_1}{\pi\Delta}\int_{-a}^{a}\frac{\sigma(s)}{s-x}ds - \frac{l_2}{\pi\Delta\vartheta_2^{(2)}}\int_{-a}^{a}\frac{u'(s)}{s-x}ds,\\
u_2'(x,0) &= -\frac{l_0}{\Delta\vartheta_2^{(2)}}u'(x) + \frac{d_0}{\Delta}\sigma(x) - \frac{d_1}{\pi\Delta}\int_{-a}^{a}\frac{\tau(s)}{s-x}ds + \frac{l_2}{\pi\Delta\vartheta_2^{(2)}}\int_{-a}^{a}\frac{v'(s)}{s-x}ds,
\end{aligned} \tag{2}$$

where

$$\begin{aligned}
&\sigma(x) = \sigma_y^{(1)}(x,0) - \sigma_y^{(2)}(x,0) = -\sigma_y^{(2)}(x,0),\\
&\tau(x) = \tau_{xy}^{(1)}(x,0) - \tau_{xy}^{(2)}(x,0) = -\tau_{xy}^{(2)}(x,0) = -\rho\sigma_y^{(2)}(x,0) = \rho\sigma(x),\\
&u^{(1)}(x,0) - u^{(2)}(x,0) = \frac{u(x)}{\theta_2^{(2)}}, \quad v^{(1)}(x,0) - v^{(2)}(x,0) = \frac{v(x)}{\theta_2^{(2)}},\\
&l_0 = \vartheta_2^{(1)}(\vartheta_2^{(1)} + \vartheta_2^{(2)}) - \vartheta_1^{(1)}(\vartheta_1^{(1)} - \vartheta_1^{(2)}), \quad l_2 = \vartheta_1^{(1)}\vartheta_2^{(2)} + \vartheta_2^{(1)}\vartheta_1^{(2)},\\
&l_1 = 2(\vartheta_2^{(2)})^{-1}(\vartheta_1^{(2)}l_0 - \vartheta_2^{(2)}l_2), \quad l_3 = 2(\vartheta_2^{(2)})^{-1}(\vartheta_1^{(2)}l_2 - \vartheta_2^{(2)}l_0),\\
&d_0 = \frac{\vartheta_1^{(1)} - \vartheta_1^{(2)}}{2}, \quad d_1 = \frac{\vartheta_2^{(1)} + \vartheta_2^{(2)}}{2}, \quad \vartheta_1^{(j)} = \frac{\mu_j}{2+\alpha_j}, \quad \vartheta_2^{(j)} = \frac{(1+\alpha_j)\mu_j}{2+\alpha_j},\\
&\Delta = (\vartheta_2^{(1)} + \theta_2^{(2)})^2 - (\vartheta_1^{(1)} - \vartheta_1^{(2)})^2, \quad \alpha_j = \frac{1}{1-2\nu_j} \quad (j=1,2).
\end{aligned}$$

We use representations (2), which satisfies the conditions (1b) on the crack edges, previously differentiated the last of them. As a result, to determine the unknown contact pressure $\sigma(x)$, shear stress $\tau(x)$ and dislocation displacement functions $u'(x)$ and $v'(x)$, the following governing system of singular integral equations are obtained:

$$\begin{aligned}
&\frac{l_0}{\Delta}\sigma(x) + \frac{l_1}{\Delta}u'(x) + \frac{l_2}{\pi\Delta}\int_{-a}^{a}\frac{\tau(s)}{s-x}ds - \frac{l_3}{\pi\Delta}\int_{-a}^{a}\frac{v'(s)}{s-x}ds = 0,\\
&\frac{l_0}{\Delta}\tau(x) - \frac{l_1}{\Delta}v'(x) - \frac{l_2}{\pi\Delta}\int_{-a}^{a}\frac{\sigma(s)}{s-x}ds - \frac{l_3}{\pi\Delta}\int_{-a}^{a}\frac{u'(s)}{s-x}ds = 0,\\
&\frac{l_0}{\Delta\vartheta_2^{(2)}}v'(x) + \frac{d_0}{\Delta}\tau(x) + \frac{d_1}{\pi\Delta}\int_{-a}^{a}\frac{\sigma(s)}{s-x}ds + \frac{l_2}{\pi\Delta\vartheta_2^{(2)}}\int_{-a}^{a}\frac{u'(s)}{s-x}ds = 0,\\
&\tau(x) = \rho\sigma(x).
\end{aligned} \tag{3}$$

Note that in the case of a homogeneous plane $l_1 = 0$, as well as system (3) and the system of governing equations obtained in [11] are the same. To construct an unambiguous solution for system (3), there is need to use the conditions for the equilibrium of the stamp and the continuity of displacements at the end points of the crack:

$$\int_{-a}^{a}\sigma(x)dx = P_0, \quad \int_{-a}^{a}v'(x)dx = 0, \quad \int_{-a}^{a}u'(x)dx = 0. \tag{4}$$

Thus, the solution of the problem is reduced to the solution of the system of singular integral equations (3) under conditions (4).

3. Solution of the system of governing equations

To construct a solution of system (3) under conditions (4), we substitute the value $\tau(x)$ from the last equation (3) into the first three equations and bring them to the canonical form.

$$\begin{aligned} &\sigma(x) + \frac{a_1}{\pi}\int_{-a}^{a}\frac{\sigma(s)}{s-x}ds + \frac{a_2}{\pi}\int_{-a}^{a}\frac{u'(s)}{s-x}ds = 0,\\ &v'(x) + \frac{b_1}{\pi}\int_{-a}^{a}\frac{\sigma(s)}{s-x}ds + \frac{b_2}{\pi}\int_{-a}^{a}\frac{u'(s)}{s-x}ds = 0,\\ &u'(x) + \frac{c_1}{\pi}\int_{-a}^{a}\frac{\sigma(s)}{s-x}ds + \frac{c_2}{\pi}\int_{-a}^{a}\frac{u'(s)}{s-x}ds + \frac{c_3}{\pi}\int_{-a}^{a}\frac{v'(s)}{s-x}ds = 0. \end{aligned} \tag{5}$$

here

$$a_1 = -\frac{\vartheta_1^{(1)}\vartheta_2^{(2)}}{\rho\vartheta},\quad a_2 = \frac{2[(\vartheta_2^{(1)})^2 - (\vartheta_1^{(1)})^2]}{\rho\vartheta},\quad b_1 = \frac{\vartheta_2^{(1)}\vartheta_2^{(2)}}{2\vartheta},\quad b_2 = \frac{\vartheta_2^{(1)}\vartheta_1^{(2)}}{\vartheta},$$

$$c_1 = \frac{\rho l_2 - a_1 l_0}{l_1},\quad c_2 = -\frac{a_2 l_0}{l_1},\quad c_3 = -\frac{l_3}{l_1},\quad \vartheta = (\vartheta_2^{(1)})^2 - (\vartheta_1^{(1)})^2 + \vartheta_1^{(1)}\vartheta_1^{(2)}.$$

Note that the representation (5) of the governing system is inapplicable only in the case when $l_1 = 0$, which corresponds to the cases of the homogeneous problem considered in [11]. Let us reduce the solution of system (5) to the solution of three independent singular integral equations. For this purpose, we multiply the second and third of the equations (5), respectively, by the unknown constants $n \neq 0$ and $m \neq 0$ and sum up with the first equation (5):

$$\sigma(x) + nv'(x) + mu'(x)$$

$$+\frac{a_1 + nb_1 + mc_1}{\pi}\int_{-a}^{a}\frac{\sigma(s) + \dfrac{mc_3 v'(s)}{a_1 + nb_1 + mc_1} + \dfrac{(a_2 + nb_2 + mc_2)u'(s)}{a_1 + nb_1 + mc_1}}{s-x}ds = 0. \tag{6}$$

Further, the following equalities must hold:

$$\frac{mc_3}{a_1 + nb_1 + mc_1} = n,\quad \frac{a_2 + nb_2 + mc_2}{a_1 + nb_1 + mc_1} = m. \tag{7}$$

Hence, introducing new variables by formulas $m/n = x$, $n = y$ and excluding y from (7), to determine x the following third-order equation is obtained:

$$c_3^2 x^3 - c_3(a_1 + c_2)x^2 - (c_3 b_2 - a_1 c_2 + a_2 c_1)x + (a_1 b_2 - a_2 b_1) = 0. \tag{8}$$

Where

$$n = y = \frac{c_3 x - a_1}{c_1 x + b_1},\quad m = nx = x\frac{c_3 x - a_1}{c_1 x + b_1}. \tag{9}$$

It is not hard to check that

$$a_1 b_2 - a_2 b_1 = -\frac{\vartheta_2^{(1)}\vartheta_2^{(2)}}{\rho\vartheta} \neq 0.$$

Hence (8) has no trivial roots. Consider the case when (8) has three different roots. In this case, introducing the functions

$$\varphi_j(x) = \sigma(x) + n_j v'(x) + m_j u'(x)\quad (j = 1, 2, 3), \tag{10}$$

where n_j and m_j $(j = 1, 2, 3)$ for each solution x_j of equation (8) are determined by formulas (9), the solution of system (5) is reduced to the solution of the following three independent singular integral equations:

$$\varphi_j(x) + \frac{q_j}{\pi} \int_{-a}^{a} \frac{\varphi_j(s)}{s - x} ds = 0 \quad (-a < x < a,\ q_j = c_3 x_j,\ j = 1, 2, 3). \tag{11}$$

The equations should be considered under condition

$$\int_{-a}^{a} \varphi_j(x) dx = P_0 \quad (j = 1, 2, 3). \tag{12}$$

The solutions of (11) has the following form [4]

$$\begin{gathered}\varphi_j(x) = C_j X_j^+(x), \\ X_j(z) = (z + a)^{-\gamma_j}(z - a)^{\gamma_j - 1}, \quad \gamma_j = \frac{1}{2\pi i} \ln|G_j| + \frac{\vartheta_j}{2\pi} = \alpha_j - i\beta_j, \\ 0 < \vartheta_j = \arg G_j < 2\pi, \quad G_j = \frac{1 - iq_j}{1 + iq_j},\end{gathered} \tag{13}$$

where function

$$X_j^+(x) = -\sqrt{G_j}(a + x)^{-\gamma_j}(a - x)^{\gamma_j - 1} = -\sqrt{G_j}\omega_j(x)$$

is the value of function $X_j(z)$ on the upper edge of interval $(-a, a)$.

Using (13), considering conditions (12) and the value of the integral [13]

$$\int_{-a}^{a} (x + a)^{-\gamma}(a - x)^{\gamma - 1} dx = \frac{\pi}{\sin(\pi\gamma)},$$

we get

$$C_j = -P_0 \frac{\sin(\pi\gamma_j)}{\pi\sqrt{G_j}} \quad (j = 1, 2, 3).$$

Therefore

$$\varphi_j(x) = \frac{P_0 \sin(\pi\gamma_j)}{\pi(a + x)^{\gamma_j}(a - x)^{1 - \gamma_j}}. \tag{14}$$

To determine the normal contact pressure and dislocation displacement components, substituting the value of functions $\varphi_j(x)$ from (13) into relation (9), a system of algebraic equations is obtained. After solving of system, we get

$$\sigma(x) = \sum_{j=1}^{3} \frac{P_0 A_j \sin(\pi\gamma_j)}{\pi(x + a)^{\gamma_j}(a - x)^{1 - \gamma_j}},$$

$$v'(x) = \sum_{j=1}^{3} \frac{P_0 B_j \sin(\pi\gamma_j)}{\pi(x + a)^{\gamma_j}(a - x)^{1 - \gamma_j}}, \quad u'(x) = \sum_{j=1}^{3} \frac{P_0 C_j \sin(\pi\gamma_j)}{\pi(x + a)^{\gamma_j}(a - x)^{1 - \gamma_j}}.$$

Here

$$A_1 = \frac{m_2 n_3 - m_3 n_2}{\Delta}, \quad A_2 = -\frac{m_3 n_1 - m_1 n_3}{\Delta}, \quad A_3 = \frac{m_2 n_1 - m_1 n_2}{\Delta}, \quad B_1 = \frac{m_2 - m_3}{\Delta},$$

$$B_2 = \frac{m_1 - m_3}{\Delta}, \quad B_3 = \frac{m_1 - m_3}{\Delta}, \quad C_1 = \frac{n_3 - n_2}{\Delta}, \quad C_2 = \frac{n_1 - n_3}{\Delta}, \quad C_1 = \frac{n_2 - n_1}{\Delta},$$

$$\Delta = (m_2 n_3 - m_3 n_2) - (m_3 n_1 - m_1 n_3) + (m_2 n_1 - m_1 n_2).$$

It is clear that in the case when (8) has three different real roots, then both $|G_j| = 1$ $(j = 1, 2, 3)$ the contact pressure $\sigma(x)$ and the components of the dislocation of the points of the crack edges $v'(x)$ and $u'(x)$ at the end points of the crack will have an exponential singularity without oscillations with three different exponents $\gamma_j = \vartheta_j/2\pi$. In the case when (8) has one real and two complex conjugate roots, these functions at the end points of the crack will have one exponential singularity without oscillation and two with oscillation.

Let us also determine the contact stresses outside the crack. For this purpose, we use the first two equations (2). Substituting into these equations the values of the contact pressure and dislocation components of the points of the crack edges, and using the values of the integral [13]

$$\int_{-a}^{a} \frac{(s+a)^{-\gamma}(a-s)^{\gamma-1}}{s-x} ds = \frac{\pi}{(a-x)\sin[\pi(1-\gamma)]} \left|\frac{a+x}{a-x}\right|^{-\gamma} \quad (|x| > a),$$

after some calculations we get:

$$\sigma_y^{(1)}(x,0) = \sigma_y^{(2)}(x,0) = -\frac{P_0}{\pi\Delta} \sum_{j=1}^{3} \frac{(\rho l_2 A_j - l_3 B_j)\mathrm{sgn}x}{|a+x|^{\gamma_j}|a-x|^{1-\gamma_j}} \quad (|x| > a, \ j = 1, 2),$$

$$\tau_{xy}^{(1)}(x,0) = \tau_{xy}^{(2)}(x,0) = \frac{P_0}{\pi\Delta} \sum_{j=1}^{3} \frac{(l_2 A_j + l_3 C_j)\mathrm{sgn}x}{|a+x|^{\gamma_j}|a-x|^{1-\gamma_j}} \quad (|x| > a, \ j = 1, 2).$$

Hence, it can be seen that the contact stresses outside the crack at the junction line of the heterogeneous half-planes at the end points of the crack have the same singularities as the contact pressure. Without referring to the case when equation (8) has multiple roots, we only note that in this case the number of equations (11) decreases. This does not allow us to determine the unknown functions immediately. However, it is possible to determine one of the unknown functions through the other two, if (8) has a threefold root, or two of the unknown functions through one of them, if (8) has one simple and one double root. Then, using, respectively, two or one of the equations (5), the solution of the problem is reduced to the solution of a system of two heterogeneous singular integral equations, in the first case, and one heterogeneous singular integral equation in the second case. Numerical calculations show that when the value of the friction coefficient ρ is less than 0.6, the equation (8) has three real roots for all values of the elastic constants of heterogeneous half-planes,.

Conclusion

Thus, using the method of singular integral equations, an exact solution for the problem of contact interaction of a piecewise-homogeneous elastic plane with a finite interfacial crack and an absolutely rigid stamp is constructed taking into account dry friction. Simple formulas are obtained both for determining the contact stresses under the stamp and outside the crack, and for the components of the dislocation of the points of the crack edges. It is shown that in the case when the characteristic cubic equation has three real roots, both the contact pressure and the normal and tangential contact stresses outside the crack at the end points of the crack have an exponential singularity without oscillation with three different exponents. The case when the characteristic cubic equation has one real and two complex conjugate roots is also studied. It is shown that, in this case, the contact stresses, have both the exponential singularity without oscillation and an exponential singularity with oscillation.

Acknowledgments

This study was carried out within the framework of the grant 21T-2C209 (Science Committee, Ministry of Education, Science, Culture and Sports of the Republic of Armenia).

References

[1] Galin L A 1980 *Contact Problems of the Theory of Elasticity and Viscoelasticity* (Moscow: Nauka) [in Russian]
[2] Popov G Ya 1982 *Concentration of Elastic Stresses near Stamps, Cuts, Thin Inclusions, and Reinforcements* (Moscow: Nauka) [in Russian]
[3] Panasyuk V V, Savruk M P, and Datsishin A P 1976 *The Distribution of Stresses near Cracks in Plates and Shells* (Kiev: Naukova Dumka) [in Russian]
[4] Muskhelishvili N I 1966 *Some Basic Problems of the Mathematical Theory of Elasticity* (Moscow: Nauka) [in Russian]
[5] Berezhnitsky L T, Panasyuk V V, and Staschuk N G 1983 *Interaction of Rigid Linear Inclusions and Cracks in Deformed Body* (Kiev: Naukova Dumka) [in Russian]
[6] Hakobyan V N 2014 *Mixed Boundary Value Problems on Interaction of Continuum Deformable Bodies with Different Types Stress Concentrators* (Yerevan: Gitutyun) [in Russian]
[7] Hakobyan V N 1995 On one mixed problem for composite plate, weakened by crack *Mechanics — Proceedings of National Academy of Sciences of Armenia* **48** (4) 57–65
[8] Hakobyan V N 2007 The stresses near the absolutely rigid coin-type inclusion in piecewise homogeneous space In *Proceedings of International Conference Dedicated to 95th Anniversary of Academician N. Kh. Arutyunyan, Yerevan, 2007* (Erevan) pp. 45–51 [in Russian]
[9] Il'ina I I and Silvestrov V V 2005 The problem of a thin interfacial inclusion detached from the medium along one side *Mechanics of Solids* **40** (3) 153–66
[10] Hakobyan V N, Mirzoyan S E, and Dashtoyan L L 2015 Axisymmetric mixed boundary value problem for composite space with coin-shaped crack *Herald of the Bauman Moscow State Technical University, Series Natural Sciences* No. 3(60) 31–46 doi: 10.18698/1812-3368-2015-3-31-46
[11] Hakobyan V N and Hakobyan L V 2020 Contact problem for a homogeneous plane with crack *Mechanics — Proceedings of National Academy of Sciences of Armenia* **73** (4) 3–20 doi: 10.33018/73.4.1
[12] Hakobyan L V and Dashtoyan L L 2021 Contact problem for piecewise-homogeneous plane with interfacial crack In *Topical Problems of Continuum Mechanics. Proceedings of VII International Conference, October 05-08, 2021, Tsakhkadzor, Armenia* (Erevan: Gitutyun) pp. 18–22 [in Russian]
[13] Prudnikov A P, Brychkov Yu A, and Marichev O I 1981 *Integtrals and Series* (Moscow: Nauka) [in Russian]

Formation of a hybrid of electroacoustic waves in piezoelectric layered composites

A S Avetisyan, V M Khachatryan*, and M H Mkrtchyan

Institute of Mechanics of the National Academy of Sciences of Armenia, Yerevan, Republic of Armenia

E-mail: *khachvaz@gmail.com

Abstract. Investigating the reflection and refraction of electroelastic waves at a plane interface without acoustic contact between two piezoelectric half-spaces of different textures, it was found that:

(i) As a result of the incidence of a two-component electroelastic shear wave on the interface between the media, the same wave is reflected into the first half-space, and a three-component electroelastic plane strain wave is refracted into the second half-space. In the utmost cases, in layered inhomogeneous waveguide, a wave hybrid of electroacoustic waves is formed. This makes it possible to transmit along the boundary of non-acoustic contact of media a hybrid of electroacoustic surface waves of the Gulyaev-Bluestein type and Rayleigh type localized near this boundary.

(ii) As a result of the incidence of a two-component electroelastic plane strain wave on the interface between the media, the same wave is reflected into the first half-space, and a three-component electroelastic shear wave is refracted into the second half-space. This also makes it possible to organize a hybrid of electroacoustic waves of elastic shear and plane deformation of the type of periodic electroelastic Floquet waves in a periodically transversely inhomogeneous piezoelectric composite.

The conditions for the possible formation of the hybrid and the characteristics of their propagation in the layered composites of piezoelectrics of the hexagonal symmetry class $6mm$ and $\bar{6}m2$ are discussed.

Introduction

The problem of reflection and refraction of elastic waves has been discussed by a number of authors. Without going into details, let us cite the famous works [1, 2]. It is known that during the propagation of an elastic wave of the given configuration (an antiplane deformation wave, a plane deformation wave, or a three-component elastic wave), the wave of the same configuration is reflected from the interface into the first medium. The wave of such a configuration is refracted into the second medium, which is allowed by the anisotropy of that medium. For example, in [3], the reflection and refraction of a plane strain wave at the interface isotropic with the crystalline homogeneous medium of triclinic symmetry was investigated. It depends both on the anisotropy of the adjoining materials and on the type of mechanical bond at the interface between the media.

In modern high-precision electrical and radio engineering, layered-inhomogeneous piezoelectric composite waveguides are widely used as transducers, filters or resonators for signals of electroacoustic waves.

Taking into account the various properties of media (related physical and mechanical properties of the medium), the problems of reflection and refraction of waves at the common interface between media with structural inhomogeneity have been studied by many authors. In particular, you can celebrate the work [4], where the reflection and transmission of waves at an interface of piezoelectric (ALN & PZT-5T) half-spaces with microstructures are researched. With the consideration of the flexoelectric effect, the dispersion feature of plane waves propagating in infinite piezoelectric medium and the reflection amplitude ratio at a mechanically traction-free and dielectrically charge-free surface of semi-infinite piezoelectric half-space are studied in paper [5]. In paper [6] the shear wave propagation through the symmetrically and asymmetrically arranged bi-material layer sandwiched between two semi-spaces is studied.

An important role in the problem of reflection and refraction of waves in layered structures is played also by non-standard boundary conditions at the interfaces between media. We meet such conditions in problems of electro-magneto-elasticity in the case of non-acoustic contact between piezoelectric bodies.

It was shown in [7] that, depending on the crystallographic symmetry of piezoelectrics in their sagittal planes, it is possible to excite two-component electroactive shear waves, or three-component electroactive plane deformation waves separated from each other. When an electroacoustic wave of one of these types propagates in the layered piezoelectric composite, with the non-acoustic contact between adjacent layers, part of the wave energy through the accompanying oscillations of the electric field seep into the adjacent layer, exciting an electroacoustic wave of another type in it.

Studies show that joint (simultaneous) propagation of these separated waves (hybrid) is possible in layered-inhomogeneous piezoelectric composites, with non-acoustic contact between adjacent layers [8–10]. Taking into account the possibility of excitation and propagation both of the electroactive elastic shear and electroactive elastic plane deformation waves in separate piezoelectric crystals, a discussion of possible layered composites physical model allowing the formation of a hybrid of these acoustic waves is proposed.

1. Basic relations and physical model

Consider the propagation of electroactive waves in a composite piezoelectric space. The space consists of two transversely isotropic piezoelectric half-spaces $\Omega_1(x, y, z)$ and $\Omega_2(x, y, z)$ of the $6mm$ class and $\bar{6}m2$ class of hexagonal symmetry respectively, which are in non-acoustic contact. The crystallographic sections and orientations of the crystallographic axes of these materials are compared with the Cartesian coordinate system $0xyz$ so that in the coordinate plane $x0y$ of neighboring half-spaces $\Omega_1(x, y, z)$ and $\Omega_2(x, y, z)$ there are separated electroactive waves of antiplane and plane deformations, respectively [7]. The axis $0y$ is directed vertically upwards. We will consider the formation of waves in the coordinate plane $x0y$. For all characteristics of the propagating wave fields in the plane $x0y$, the following is observed $\partial[*]/\partial z \equiv 0$.

An electroactive antiplane deformation state is possible in the sagittal plane $x_1 0 x_2$ of the $6mm$ class piezocrystal, combined with the coordinate plane $x0y$. The sixth-order inversion axis $\vec{p}_6$ of symmetry of the hexagonal piezocrystal is directed along the coordinate axis $0z$.

The non-zero components of the tensor of mechanical stresses and the electric displacement vector on the coordinate plane $x0y$ in the case of antiplane deformation for the piezoelectric crystals of the $6mm$ class, are represented as [7]

$$\sigma_{zx}^{(1)}(x,y,t)=c_{44}^{(1)}w_{1,x}(x,y,t)+e_{15}^{(1)}\varphi_{1,x}(x,y,t),\quad \sigma_{zy}^{(1)}(x,y,t)=c_{44}^{(1)}w_{1,y}(x,y,t)+e_{15}^{(1)}\varphi_{1,y}(x,y,t), \quad (1)$$

$$D_x^{(1)}(x,y,t)=e_{15}^{(1)}w_{1,x}(x,y,t)-\varepsilon_{11}^{(1)}\varphi_{1,x}(x,y,t),\quad D_y^{(1)}(x,y,t)=e_{15}^{(1)}w_{1,y}(x,y,t)-\varepsilon_{11}^{(1)}\varphi_{1,y}(x,y,t). \quad (2)$$

The quasistatic equations of the electroactive antiplane stress-strain state describing the separate excitation and propagation of electro-elastic shear waves of type SH in this band are written as

$$w_{1,xx}(x,y,t) + w_{1,yy}(x,y,t) = \tilde{C}_{1t}^{-2}\ddot{w}_1(x,y,t),$$
$$\varphi_{1,xx}(x,y,t) + \varphi_{1,yy}(x,y,t) = \frac{e_{15}^{(1)}}{\varepsilon_{11}^{(1)}}[w_{1,xx}(x,y,t) + w_{1,yy}(x,y,t)]. \tag{3}$$

where $\tilde{C}_{1t} = \sqrt{\tilde{c}_{44}^{(1)}/\rho_1}$ is the velocity of the electroelastic shear wave, $\tilde{c}_{44}^{(1)} = c_{44}^{(1)}(1+\chi_1^2)$ is the reduced shear stiffness, $c_{44}^{(1)}$ is the shear stiffness, $\chi_1^2 = (e_{15}^{(1)})^2/(c_{44}^{(1)}\varepsilon_{11}^{(1)})$ is the coefficient of electro-mechanical coupling, $e_{15}^{(1)}$ is the piezoelectric module, $\varepsilon_{11}^{(1)}$ is the relative dielectric constant, ρ_1 is the density of the $6mm$ class piezocrystal.

The electroactive plane stress-strain state is possible in the sagittal plane $x_3 0 x_1$ of the $\bar{6}m2$ class piezocrystal, combined with the coordinate plane $x0y$. The sixth-order inversion axis $\vec{p}_{\bar{6}}$ of symmetry of the hexagonal piezocrystal is directed along the coordinate axis $0y$.

The non-zero components of the tensor of mechanical stresses and the electric displacement vector on the coordinate plane $x0y$ in the case of the plane stress-strain state for the piezoelectric crystals of the $\bar{6}m2$ class, are represented in the form [7]

$$\sigma_{xx}^{(2)}(x,y,t) = c_{11}^{(2)}u_{2,x}(x,y,t) + c_{13}^{(2)}v_{2,y}(x,y,t) + e_{11}^{(2)}\varphi_{2,x}(x,y,t),$$
$$\sigma_{yy}^{(2)}(x,y) = c_{13}^{(2)}u_{2,x}(x,y,t) + c_{33}^{(2)}v_{2,y}(x,y,t), \tag{4}$$
$$\sigma_{xy}^{(2)}(x,y) = c_{44}^{(2)}v_{2,x}(x,y,t) + c_{44}^{(2)}u_{2,y}(x,y,t),$$
$$D_x^{(2)}(x,y,t) = e_{11}^{(2)}u_{2,x}(x,y,t) - \varepsilon_{11}^{(2)}\varphi_{2,x}(x,y,t),$$
$$D_y^{(2)}(x,y,t) = -\varepsilon_{33}^{(2)}\varphi_{2,y}(x,y,t). \tag{5}$$

Taking into account the compatibility conditions for the axial stress $\sigma_{zz}^{(2)}(x,y,t) \equiv 0$ and the third component of the displacement of the electric field $D_z^{(2)}(x,y,t) \equiv 0$ in the formulation of two-dimensional problem of electro elasticity [7], the quasistatic equations of the electroactive plane stress-strain state, with respect to elastic displacements $u_2(x,y,t)$, $v_2(x,y,t)$ and the potential of the electric field $\varphi_2(x,y,t)$ are written in the form

$$(c_{11}^{(2)} - \vartheta_2 c_{12}^{(2)})u_{2,xx}(x,y,t) + c_{44}^{(2)}u_{2,yy}(x,y,t) - e_{11}^{(2)}(1+\vartheta_2)\varphi_{2,xx}(x,y,t) = \rho_2\ddot{u}_2(x,y,t), \tag{6}$$
$$c_{44}^{(2)}v_{2,xx}(x,y,t) + (c_{33}^{(2)} - \vartheta_1 c_{13}^{(2)})v_{2,yy}(x,y,t) - e_{11}^{(2)}\vartheta_1\varphi_{2,xy}(x,y,t) = \rho_2\ddot{v}_2(x,y,t), \tag{7}$$
$$\varepsilon_{11}^{(2)}\varphi_{2,xx}(x,y,t) + \varepsilon_{33}^{(2)}\varphi_{2,yy}(x,y,t) - e_{11}^{(2)}u_{2,xx}(x,y,t) = 0. \tag{8}$$

In relations (4)–(8) $c_{11}^{(2)}$, $c_{12}^{(2)}$, $c_{44}^{(2)}$, $c_{13}^{(2)}$, and $c_{33}^{(2)}$ are the elastic stiffnesses, $e_{11}^{(2)}$ is the piezoelectric modulus, $\varepsilon_{11}^{(2)}$ and $\varepsilon_{33}^{(2)}$ are the relative permittivity, and ρ_2 is the density of the piezoelectric crystal. The introduced dimensionless coefficients $\vartheta_1 = (c_{13}^{(2)} + c_{44}^{(2)})/c_{12}^{(2)}$ and $\vartheta_2 = (c_{13}^{(2)} + c_{44}^{(2)})/c_{13}^{(2)}$ characterize the anisotropy of the piezoelectric of the class $\bar{6}m2$.

When the piezoelectric layers in the composite are in non-acoustic contact, the conditions of the mechanically free surface for both piezoelectric half-spaces are satisfied on the surface $y = 0$

$$[c_{44}^{(1)}w_{1,y}(x,y,t) + e_{15}^{(1)}\varphi_{1,y}(x,y,t)]_{y=0} = 0, \tag{9}$$
$$[c_{13}^{(2)}u_{2,x}(x,y,t) + c_{33}^{(2)}v_{2,y}(x,y,t)]_{y=0} = 0, \tag{10}$$
$$[u_{2,y}(x,y,t) + v_{2,x}(x,y,t)]_{y=0} = 0. \tag{11}$$

The condition (10) is obtained by taking into account the second equation of material relations (4) and the compatibility condition for mechanical stresses $\sigma_{zz}^{(2)}(x,y,t) \equiv 0$.

The conditions for the conjugation of the electric field on the surfaces of the crack $y = 0$ between the neighboring half-spaces, by taking into account the zero gap width are written as

$$[\varphi_1(x,y,t) - \varphi_2(x,y,t)]_{y=0} = 0, \tag{12}$$

$$[e_{15}^{(1)} w_{1,y}(x,y,t) - \varepsilon_{11}^{(1)} \varphi_{1,y}(x,y,t) + \varepsilon_{33}^{(2)} \varphi_{2,y}(x,y,t)]_{y=0} = 0. \tag{13}$$

2. Reflection and transmission of the electroelastic *SH* wave at the non-acoustic surface between the piezoelectrics of the symmetry classes $6mm$ and $\bar{6}m2$ [8, 9]

The electroactive elastic *SH* wave incident on the non-acoustic contact interface of two piezoelectrics, will generate the reflected elastic *SH* wave, with the accompanying potential of electric field in the lower half-space, as well as the refracted electroactive elastic *P* and refracted electroactive elastic *SV* waves with the accompanying potential of electric field in the upper half-space, as shown in figure 1.

The solution of equation (3) as the elastic displacement $W^{(1)}(x,y,t)$ and the accompanying potential of electric field $\varphi^{(1)}(x,y,t)$ for the incident *SH* electroelastic wave can be presented as

$$\begin{aligned} &W_{\mathrm{i}}^{(1)}(x,y,t) = W_{0\mathrm{i}} \exp[i(k_{\mathrm{i}x}x + k_{\mathrm{i}y}y - \tilde{C}_{1t}t)], \\ &\Phi_{\mathrm{i}}^{(1)}(x,y,t) = \frac{e_{15}^{(1)}}{\varepsilon_{11}^{(1)}} W_{0\mathrm{i}} \exp[i(k_{\mathrm{i}x}x + k_{\mathrm{i}y}y - \tilde{C}_{1t}t)], \end{aligned} \tag{14}$$

where $k_{\mathrm{i}x} = k_{SH}\sin\theta_0$, $k_{\mathrm{i}y} = k_{SH}\cos\theta_0$, k_{SH} is the wave number and θ_0 is the angle of incidence.

Due to the reflection from the surface of non-acoustic contact in the half-space of the piezoelectric for the elastic displacement and the accompanying potential of electric field we will have

$$\begin{aligned} &W_{\mathrm{rl}}^{(1)}(x,y,t) = W_{\mathrm{rl}} \exp[i(k_{\mathrm{rl}x}x - k_{\mathrm{rl}y}y - \tilde{C}_{1t}t)], \\ &\Phi_{\mathrm{rl}}^{(1)}(x,y,t) = \varphi_{\mathrm{rl}} \exp[i(k_{\mathrm{rl}x}x - k_{\mathrm{rl}y}y - \tilde{C}_{1t}t)] + \frac{e_{15}^{(1)}}{\varepsilon_{11}^{(1)}} W_{\mathrm{rl}} \exp[i(k_{\mathrm{rl}x}x - k_{\mathrm{rl}y}y - \tilde{C}_{1t}t)], \end{aligned} \tag{15}$$

where $k_{\mathrm{rl}x} = k_{SH}\sin\theta_1$, $k_{\mathrm{rl}y} = k_{SH}\cos\theta_1$, k_{SH} is the wave number in the $6mm$ class piezoelectric half-space and θ_1 is the angle of reflecting wave.

Then, the elastic *SH* wave, with the accompanying potential of electric field in the lower half-space is formed as

$$\begin{aligned} &w_1(x,y,t) = W_{\mathrm{i}} \exp[i(k_{\mathrm{i}x}x + k_{\mathrm{i}y}y - \tilde{C}_{1t}t)] + W_{\mathrm{rl}} \exp[i(k_{\mathrm{rl}x}x - k_{\mathrm{rl}y}y - \tilde{C}_{1t}t)], \\ &\varphi_1(x,y,t) = \varphi_{\mathrm{rl}} \exp[i(k_{\mathrm{rl}x}x - k_{\mathrm{rl}y}y - \tilde{C}_{1t}t)] + \\ &\quad + \frac{e_{15}^{(1)}}{\varepsilon_{11}^{(1)}} \{W_{ii} \exp[i(k_{\mathrm{i}x}x + k_{\mathrm{i}y}y - \tilde{C}_{1t}t)] + W_{\mathrm{rl}} \exp[i(k_{\mathrm{rl}x}x - k_{\mathrm{rl}y}y - \tilde{C}_{1t}t)]\}. \end{aligned} \tag{16}$$

Through the seepage of the accompanying electric field from half-space of the piezoelectric $6mm$ into the half-space of the piezoelectric $\bar{6}m2$, an electroactive elastic wave of the plane deformation $\{u_{rr}^{(2)}(x,y,t), v_{rr}^{(2)}(x,y,t), 0, \varphi_{rr}^{(2)}(x,y,t)\}$ is formed

$$\begin{aligned} &u_{rr}^{(2)}(x,y,t) = \alpha_u\{\sin\theta_3 U_{rr} \exp[i(k_{Px}x + k_{Py}y - \tilde{C}_{2l}t)] + \sin\theta_2 V_{rr} \exp[i(k_{SVx}x + k_{SVy}y - \tilde{C}_{2t}t)]\}, \\ &v_{rr}^{(2)}(x,y,t) = \alpha_v\{\cos\theta_3 U_{rr} \exp[i(k_{Px}x + k_{Py}y - \tilde{C}_{2l}t)] + \cos\theta_2 V_{rr} \exp[i(k_{SVx}x + k_{SVy}y - \tilde{C}_{2t}t)]\}, \end{aligned} \tag{17}$$

$$\varphi_{rr}^{(2)}(x,y,t) = \varphi_{rr}\{\exp[i(k_{Px}x + k_{Py}y - \tilde{C}_{2l}t)] + \exp[i(k_{SVx}x + k_{SVy}y - \tilde{C}_{2t}t)]\}, \tag{18}$$

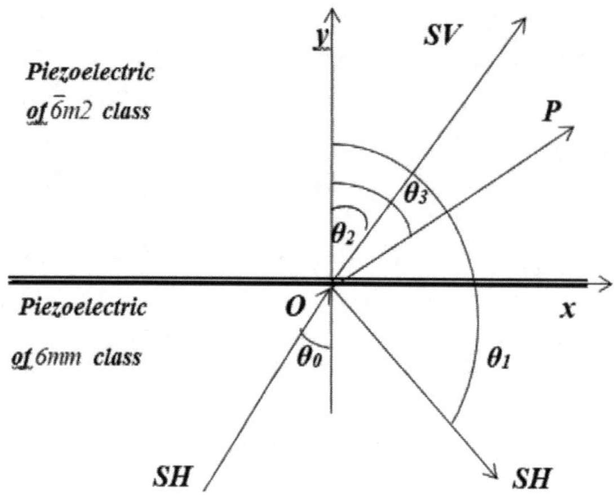

Figure 1. The physical model of the reflection and refraction of an electroelastic SH wave at the interface of two piezoelectrics with non-acoustic contact (θ_1 is angle of reflecting wave, θ_2 and θ_3 are angles of refracting wave.

where $k_{SVx} = k_{SV} \sin\theta_2$, $k_{SVy} = k_{SV} \cos\theta_2$, $k_{Px} = k_P \sin\theta_3$, $k_{Py} = k_P \cos\theta_3$, k_{SV} and k_P are the wave numbers, θ_2 and θ_3 are the refraction angles for the SV type and the P type waves in the second half-space respectively. Indicators of the natural vibration modes $p_u(\omega/k_2)$, $p_v(\omega/k_2)$, and $p_\varphi(\omega/k_2)$ are the three real solutions of the characteristic equation

$$(p^2 - k_2^2)\left[\chi_{16}^2 p^2 \begin{pmatrix} ik_2(1-\theta_{61}) & 1 \\ p^2 - k_2^2\alpha_{2l}^2 & ik_2 \end{pmatrix} - \begin{pmatrix} p^2 - k_2^2\alpha_{2t}^2 & ik_2(\theta_{16}-1)p \\ ik_2(1-\theta_{61})p & p^2 - k_2^2\alpha_{2l}^2 \end{pmatrix}\right] = 0. \quad (19)$$

Here $\chi_{16}^2 = e_{11}^{(2)}/(c_{66}^{(2)}\varepsilon_{11}^{(2)})$ is the electromechanical coupling coefficient for the second piezoelectric, $\alpha_{2l} = \sqrt{\theta_{61} - [\omega^2/k_2^2(\omega)]C_{2l}^{-2}}$ and $\alpha_{2t}(\omega) = \sqrt{\theta_{16} - [\omega^2/k_2^2(\omega)]C_{2t}^{-2}}$ are the wave modes parameters. The multipliers $\alpha_u = e_{11}^2(1+\vartheta_2)/(c_{11}^{(2)} - \vartheta_2 c_{12}^{(2)})$ and $\alpha_v = e_{11}^{(2)}\vartheta_1/(c_{33}^{(2)} - \vartheta_1 c_{13}^{(2)})$ are described the inverse (direct) piezoelectric effect in the $\bar{6}m2$ class piezoelectric half-space.

In non-acoustic contact surface $y = 0$ between the neighboring half-spaces, both the conditions of the mechanically free surfaces for piezoelectric half-space (9)–(11) and the conditions for the conjugation of the electric field (12) and (13) are satisfied.

Substituting the values of the wave components $w_1(x,y,t)$, $\varphi_1(x,y,t)$, $u_{rr}^{(2)}(x,y,t)$, $v_{rr}^{(2)}(x,y,t)$ and $\varphi_{rr}^{(2)}(x,y,t)$ into the surface conditions (9)–(13), we obtain a system of five linear algebraic equations with respect to the amplitudes of the reflected and refracted electroelastic wave components $(W_{\rm rl}/W_{\rm i})$, $(\varphi_{\rm rl}/W_{\rm i})$, $(U_{rr}/W_{\rm i})$, $(V_{rr}/W_{\rm i})$, and $(\varphi_{rr}/W_{\rm i})$:

$$(W_{\rm rl}/W_{\rm i}) + \frac{e_{15}^{(1)}}{\tilde{c}_{44}^{(1)}}(\varphi_{\rm rl}/W_{\rm i}) = 1, \tag{20}$$

$$\begin{gathered}\tilde{C}_{2t}\sin\theta_2[\alpha_u c_{13}^{(2)}\sin^2\theta_3 + \alpha_v c_{33}^{(2)}\cos^2\theta_3](U_{rr}/W_{\rm i}) \\ + \tilde{C}_{2l}\sin\theta_3[\alpha_u c_{13}^{(2)}\sin^2\theta_2 + \alpha_v c_{33}^{(2)}\cos^2\theta_2](V_{rr}/W_{\rm i}) = 0, \\ \tilde{C}_{2t}\sin\theta_2\sin\theta_3\cos\theta_3(U_{rr}/W_{\rm i}) + \tilde{C}_{2l}\sin\theta_3\sin\theta_2\cos\theta_2(V_{rr}/W_{\rm i}) = 0,\end{gathered} \tag{21}$$

$$\begin{gathered}\frac{e_{15}^{(1)}}{\varepsilon_{11}^{(1)}}(W_{\rm rl}/W_{\rm i}) + (\varphi_{\rm rl}/W_{\rm i}) - 2(\varphi_{rr}/W_{\rm i}) = -\frac{e_{15}^{(1)}}{\varepsilon_{11}^{(1)}}, \\ \cos\theta_0\sin\theta_2\sin\theta_3\varepsilon_{11}^{(1)}\tilde{C}_{2t}\tilde{C}_{2l}(\varphi_{\rm rl}/W_{\rm i}) \\ + \varepsilon_{33}^{(2)}\tilde{C}_{1t}\sin\theta_0[\sin\theta_2\cos\theta_3\tilde{C}_{2t} + \cos\theta_2\sin\theta_3\tilde{C}_{2l}](\varphi_{rr}/W_{\rm i}) = 0.\end{gathered} \tag{22}$$

The resulting system is solvable in the case of known identities

$$\frac{\omega(k)}{\tilde{C}_{1t}} = k_{SH}\sin\theta_0 = k_{\rm rl}\sin(\pi-\theta_1), \quad \frac{\omega(k)}{\tilde{C}_{2t}} = k_{SV}\sin\theta_2, \quad \frac{\omega(k)}{\tilde{C}_{2l}} = k_P\sin\theta_3, \tag{23}$$

where $\omega(k)$ is the circular frequency, and the wave numbers satisfy the following conditions

$$k_{\rm rl} = k_{SH} \quad \Rightarrow \quad \theta_1 = \pi - \theta_0, \quad \sin\theta_2 = \frac{\tilde{C}_{2t}}{\tilde{C}_{1t}}\sin\theta_0, \quad \sin\theta_3 = \frac{\tilde{C}_{2l}}{\tilde{C}_{1t}}\sin\theta_0. \tag{24}$$

Here k_* is the apparent wave number of the electroelastic wave process, $\tilde{C}_{1t} = \sqrt{c_{44}^{(1)}(1+\chi_1^2)/\rho_1}$ is the velocity of the electroelastic SH type wave in the piezoelectric of $6mm$ class, $\tilde{C}_{2l} = \sqrt{[1-\vartheta_{12}(c_{13}^{(2)}/c_{11}^{(2)}) - \chi_2^2(1+\vartheta_{12})](c_{11}^{(2)}/\rho_2)}$ is the speed of P type electroactive wave and $\tilde{C}_{2t} = \sqrt{c_{44}^{(2)}/\rho_2}$ is the speed of SV type wave in the piezoelectric of the $\bar{6}m2$ class.

Solving the system of equations (20)–(22), for the amplitudes of reflected components of the electroelastic SH type wave and the refracted (transmitted) components of the electroelastic P and SV type wave we obtain, respectively

$$(W_{\rm rl}/W_{\rm i}) = \frac{\tilde{\chi}_1^2}{1-\tilde{\chi}_1^2}(1+A_{rr}), \quad (\varphi_{\rm rl}/W_{\rm i}) = -\frac{e_{15}^{(1)}}{\varepsilon_{11}^{(1)}}\frac{\tilde{\chi}_1^2}{1-\tilde{\chi}_1^2}(1+A_{rr}), \quad (\varphi_{rr}/W_{\rm i}) = \frac{e_{15}^{(1)}}{\varepsilon_{11}^{(1)}}B_{rr}, \tag{25}$$

$$\begin{gathered}(U_{rr}/W_{\rm i}) = \frac{e_{11}^{(2)}e_{15}^{(1)}}{\varepsilon_{11}^{(1)}}\left[\frac{(1+\vartheta_2)(c_{33}^{(2)}-\vartheta_1 c_{13}^{(2)})}{(c_{11}^{(2)}-\vartheta_2 c_{12}^{(2)})}B_{rr}\sin\theta_3 + \frac{\vartheta_1(c_{11}^{(2)}-\vartheta_2 c_{12}^{(2)})}{(c_{33}^{(2)}-\vartheta_1 c_{13}^{(2)})}B_{rr}\cos\theta_3\right], \\ (V_{rr}/W_{\rm i}) = \frac{e_{11}^{(2)}e_{15}^{(1)}}{\varepsilon_{11}^{(1)}}\left[\frac{(1+\vartheta_2)(c_{33}^{(2)}-\vartheta_1 c_{13}^{(2)})}{(c_{11}^{(2)}-\vartheta_2 c_{12}^{(2)})}B_{rr}\sin\theta_2 + \frac{\vartheta_1(c_{11}^{(2)}-\vartheta_2 c_{12}^{(2)})}{(c_{33}^{(2)}-\vartheta_1 c_{13}^{(2)})}B_{rr}\cos\theta_2\right].\end{gathered} \tag{26}$$

In solutions (25) and (26), the following designations of dimensionless quantities due to the anisotropy of the piezoelectric of the $6m2$ class are introduced

$$A_{rr} = \frac{(1-\tilde{\chi}_1^2)\varepsilon_{33}^{(2)}\tilde{C}_{1t}\sin\theta_0[\tilde{C}_{2t}\sin\theta_2\cos\theta_3 + \tilde{C}_{2l}\cos\theta_2\sin\theta_3]}{(1-\tilde{\chi}_1^2)\varepsilon_{33}^{(2)}\tilde{C}_{1t}\sin\theta_0[\tilde{C}_{2t}\sin\theta_2\cos\theta_3 + \tilde{C}_{2l}\cos\theta_2\sin\theta_3] + 2\varepsilon_{11}^{(1)}\tilde{C}_{2t}\tilde{C}_{2l}\cos\theta_0\sin\theta_2\sin\theta_3},$$

$$B_{rr} = \frac{2\varepsilon_{11}^{(1)}\tilde{C}_{2t}\tilde{C}_{2l}\cos\theta_0\sin\theta_2\sin\theta_3}{(1-\tilde{\chi}_1^2)\varepsilon_{33}^{(2)}\tilde{C}_{1t}\sin\theta_0[\tilde{C}_{2t}\sin\theta_2\cos\theta_3 + \tilde{C}_{2l}\cos\theta_2\sin\theta_3] + 2\varepsilon_{11}^{(1)}\tilde{C}_{2t}\tilde{C}_{2l}\cos\theta_0\sin\theta_2\sin\theta_3}.$$

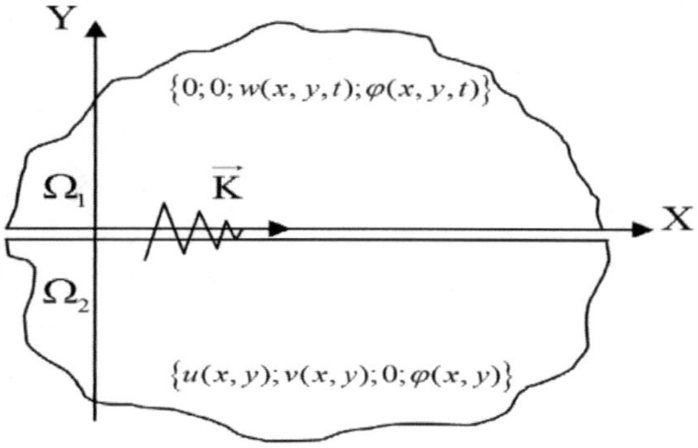

Figure 2. Scheme of signal propagation of a normal electroelastic wave in adjoining piezoelectric half-spaces without acoustic contact.

Nonzero solutions of refracted elastic waves of type P and type SV (25) exist taking into account the relationship between the angles of their unit vectors

$$\frac{\tan\theta_3}{\tan\theta_2} = \frac{\tan^2\theta_3 + [\vartheta_1 c_{33}^{(2)}(c_{11}^{(2)} - \vartheta_2 c_{12}^{(2)})]/[(1+\vartheta_2)c_{13}^{(2)}(c_{33}^{(2)} - \vartheta_1 c_{13}^{(2)})]}{\tan^2\theta_2 + [\vartheta_1 c_{33}^{(2)}(c_{11}^{(2)} - \vartheta_2 c_{12}^{(2)})]/[(1+\vartheta_2)c_{13}^{(2)}(c_{33}^{(2)} - \vartheta_1 c_{13}^{(2)})]}. \quad (27)$$

The values of the amplitude coefficients $A_{rr}(\theta_0)$ and $B_{rr}(\theta_0)$ in the margin cases of the angle of incidence $\theta_0 = 0$ and $\theta_0 = \pi/2$ are easy to obtain from their expressions.

In the case, when the electroelastic SH wave falls on the piezoelectric interface perpendicularly $\theta_0 = 0$, the same electroelastic shear wave is reflected, as well as the accompanying oscillations of the electric field seep into the second piezoelectric half-space. The angles and amplitudes of the reflected and refracted waves in this case are equal to, respectively

$$\theta_1 = \pi, \quad \theta_2 = \theta_3 = 0,$$
$$W_{\rm rl} = \frac{\tilde{\chi}_1^2}{1-\tilde{\chi}_1^2} W_{\rm i}, \quad \varphi_{\rm rl} = -\frac{e_{15}^{(1)}}{\varepsilon_{11}^{(1)}} \frac{\tilde{\chi}_1^2}{1-\tilde{\chi}_1^2} W_{\rm i}, \quad \varphi_{rr} = \frac{e_{15}^{(1)}}{\varepsilon_{11}^{(1)}} W_{\rm i}, \quad (28)$$

In the case when electroactive elastic SH wave and its accompanying electrical vibrations propagate along the interface of piezoelectric media $\theta_0 = \pi/2$, the same accompanying electrical vibrations are induced in the second piezoelectric half-space. The angles and amplitudes of the reflected and refracted waves in this case are equal to, respectively

$$\theta_1 = \frac{3\pi}{2}, \quad \theta_2 = \theta_3 = \frac{\pi}{2},$$
$$W_{\rm rl} = 0, \quad \varphi_{\rm rl} = 0, \quad \varphi_{rr} = \frac{e_{15}^{(1)}}{\varepsilon_{11}^{(1)}} \left[1 + \tilde{C}_{1t}(1-\tilde{\chi}_1^2)\varepsilon_{33}^{(2)}/2\varepsilon_{11}^{(1)} \left(\frac{\tilde{C}_{2t}}{\tilde{C}_{2l}} + \frac{\tilde{C}_{2l}}{\tilde{C}_{2t}}\right)\right]^{-1} W_{\rm i}. \quad (29)$$

3. Propagation of a hybrid of electroactive unidirectional elastic waves in a piezoelectric composite [10]

The propagation of high-frequency electroelastic waves in a two-layer piezoelectric body, consisting of piezoelectrics of the $6mm$ class and the $\bar{6}m2$ class, related to the Cartesian coordinate system $0xyz$ is studied. The crystallographic sections and orientations of the crystallographic axes of the strip materials are compared with the Cartesian coordinate system $0xyz$ so that in the coordinate plane $x0y$ of the adjacent layers $\Omega_1(x,y)$ and $\Omega_2(x,y)$ there are separate electroactive waves of antiplane and plane deformations, respectively (figure 2).

In the case when the high-frequency electroacoustic SH wave signal and its accompanying electrical vibrations $\{0, 0, w_1(x,y,t), \varphi_1(x,y,t)\}$ propagate along the interface of piezoelectric media, the equations of electroelasticity (3) are solved, together with the boundary conditions on the mechanically free surface (9) with the damping conditions deep into half-space $\Omega_1(x,y,t)$ from the common surface of non-acoustic contact

$$\lim_{y\to\infty} w_1(x,y,t)=0,\quad \lim_{y\to\infty}\varphi_1(x,y,t)=0. \tag{30}$$

Due to the conjugation conditions (12) and (13) of the electric field on the surface $y=0$, the oscillations of the electric field $\varphi_1(x,y,t)$, accompanying the electroacoustic SH wave signal in the layers $\Omega_1(x,y)$ leak through the vacuum gap into piezoelectric layer $\Omega_2(x,y)$. In the second medium, another type of electroelastic wave is generated $\{u_2(x,y,t),v_2(x,y,t),0,\varphi_2(x,y,t)\}$.

The equations system of electroelasticity (6)–(8) in the half-space $\Omega_2(x,y)$ are solved, together with the boundary conditions on the mechanically free surface (10) and (11) with the damping conditions deep into half-space $\Omega_2(x,y)$ from the common surface of non-acoustic contact

$$\lim_{y\to-\infty} u_2(x,y,t)=0,\quad \lim_{y\to-\infty} v_2(x,y,t)=0,\quad \lim_{y\to-\infty}\varphi_2(x,y,t)=0. \tag{31}$$

From the formulating mixed boundary-value problem, it is obvious that the unidirectional elastic waves of electroactive shear in the layer $\Omega_1(x,y)$ and plane deformation in the layer $\Omega_2(x,y)$ as solutions of the system of quasistatic equations of the electroelasticity can by presents in the form

$$\begin{aligned} w_1(x,y,t)&=A_1\exp(-k\alpha_{1w}(\omega,k)y)\exp[i(kx-\omega t)]\\ \varphi_1(x,y,t)&=\left[C_1\exp(-ky)+\frac{e_{15}^{(1)}}{\varepsilon_{11}^{(1)}}A_1\exp(-k\alpha_{1w}(\omega,k)y)\right]\exp[i(kx-\omega t)], \end{aligned} \tag{32}$$

$$\begin{aligned} u_2(x,y,t)&=\{A_u\exp[kq_{2u}(\omega,k)y]+a_{u\phi}C_\varphi\exp[kq_{2\varphi}(\omega,k)y]\}\exp[i(kx-\omega t)],\\ v_2(x,y,t)&=\{B_v\exp[kq_{2v}(\omega,k)]+b_{vu}A_u\exp[kq_{2u}(\omega,k)y]\\ &\quad+b_{v\varphi}C_\varphi\exp[kq_{2\varphi}(\omega,k)y]\}\exp[i(kx-\omega t)],\\ \varphi_2(x,y,t)&=\{C_\varphi\exp[kq_{2\varphi}(\omega,k)y]+c_{\varphi u}A_u\exp[kq_{2u}(\omega,k)y]\}\exp[i(kx-\omega t)]. \end{aligned} \tag{33}$$

The permissible values of the phase velocity, as in the case of Gulyaev-Bluestein wave, have the form $\eta(\omega,k)=\omega/k<\tilde{C}_{1t}$.

From the solutions (33) it is obvious that the components of the induced electroelastic wave in the second piezoelectric are damped along the depth of the second thick layer in the zone of permissible phase velocities, when

$$\eta(\omega,k)<\min\left\{\frac{C_{2t}}{\sqrt{\vartheta_3}},\tilde{C}_{1t},C_{2l*}\sqrt{1-\chi_2^2(1+\vartheta_2)\vartheta_4}\right\}\quad \text{if } c_{33}^{(2)}-\vartheta_1 c_{13}^{(2)}>0, \tag{34}$$

$$\text{or}\quad C_{2t}\sqrt{1-\vartheta_3}<\eta(\omega,k)<C_{2l*}\sqrt{1-\chi_2^2(1+\vartheta_2)\vartheta_4}\quad \text{if } c_{33}^{(2)}-\vartheta_1 c_{13}^{(2)}<0. \tag{35}$$

From the conditions for the existence of electroactive waves of the Rayleigh type (33) it follows that such waves in a medium can exist only in the case of a small value of the piezoelectric effect $\chi_2^2<[c_{11}^{(2)}c_{13}^{(2)}-c_{12}^{(2)}(c_{13}^{(2)}+c_{44}^{(2)})]/c_{44}^{(2)}[c_{13}^{(2)}+(c_{13}^{(2)}+c_{44}^{(2)})]$.

Substituting the decaying solutions (32) and (33) into the boundary conditions, we obtain the dispersion equation of electroelastic hybrid of $R\&GB$ waves in the following form

$$\frac{\alpha_{1w}-\tilde{\chi}_1^2}{\alpha_{1w}(\varepsilon_{11}^{(1)}/\varepsilon_{33}^{(2)})}=-\begin{pmatrix}\dfrac{c_{\varphi u}}{\vartheta_5+e_{11}^{(2)}c_{\varphi u}} & \dfrac{1}{e_{11}^{(2)}+\vartheta_5 a_{u\varphi}}\end{pmatrix}\Big/\begin{pmatrix}\dfrac{q_{2u}c_{\varphi u}}{\vartheta_5+e_{11}^{(2)}c_{\varphi u}} & \dfrac{q_{2\varphi}}{e_{11}^{(2)}+\vartheta_5 a_{u\varphi}}\end{pmatrix}. \tag{36}$$

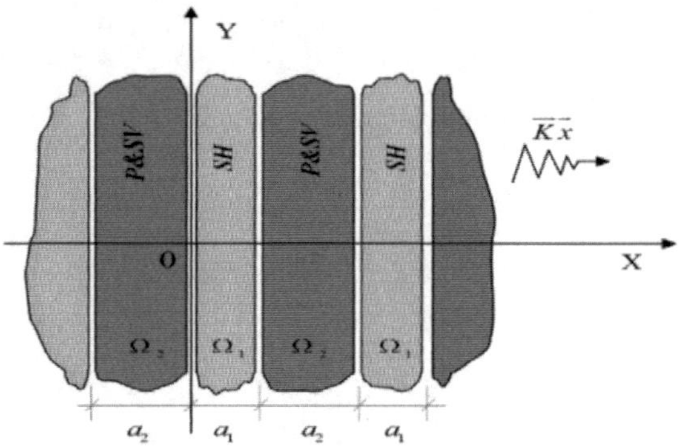

Figure 3. Transversally inhomogeneous periodic composite waveguide with non-acoustic contact between sub-layers.

From the form of dispersion equation (36) it follows that it has a solution in the case of permissible wave signal frequencies at which the dispersion function in the second piezoelectric takes on values

$$\begin{pmatrix} c_{\varphi u} & 1 \\ \vartheta_5 + e_{11}^{(2)} c_{\varphi u} & e_{11}^{(2)} + \vartheta_5 a_{u\varphi} \end{pmatrix} \Big/ \begin{pmatrix} q_{2u} c_{\varphi u} & q_{2\varphi} \\ \vartheta_5 + e_{11}^{(2)} c_{\varphi u} & e_{11}^{(2)} + \vartheta_5 a_{u\varphi} \end{pmatrix} \leq \frac{\varepsilon_{33}^{(2)}}{\varepsilon_{11}^{(1)}} (1 - \tilde{\chi}_1^2) \qquad (37)$$

Here the notation $\vartheta_5 = [(c_{13}^{(2)})^2 - c_{12}^2 c_{33}^2]/(c_{33}^{(2)} c_{44}^2)$ is introduced.

The determination of the zones of permissible phase velocities and phase function of the resulting electro-acoustic hybrid wave is carried out by selecting neighboring materials with different physicomechanical constants [10]. The formation of the electro-acoustic hybrid *R&GB* (a combination of electro-acoustic waves of the Rayleigh type and the Gulyaev-Bluestein type) is determined by the solution of dispersion equation (36) in the zone of permissible values of the phase velocity of the decaying components of the hybrid wave (35) and relation (37). For different duets of adjacent piezoelectric materials, the zones of permissible frequencies of the electro-acoustic wave hybrid are determined differently. A complete numerical analysis of the formation and propagation of a hybrid of surface electroelastic waves of the Rayleigh type and the Gulyaev–Bleustein type for a particular pair of materials is given in the article [10].

4. Propagation of a hybrid of electroactive unidirectional elastic waves in a transversally inhomogeneous periodic piezoelectric composite [11]

Let us consider the propagation of electroelastic 1D normal waves in transversally inhomogeneous periodic piezoelectric medium, which consists of alternating infinite plates $\Omega_{1n}(x,y)$ and $\Omega_{2n}(x,y)$ made of piezoelectric crystals (figure 3). The medium Linear sizes of these plates towards $0y$ and $0z$ are much larger than the wavelength of the propagating high-frequency wave. Here, $n \in \mathbb{N}^+$ is the number of the repeating cell of two sub-layers with section $\Omega_0(x,y) = \Omega_{20}(x,y) \cup \Omega_{10}(x,y)$ in $x0y$ plane, where

$$\Omega_{10}(x,y) \triangleq \{x \in [0, a_1], y \in (-\infty, +\infty)\}, \quad \Omega_{20}(x,y) \triangleq \{x \in [-a_2, 0], y \in (-\infty, +\infty)\}. \qquad (38)$$

For the sake of definiteness, we assume that in $x0y$ plane, the material of layers with sections $\Omega_{10}(x,y)$ allows separate excitation and propagation of electroactive elastic shear, while the material of layers with sections $\Omega_{20}(x,y)$ allows separate excitation and propagation of electroactive plane stress-strain state.

In the layer with section $\Omega_{10}(x,y)$, for a hybrid of electroactive unidirectional elastic waves the system of one-dimensional quasi-static equations of anti-plane electroactive deformation will

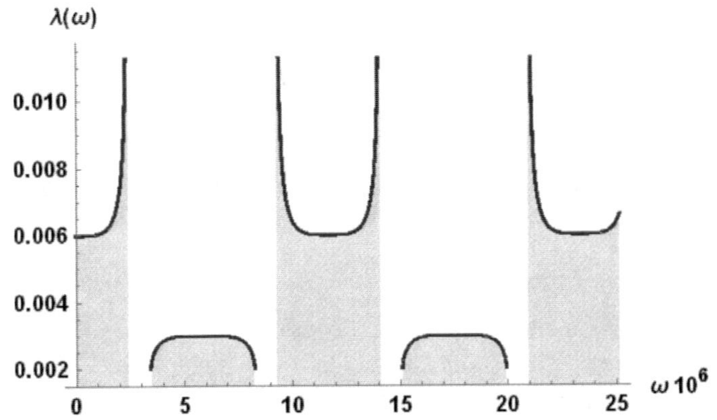

Figure 4. Zones of allowed (and forbidden) lengths $[\lambda(\omega)]$ of electroelastic hybrid wave in piezoelectric layered composite.

be solved

$$\frac{\partial^2 W_1^{(1)}}{\partial x^2} = -\frac{\omega^2}{\tilde{C}_{1t}^2} W_1^{(1)}, \quad \frac{\partial^2 \Phi_1^{(1)}}{\partial x^2} = \frac{e_{15}^{(1)}}{\varepsilon_{11}^{(1)}} \frac{\partial^2 W_1^{(1)}}{\partial x^2}. \tag{39}$$

And, in the layer with section $\Omega_{20}(x,y)$, the system of one-dimensional quasi-static equations of plane electroactive deformation will be solved

$$\frac{\partial^2 U_2(x)}{\partial x^2} = -\frac{\omega^2}{C_{2l*}^2} U_2(x), \quad \frac{\partial^2 V_2(x)}{\partial x^2} = -\frac{\omega^2}{C_{2t}^2} V_2(x), \frac{\partial^2 \Phi_2(x)}{\partial x^2} = \frac{e_{11}^{(2)}}{\varepsilon_{11}^{(2)}} \frac{\partial^2 U_2(x)}{\partial x^2}. \tag{40}$$

Here, $C_{2l*}^2 = [1 - \vartheta_{12}(c_{13}^{(2)}/c_{11}^2) - \chi_2^2(1+\vartheta_{12})](c_{11}^{(2)}/\rho_2)$ is the speed of longitudinal electroactive wave, $C_{2t}^2 = c_{44}^{(2)}/\rho_2$ is the speed of transverse elastic wave in the piezoelectric of $\bar{6}m2$ class. It is evident from the expression of the longitudinal electroactive wave speed C_{2l*}^2 that for values of the electromechanical connection coefficient satisfying $\chi_2^2 > 1 - [(c_{13}^{(2)} + c_{44}^{(2)})(c_{11}^{(2)} + c_{13}^{(2)})]/[c_{11}^{(2)}(c_{12}^{(2)} + c_{13}^{(2)} + c_{44}^{(2)})]$, the extension stiffness becomes negative.

This is possible in case of relation $c_{11}^{(2)}c_{12}^{(2)} > c_{13}^{(2)}(c_{13}^{(2)} + c_{44}^{(2)})$, because then, $0 < \chi_2^2 < 1$ always holds.

In non-acoustic contact surfaces free of mechanical stresses the electric field conjugacy conditions are satisfied:

$$\Phi_1(0,y,t) = \Phi_2(0,y,t), \quad \Phi_1(a_1,y,t) = \mu\Phi_2(-a_2,y,t), \tag{41}$$

$$e_{15}^{(1)}\frac{dW_1(0,y,t)}{dx} - \varepsilon_{11}^{(1)}\frac{d\Phi_1(0,y,t)}{dx} = e_{11}^{(2)}\frac{dU_2(0,y,t)}{dx} - \varepsilon_{11}^{(2)}\frac{d\Phi_2(0,y,t)}{dx}, \tag{42}$$

$$e_{15}^{(1)}\frac{dW_1(a_1,y,t)}{dx} - \varepsilon_{11}^{(1)}\frac{d\Phi_1(a_1,y,t)}{dx} = \mu\left[e_{11}^{(2)}\frac{dU_2(-a_2,y,t)}{dx} - \varepsilon_{11}^{(2)}\frac{d\Phi_2(-a_2,y,t)}{dx}\right]. \tag{43}$$

The following eigen lengths of vibrations correspond to wave forms in sublayers $\Omega_{10}(x,y)$ and $\Omega_{20}(x,y)$: $k(\omega)_{nw} = n\pi/a_1$ and $k(\omega)_{nu} = k(\omega)_{nv} = n\pi/a_2$.

The dispersion equation solution of propagation (frequency filtration) of hybrid of electroacoustic waves can be represented as

$$k(\omega) = \frac{1}{a_1+a_2}\arccos\frac{[1-\cos(\omega a_2/C_{2l*})]^2 + [1-\cos(\omega a_1/C_{1t})]^2}{2[1-\cos(\omega a_1/C_{1t})][1-\cos(\omega a_2/C_{2l*})]}. \tag{44}$$

In the particular case when the propagation of electroactive SH waves in piezoelectric medium with system of infinite cracks is considered, the dispersion equation for determination of phase velocity has a simple form $\cos(a_1 k) = [1+\cos^2(\omega a_1/\tilde{C}_{1t})]/[2\cos(\omega a_1/\tilde{C}_{1t})]$.

From this equation it follows that the electroactive SH wave can propagate in this case as well. The solution of this dispersion equation for wavelengths is obtained as

$$\lambda(\omega) = 2\pi a_1 \arccos^{-1}[1 + \cos^2(\omega a_1/\tilde{C}_{1t})]/[2\cos(\omega a_1/\tilde{C}_{1t})] \tag{45}$$

Despite the case of homogeneous space (when space without periodic cracks), in this case zones of forbidden frequencies occur (use figure 3).

Conclusions

The system of a selected pair of piezoelectrics with the non-acoustic contact can be the physical model of hybridization for the electroacoustic waves.

In a layered piezoelectric composite, by choosing pairs of different piezoelectrics with non-acoustic contact, a hybrid of localized electro-acoustic waves of the Rayleigh and Gulyaev-Bluestein types can be obtained near the surface of non-acoustic contact.

In a transversely periodic inhomogeneous piezoelectric space, made of piezocrystalls of $6mm$ class and $\bar{6}m2$ class of hexagonal symmetry, it is possible to propagate a hybrid of unidirectional electroactive waves of elastic shear and plane deformation with zones of permissible and forbidden frequencies (permissible and forbidden wavelengths).

Acknowledgments

The work was supported by the Science Committee of RA, in the frames of the research project No. 21T-2C130

References

[1] Achenbach J D 1976 *Wave Propagation in Elastic Solids* (New York: North Holland Publishing Company)
[2] Auld B A 1990 *Acoustic Fields and Waves in Solids.* Vol. 1 (Malabar: Krieger Publishing Company)
[3] Chattopadhyay A and Rajneesh H 2006 Reflection and refraction of waves at the interface of an isotropic medium over a highly anisotropic medium *Acta Geophysica* **54** (3) 239–49 doi: 10.2478/s11600-006-0022-y
[4] Nirwal S, Sahu S A, and Mondal S 2022 Reflection and transmission of waves at the common interface of piezoelectric half-spaces with microstructure *Applied Mathematical Modelling* **101** 533–52 doi: 10.1016/j.apm.2021.08.034
[5] Jiao F Y, Wei P J, and Li Y Q 2019 Wave propagation in piezoelectric medium with the flexoelectric effect considered *Journal of Mechanics* **35** (1) 51–63 doi: 10.1017/jmech.2017.87
[6] Ohanyan S K 2020 Shear wave reflection and transmission from symmetrically and asymmetrically arranged bimaterial layer *Mechanics — Proceedings of National Academy of Sciences of Armenia* **73** (1) 63–71 doi: 10.33018/73.1.6
[7] Avetisyan A S 2019 Two-dimensional problems of electroacoustics in homogeneous piezoelectric crystals *Mechanics — Proceedings of National Academy of Sciences of Armenia* **72** (3) 56–79 doi: 10.33018/72.3.4
[8] Avetisyan A S, Khachatryan V M, and Galichyan T A 2019 Reflection and transmission of electro elastic waves on plane nonacoustic contact interface of two piezoelectric half-spaces In *Multiferroics: Obtaining, Properties, Application: Proceedings of International Scientific-Practice Conference, Vitebsk, September 24–27, 2019.* Ed. by V V Rubanik (Vitebsk: UO "VGTU") pp. 92–5 doi: 10.26201/ISSP.2019.45.557/MFerro.37
[9] Avetisyan A S and Khachatryan V M 2019 Reflection and transmission of electro-elastic waves at plane non-acoustic contact interface of two different piezoelectric half-spaces *Topical Issues of Mechanical Engineering* **8** 370–4
[10] Avetisyan A S and Jilavyan H S 2020 Hybrid of rayleigh and gulyaev-bluestein electroacoustic waves near the inner surface of a layered piezoelectric composite*Mechanics — Proceedings of National Academy of Sciences of Armenia* **73** (2) 3–21 doi: 10.33018/73.2.1
[11] Avetisyan A S and Khachatryan V M 2020 Propagation of hybrid electroelastic waves in transversally inhomogeneous periodic piezoelectric structure *Mechanics — Proceedings of National Academy of Sciences of Armenia* **73** (1) 6–22 doi: 10.33018/73.1.1

Residual stresses calculation in a thermoelastoplastic torus after cooling

E P Dats[1,2], **V A Kovalev**[3], **and E V Murashkin**[4*]

[1]Institute of Applied Mathematics of Far Eastern Branch of Russian Academy of Sciences, Vladivostok, Russia
[2]Vladivostok State University of Economics and Service, Vladivostok, Russia
[3]Moscow City University of Management, Moscow, Russia
[4]Ishlinsky Institute for Problems in Mechanics of the Russian Academy of Sciences, Moscow, Russia

E-mail: *evmurashkin@gmail.com

Abstract. The present study deals with the boundary value problems under toroidal symmetry conditions. The residual stresses after cooling (unloading) in an elasto-plastic material are calculated. Throughout the paper the conventional Prandtl-Reuss model is generalised and used. The solution to the problem of hollow torus cooling under a temperature gradient is obtained and discussed. Analytical solutions, as an approximation of complete boundary value problem, describing residual deformations and stresses under conditions of toroidal symmetry are constructed and discussed.

Preliminary remarks

Requirement for the use of lightweight parts and structures has significantly increased in many branches of modern mechanical engineering and aircraft construction. This problem is partially solved by the use of functionally gradient materials (for example, titanium alloys) [1–5]. Functionally graded material is a class of modern materials with different properties depending on the characteristic microstructural size. In nature, functionally graded materials are bones, teeth, etc. One of the unique characteristics of functionally graded materials is their ability to adapt to specific operational loads.

Another topical problem is the quick and easy replacement of failed parts. The processes of additive manufacturing have undoubted advantages in the case of their replacement. These production methods include physical or chemical liquid gas deposition, plasma spraying, self-propagating high temperature synthesis, powder metallurgy, centrifugal casting, and laser metal deposition. The laser metal deposition process is a class of additive manufacturing processes that allows a functional part to be produced directly from a 3D computer model and possibly from the various materials.

Products produced by such methods are more economically profitable, and the production is less toxic in comparison with other technological processes. Nevertheless, products and materials obtained by the additive method often exhibit the microstructural features and are functionally graded materials.

Mathematical models of deformation of the items manufactured by the methods described above must undoubtedly take into account temperature effects. The thermoelasticity model obtained by generalising the classical Prandtl-Reuss model fully meets the requirements of modern engineering for researchers. Earlier, the authors of this research have solved a number of boundary value problems for temperature stresses calculation in the bodies with axial and central symmetry [6–18]. In this work, we will consider the problem of the residual stresses calculation under conditions of toroidal symmetry. The basis for the plastic flow calculation preceding the stage of material unloading is the results presented in publications [19–25].

1. Differential equations of thermoelastoplastic model

Transformation from Cartesian $(X,\ Y,\ Z)$ to the toroidal (r, θ, φ) coordinates is given by relations:

$$X = \Omega \cos\varphi, \quad Y = \Omega \sin\varphi, \quad Z = R_0 \cos\theta, \quad \Omega = R_0 + r\sin\theta, \tag{1}$$

where R_0 is the major radius of the torus, $r \in [r_1, r_2]$, r_1 and r_2 are the inner and outer radius of the toroidal surface. The center of the torus corresponds to the origin of the Cartesian coordinates and the center of the toroidal system is located on the generatrix of the torus.

Strain tensor components are the sum of the thermoelastic e_{ij} and the plastic p_{ij} parts.

$$d_{ij} = e_{ij} + p_{ij} \tag{2}$$

The strain tensor components depend on the displacement vector u_i by following equations

$$\begin{gathered} d_{\theta\theta} = \frac{u_{\theta,\theta}}{r} + \frac{u_r}{r}, \quad d_{\varphi\varphi} = \frac{u_r \sin\theta + u_\theta \cos\theta}{\Omega} + \frac{u_{\varphi,\varphi}}{\Omega}, \quad d_{r\theta} = \frac{1}{2}\left(\frac{u_{r,\theta}}{r} + u_{\theta,r} - \frac{u_\theta}{r}\right), \\ d_{rr} = u_{r,r}, \quad d_{r\varphi} = \frac{1}{2}\left(\frac{u_{r,\varphi}}{\Omega} + u_{\varphi,r} - \frac{u_\varphi \sin\theta}{\Omega}\right), \quad d_{\theta\varphi} = \frac{1}{2}\left(\frac{u_{\theta,\varphi}}{\Omega} + u_{\varphi,\theta} - \frac{u_\varphi \cos\theta}{\Omega}\right). \end{gathered} \tag{3}$$

There comma denotes the partial derivatives with respect to the corresponding spatial coordinate.

In the toroidal coordinate net the equilibrium equations take the form

$$\begin{gathered} \sigma_{rr,r} + \frac{\sigma_{r\theta,\theta}}{r} + \frac{\sigma_{r\varphi,\varphi}}{\Omega} + \frac{\sigma_{rr} - \sigma_{\theta\theta}}{r} + \frac{\sin\theta}{\Omega}(\sigma_{rr} - \sigma_{\varphi\varphi} + \cot\theta\sigma_{r\theta}) = 0, \\ \sigma_{r\theta,r} + \frac{\sigma_{\theta\theta,\theta}}{r} + \frac{\sigma_{\theta\varphi,\varphi}}{\Omega} + \frac{2\sigma_{r\theta}}{r} + \frac{\sin\theta}{\Omega}(\sigma_{r\theta} + \cot\theta(\sigma_{\theta\theta} - \sigma_{\varphi\varphi})) = 0, \\ \sigma_{r\varphi,r} + \frac{\sigma_{\theta\varphi,\theta}}{r} + \frac{\sigma_{\varphi\varphi,\varphi}}{\Omega} + \frac{\sigma_{r\varphi}}{r} + \frac{2\sin\theta}{\Omega}(\sigma_{r\varphi} + \cot\theta\,\sigma_{\theta\varphi}) = 0. \end{gathered} \tag{4}$$

The constitutive equations of the thermoelastic continuum can be assumed in the form of the Duhamel-Neumann's law:

$$\sigma_{ij} = \lambda\delta_{ij}\mathrm{tr}e_{ij} - \alpha\delta_{ij}(3\lambda + 2\mu)(T - T_0) + 2\mu e_{ij}, \tag{5}$$

where δ_{ij} is Kronecker delta, λ, μ are Lame constants, α is coefficient of the linear thermal expansion, $(T - T_0)$ is the difference between the initial T_0 and the current temperature T.

Note that in the further consideration, we will neglect the influence of deformation processes on the change in the temperature field. In the toroidal coordinates the heat equation reads by

$$T_{,rr} + \frac{(R_0 + 2r\sin\theta)T_{,r}}{r(R_0 + r\sin\theta)} + \frac{T_{,\theta\theta}}{r^2} + \frac{\cos\theta\, T_{,\theta}}{r(R_0 + r\sin\theta)} + \frac{T_{,\varphi\varphi}}{(R_0 + r\sin\theta)^2} = \frac{1}{\kappa}\frac{\partial T}{\partial t}. \tag{6}$$

Under the given boundary conditions and known distributions of irreversible deformations p_{ij}, the system of equations (3)–(6) specifies the evolution of the stress-strain state with thermal impact of the toroidal solid in the toroidal coordinates.

2. Statement of boundary value problem

Consider hollow torus with the major radius R_0 and $r_1 < r < r_2$. Thermal influence is given by axisymmetric (according to Z-axis) temperature distribution. In this case, the stress-strain state does not depend on angular coordinate φ. Thus, the following components of the displacement vector, the strain tensor and the stress tensor will be equal to zero:

$$u_\varphi = 0, \quad d_{r\varphi} = d_{\theta\varphi} = 0, \quad \sigma_{r\varphi} = \sigma_{\theta\varphi} = 0. \tag{7}$$

On the outer surface $r = r_2$, we define the state of free thermal expansion according to the boundary conditions

$$\sigma_{rr}(r_1, \theta) = 0, \quad \sigma_{r\theta}(r_1, \theta) = 0, \quad \sigma_{rr}(r_2, \theta) = 0, \quad \sigma_{r\theta}(r_2, \theta) = 0. \tag{8}$$

Consider the solution of the stationary heat equation (6) with the boundary conditions:

$$T(r_1, \theta) = T_k, \quad T(r_2, \theta) = T_0. \tag{9}$$

Numerical analysis of solutions to the heat equation showed that the calculated temperature distribution depends significantly on the geometry of the torus and for small values of the parameter $\epsilon = r_2/R_0$ it can be described by a function that depends only on the radial coordinate. As $\epsilon = r_2/R_0$ tends to zero, the toroidal symmetry turns into cylindrical, which allows one-dimensional analytical solutions to be taken with a sufficient degree of accuracy in the approximation of the hypothesis of generalized plane deformation. With this approach, it is important to determine the admissible finite values of the parameter ϵ, for which the cylindrical solutions will satisfactorily describe two-dimensional numerical solutions in toroidal coordinates.

The stationary heat conduction equation at $\epsilon = 0$ (under conditions of axial symmetry) takes the form:

$$T_{,r} + rT_{,rr} = 0. \tag{10}$$

Numerical experiments have shown that the maximum deviation of the analytical solution of the equation (10) from the numerical solution of the equation (6) is less than 2% for $\epsilon = 0.1$ and $r_1/r_2 = 0.4$. Therefore, with a sufficiently high degree of accuracy, the temperature distribution at $\epsilon < 0.1$ can be considered for the one-dimensional case.

For $\epsilon = 0$ we have the equilibrium equations and the relations for the strains:

$$\begin{gathered}\sigma_{rr,r} + \frac{\sigma_{r\theta,\theta}}{r} + \frac{\sigma_{rr} - \sigma_{\theta\theta}}{r} = 0, \quad \sigma_{r\theta,r} + \frac{\sigma_{\theta\theta,\theta}}{r} + \frac{2\sigma_{r\theta}}{r} = 0. \\ d_{rr} = F_{,r} \quad d_{\varphi\varphi} = C, \quad d_{\theta\theta} = \frac{F}{r}, \quad d_{r\theta} = 0,\end{gathered} \tag{11}$$

where $F(r)$ is the unknown function, C is the unknown constant. In this case, the components of the displacement vector can be represented as:

$$u_r(r, \theta) = F(r) + R_0 C \sin\theta, \quad u_\theta(r, \theta) = R_0 C \cos\theta. \tag{12}$$

The form of the function $F(r)$ depends on the stress-strain state and it is determined by taking into account the presence or absence of plastic flow in a given area of the material.

3. Stress-strain state under accumulated irreversible strains

As early shown in papers [19–21], with free thermal expansion, the temperature gradient specified by the conditions (9) leads to the appearance of several regions of deformation in the material: two regions of plastic flow corresponding to the edge and face of the Tresca prism, and a region

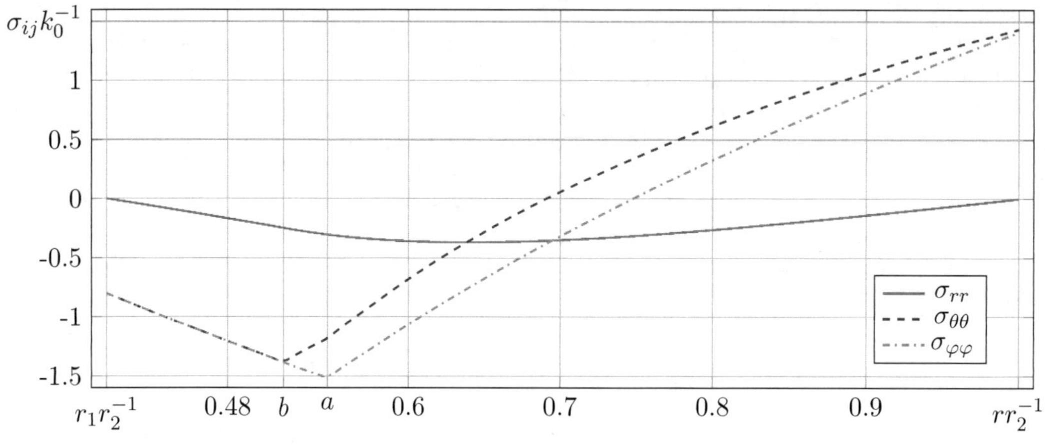

Figure 1. Thermal stresses in the toroidal coordinates, $r_1r_2^{-1} = 0.4$, $\epsilon = 0$, $br_2^{-1} = 0.517$, $ar_2^{-1} = 0.547$.

of thermoelastic deformation. Solutions in each area of a specific area are given in [19–21] and differ from each other by the form of the function $F(r)$. Distributions of the thermal stresses in the case of plastic flow are shown on the figure 1.

Consider the cooling process of the torus from the stress-strain state showing on figure 1 to the state when the temperature field returns to the initial distribution ($T = T_0$). In this case, after the plastic flow, the unloading process will begin, characterised by thermoelastic deformation taking into account the accumulated irreversible deformations. Expressing one of the components of plastic deformations in terms of two others ($p_{\varphi\varphi} = -p_{rr} - p_{\theta\theta}$), we obtain the final equations for stresses forming in the material during unloading:

$$\sigma_{rr} = \frac{2\mu}{\eta^2}\int_{r_1}^{r}\frac{p_{rr}(\rho) - p_{\theta\theta}(\rho)}{\rho}d\rho - \frac{2\mu^2}{(\lambda+2\mu)r^2}\int_{r_1}^{r}\rho[p_{rr}(\rho) + p_{\theta\theta}(\rho)]d\rho + \frac{Q}{r^2} + P,$$
$$\sigma_{\theta\theta} = [r\sigma_{rr}(r)]_{,r}, \quad \sigma_{\varphi\varphi} = \mu\gamma[p_{rr}(r) + p_{\theta\theta}(r)] + \frac{\lambda\sigma_{rr}(r) + \lambda\sigma_{\theta\theta}(r)}{2(\lambda+\mu)}. \tag{13}$$

Here, P, Q are integration constants.

According to the plasticity conditions, plastic deformations [19–21] can be represented as:

$$p_{rr} = \begin{cases} p_{rr}^{*}, & r_1 \le r \le b, \\ p_{rr}^{**}, & b \le r \le a, \\ 0, & r_1 \le a \le r_2, \end{cases} \qquad p_{\theta\theta} = \begin{cases} p_{\theta\theta}^{*}, & r_1 \le r \le b, \\ 0, & b \le r \le a, \\ 0, & r_1 \le a \le r_2. \end{cases} \tag{14}$$

The plastic deformations are described by the formulas in the domain $r_1 \le r \le b$:

$$p_{rr}^{*} = -\frac{C}{2} + \frac{D}{r^2} - \omega\int_{r_1}^{r}\frac{k(\rho)}{\rho}d\rho + \frac{\omega}{r^2}\int_{r_1}^{r}k(\rho)\rho d\rho - \frac{3}{r^2}\int_{r_1}^{r}\Delta(\rho)\rho d\rho - 2\frac{k}{\mu\gamma} + 2\Delta,$$
$$p_{\theta\theta}^{*} = -\frac{C}{2} - \frac{D}{r^2} - \omega\int_{r_1}^{r}\frac{k(\rho)}{\rho}d\rho - \frac{\omega}{r^2}\int_{r_1}^{r}k(\rho)\rho d\rho + \frac{3}{r^2}\int_{r_1}^{r}\Delta(\rho)\rho d\rho + \frac{k}{\mu\gamma} - \Delta, \tag{15}$$
$$p_{\varphi\varphi}^{*} = C + 2\omega\int_{r_1}^{r}\frac{k(\rho)}{\rho}d\rho + \frac{k}{\mu\gamma} - \Delta, \quad \omega = \frac{1}{3\lambda+2\mu},$$

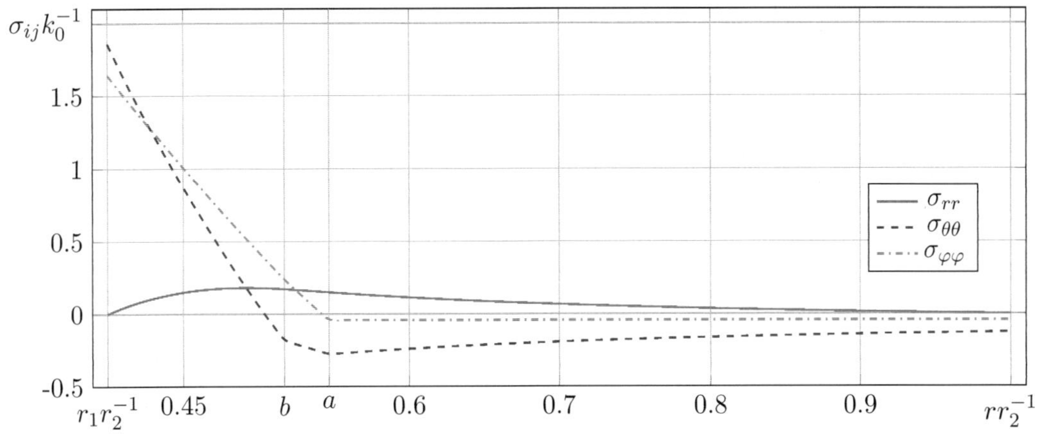

Figure 2. Residual stresses, $r_1r_2^{-1} = 0.4$, $\epsilon = 0$, $br_2^{-1} = 0.517$, $ar_2^{-1} = 0.547$.

and in the domain $b \leq r \leq a$:

$$F^{**}(r) = \frac{\psi}{2\eta}\left[\frac{\eta+1}{r^\eta}\int_{r_1}^{r}\Delta(\rho)\rho^\eta d\rho + (\eta-1)r^\eta\int_{r_1}^{r}\frac{\Delta(\rho)}{\rho^\eta}d\rho\right] + rC$$
$$-\frac{1}{2(\lambda+\mu)}\left[\frac{1}{r^\eta}\int_{r_1}^{r}k(\rho)\rho^\eta d\rho + r^\eta\int_{r_1}^{r}\frac{k(\rho)}{\rho^\eta}d\rho\right] + Mr^\eta + \frac{N}{r^\eta}, \qquad (16)$$
$$p_{rr}^{**} = \frac{1}{2}\left(F_{,r}^{**} - C - \frac{k}{\mu}\right), \quad p_{\varphi\varphi}^{**} = \frac{1}{2}\left(C + \frac{k}{\mu} - F_{,r}^{**}\right), \quad \psi = \frac{3\lambda+2\mu}{\lambda+\mu},$$

where C, D are the constants, Δ is the difference between the maximum temperature in the each point of the material and the initial temperature.

Equations (12)–(16) determine the stress-strain state under conditions of elastic unloading of the material. Distributions of the residual stresses after torus cooling are shown on the figure 2. Note that, in this case, the zones of plastic deformation are not sufficient for the occurrence of repeated plastic flow.

4. Conclusion

The paper is devoted to the boundary value problems under toroidal symmetry conditions. The residual stresses after cooling (unloading) in an elasto-plastic material have been calculated. Throughout the paper the conventional Prandtl-Reuss model has been generalised and used. The solution to the problem of hollow torus cooling under a temperature gradient has been obtained and discussed. Analytical solutions, as an approximation of complete boundary value problem, describing residual deformations and stresses under conditions of toroidal symmetry have constructed and discussed. The present analytical solutions can be applied to the calculation thick-walled long constructions manifesting symmetry close to tororidal one.

Acknowledgments

This work was carried out within the framework of a state assignment (state registration No. AAAA-A20-120011690132-4) and with the support of the Russian Foundation for Basic Research (project No. 20-01-00666).

References

[1] Mahamood R, Akinlabi S A, Shatalov M, et al. 2019 Additive manufacturing / 3D printing technology: A review *Annals of "Dunarea de Jos" University of Galati. Fascicle XII, Welding Equipment and Technology* **30** 51-8 doi: 10.35219/awet.2019.07

[2] Lasisi A, Fatoba O S, Akinlabi S A, et al. 2020 Experimental investigation of laser metal deposited AlCuTi coatings on Ti6Al4V alloy In *Advances in Manufacturing Engineering. Lecture Notes in Mechanical Engineering* (Singapore: Springer) pp. 515-22 doi: 10.1007/978-981-15-5753-8_47
[3] Lasisi A, Fatoba O S, Akinlabi S A, et al. 2020 Effect of process parameters on the hardness property of laser metal deposited AlCuTi coatingson Ti6Al4V alloy In *Advances in Manufacturing Engineering. Lecture Notes in Mechanical Engineering* (Singapore: Springer) pp. 523–9 doi: 10.1007/978-981-15-5753-8_48
[4] Akinlabi E T, Soliu G A, Mahamood R M, et al. 2020 Laser metal deposition of titanium composites: A review In *Advances in Manufacturing Engineering. Lecture Notes in Mechanical Engineering* (Singapore: Springer) pp. 555-64 doi: 10.1007/978-981-15-5753-8_51
[5] Naidoo L C, Fatoba O S, Akinlabi S A, et al. 2020 Material characterization and corrosion behavior of hybrid coating TiAlSiCu / Ti6Al-4V composite *Materialwissenschaft und Werkstofftechnik* **51** (6) 766-73 doi: 10.1002/mawe.202000019
[6] Dats E P, Murashkin E V, Tkacheva A V, and Shcherbatyuk G A 2018 Thermal stresses in an elastoplastic tube depending on the choice of yield conditions *Mechanics of Soilds* **53** (1) 23-32 doi: 10.3103/S002565441801003X
[7] Murashkin E V, Dats E P, and Klindukhov V V 2017 Numerical analysis of the elastic-plastic boundaries in the thermal stresses theory frameworks *Journal of Physics: Conference Series* **937** 012030 doi: 10.1088/1742-6596/937/1/012030
[8] Murashkin E V, Dats E P, and Stadnik N E 2017 Piecewise linear yield criteria in the problems of thermoelasticity *IAENG International Journal of Applied Mathematics* **47** (3) 261–4
[9] Dats E, Murashkin E, and Gupta N K 2017 On yield criterion choice in thermoelastoplastic problems *Procedia IUTAM* **23** 187–200 doi: 10.1016/J.PIUTAM.2017.06.020
[10] Mack W 1993 Thermal assembly of an elastic-plastic hub and a solid shaft *Archive of Applied Mechanics* **63** (1) 42–50 doi: 10.1007/BF00787908
[11] Burenin A A, Dats E P, and Murashkin E V 2014 Formation of the residual stress field under local thermal actions *Mechanics of Solids* **49** (2) 218–24 doi: 10.3103/S0025654414020113
[12] Dats E P, Murashkin E V, and Stadnik N E 2017 On heating of thin circular elastic-plastic plate with the yield stress depending on temperature *Procedia Engineering* **173** 891–6 doi: 10.1016/j.proeng.2016.12.134
[13] Dats E P, Murashkin E V, and Stadnik N E 2017 On a multi-physics modelling framework for thermo-elastic-plastic materials processing *Procedia Manufacturing* **7** 427–34 doi: 10.1016/j.promfg.2016.12.025
[14] Murashkin E and Dats E 2017 Thermoelastoplastic deformation of a multilayer ball *Mechanics of Solids***52** (5) 495–500 doi: 10.3103/S0025654417050041
[15] Burenin A A, Murashkin E V, and Dats E P 2018 Residual stresses in am fabricated ball during a heating process *AIP Conference Proceedings* **1959** 070008 doi: 10.1063/1.5034683
[16] Stadnik N and Dats E 2018 Continuum mathematical modelling of pathological growth of blood vessels *Journal of Physics: Conference Series* **991** 012075 doi: 10.1088/1742-6596/991/1/012075
[17] Murashkin E and Dats E Coupled thermal stresses analysis in the composite elastic-plastic cylinder *Journal of Physics: Conference Series* **991** 012060 doi: 10.1088/1742-6596/991/1/012060
[18] Akinlabi E T, Dats E P, and Murashkin E V 2020 Thermoelasticplastic deformation of a functionally graded spherical layer *Journal of Physics: Conference Series* **1474** 012002 doi: 10.1088/1742-6596/1474/1/012002
[19] Orçan Y 1995 Residual stresses and secondary plastic flow in a heat generating elastic-plastic cylinder with free ends *International Journal of Engineering Science* **33** (12) 1689–98 doi: 10.1016/0020-7225(95)00032-S
[20] Murashkin E and Dats E 2019 Thermal stresses computation in donut *Engineering Letters* **27** (3) 568–71
[21] Dats E and Murashkin E 2019 Thermal stresses under conditions of toroidal symmetry *Vestnik Chuvashskogo Gosudarstvennogo Pedagogicheskogo Universiteta. Seriya: Mekhanika Predel'nogo Sostoyaniya* No. 2(40) 57–70 doi: 10.26293/chgpu.2019.40.2.006
[22] Murashkin E and Dats E 2019 Thermal stresses computation under toroidal symmetry conditions *AIP Conference Proceedings* **2116** (1) 380012 doi: 10.1063/1.5114393
[23] Gorshkov S A, Dats E P, and Murashkin E V 2020 An approach of simulation of the processes of temperature stresses formation by OpenFOAM *Vestnik Chuvashskogo Gosudarstvennogo Pedagogicheskogo Universiteta. Seriya: Mekhanika Predel'nogo Sostoyaniya* No. 1(43) 131–41 doi: 10.37972/chgpu.2020.43.1.014
[24] Dats E P, Murashkon E V, Bururuev A M, et al. 2020 Thermoelastic plastic deformation of a functional gradient material under conditions of central symmetry *Vestnik Chuvashskogo Gosudarstvennogo Pedagogicheskogo Universiteta. Seriya: Mekhanika Predel'nogo Sostoyaniya* No. 3(45) 238–46 doi: 10.37972/chgpu.2020.46.88.027
[25] Dats E P, Murashkon E V, Bururuev A M, et al. 2021 Calculation of residual stresses in the state of elastic unloading of a preheated inhomogeneous thermoelastoplastic material under conditions of toroidal symmetry *Vestnik Chuvashskogo Gosudarstvennogo Pedagogicheskogo Universiteta. Seriya: Mekhanika Predel'nogo Sostoyaniya* No. 1(47) 105–13 doi: 10.37972/chgpu.2021.1.47.011

Exact and beam solutions for a narrow clamped rectangle

M D Kovalenko[1], I V Menshova[2,3*], A P Kerzhaev[2], and T D Shulyakovskaya[4]

[1]Institute of Applied Mechanics of Russian Academy of Sciences, Moscow, Russia
[2]Institute of Earthquake Prediction Theory and Mathematical Geophysics, Russian Academy of Sciences, Moscow, Russia
[3]Bauman Moscow State Technical University, Moscow, Russia
[4]Schmidt Institute of Physics of the Earth, Russian Academy of Sciences, Moscow, Russia

E-mail: *menshovairina@yandex.ru

Abstract. The paper presents formulas describing an exact solution to the boundary value problem of the theory of elasticity in a rectangle in which the horizontal sides are rigidly clamped, and normal and tangential stresses are given on the vertical ones. Only an odd-symmetric deformation of the rectangle with respect to the horizontal axis of symmetry and an even-symmetric deformation of the rectangle with respect to the vertical axis of symmetry are considered. The paper is based on the previously obtained solutions for a free half-strip and a free rectangle.

1. Introduction

In the article [1], an exact solution to the boundary value problem of the theory of elasticity was obtained for a rectangle in which two opposite sides are free, and normal or tangential loads act on the other two. All cases of symmetry with respect to the central axes of the rectangle were considered. The formulas obtained in [1] can be used in the considered problem for a rectangle in which the vertical sides are rigidly clamped, and normal stresses are applied to the horizontal ones (an odd-symmetric deformation with respect to the horizontal axis and an even-symmetric deformation with respect to the vertical one). In this case, the eigenvalues, Papkovich–Fadle eigenfunctions, normalizing factors and Lagrange coefficients will be different. For a quite narrow rectangle, the obtained solution is compared with the solution for a beam with rigidly clamped ends.

2. Solution to the boundary value problem

Consider the rectangle $\{P : |x| \leq d, |y| \leq 1\}$ shown in figure 1 in which the vertical sides $y = \pm 1$ are rigidly clamped, and the stresses

$$\sigma_x(\pm d, y) = \mp\sigma(y), \quad \tau_{xy}(\pm d, y) = 0 \tag{1}$$

are given on the horizontal sides $x = \pm d$.

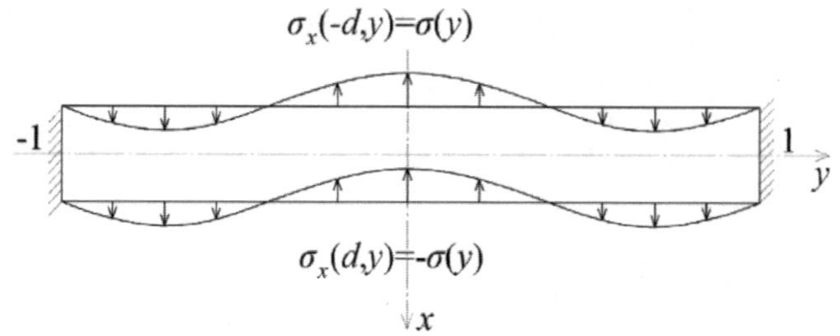

Figure 1. Scheme of the boundary value problem.

Due to the symmetry of the deformation, $\sigma(y)$ is an even function. The solution to the problem is described by the following formulas [1]:

$$\begin{aligned}
U(x,y) &= -U^1(x,y) - \sum_{k=2}^{\infty}\left\{2\mathrm{Re}\left[\sigma_k\lambda_k\frac{\xi(\lambda_k,y)}{\lambda_k M_k}\right]\frac{\mathrm{Im}[\bar\lambda_k\cosh(\bar\lambda_k d)\cosh(\lambda_k x)]}{\mathrm{Im}[\bar\lambda_k\cosh(\bar\lambda_k d)\sinh(\lambda_k d)]}\right\},\\
V(x,y) &= V^1(x,y) + \sum_{k=2}^{\infty}\left\{2\mathrm{Re}\left[\sigma_k\frac{\chi(\lambda_k,y)}{M_k}\right]\frac{\mathrm{Im}[\bar\lambda_k\cosh(\bar\lambda_k d)\sinh(\lambda_k x)]}{\mathrm{Im}[\bar\lambda_k\cosh(\bar\lambda_k d)\sinh(\lambda_k d)]}\right\},\\
\sigma_x(x,y) &= -\sigma_x^1(x,y) - \sum_{k=2}^{\infty}\left\{2\mathrm{Re}\left[\sigma_k\frac{s_x(\lambda_k,y)}{M_k}\right]\frac{\mathrm{Im}[\bar\lambda_k\cosh(\bar\lambda_k d)\sinh(\lambda_k x)]}{\mathrm{Im}[\bar\lambda_k\cosh(\bar\lambda_k d)\sinh(\lambda_k d)]}\right\},\\
\sigma_y(x,y) &= -\sigma_y^1(x,y) - \sum_{k=2}^{\infty}\left\{2\mathrm{Re}\left[\sigma_k\lambda_k\bar\lambda_k\frac{s_y(\lambda_k,y)}{\lambda_k^2 M_k}\right]\frac{\mathrm{Im}[\lambda_k\cosh(\bar\lambda_k d)\sinh(\lambda_k x)]}{\mathrm{Im}[\bar\lambda_k\cosh(\bar\lambda_k d)\sinh(\lambda_k d)]}\right\},\\
\tau_{xy}(x,y) &= -\tau_{xy}^1(x,y) - \sum_{k=2}^{\infty}\left\{2\mathrm{Re}\left[\sigma_k\lambda_k\bar\lambda_k\frac{t_{xy}(\lambda_k,y)}{\lambda_k M_k}\right]\frac{\mathrm{Im}[\cosh(\bar\lambda_k d)\cosh(\lambda_k x)]}{\mathrm{Im}[\bar\lambda_k\cosh(\bar\lambda_k d)\sinh(\lambda_k d)]}\right\}.
\end{aligned} \tag{2}$$

Here λ_k, $\bar\lambda_k$ $(k = 2,3,\ldots)$ are complex conjugate eigenvalues that are the roots of the transcendental characteristic equation

$$L(\lambda) = \frac{(3-\nu)\sin(2\lambda)}{8\lambda} - \frac{1+\nu}{4} = 0, \tag{3}$$

ν is Poisson's ratio (in calculations $\nu = 1/3$). The true displacements along the vertical x axis and horizontal y axis multiplied by the shear modulus are denoted by $U(x,y)$ and $V(x,y)$, respectively. The Papkovich–Fadle eigenfunctions are

$$\begin{aligned}
\xi(\lambda_k,y) &= \left(\frac{3-\nu}{4\lambda_k}\sin\lambda_k - \frac{1+\nu}{4}\cos\lambda_k\right)\cos(\lambda_k y) - \frac{1+\nu}{4}y\sin\lambda_k\sin(\lambda_k y),\\
\chi(\lambda_k,y) &= \frac{1+\nu}{4}[\cos\lambda_k\sin(\lambda_k y) - y\sin\lambda_k\cos(\lambda_k y)],\\
s_x(\lambda_k,y) &= \left(\frac{3+\nu}{2}\sin\lambda_k - \frac{1+\nu}{2}\lambda_k\cos\lambda_k\right)\cos(\lambda_k y) - \frac{1+\nu}{2}\lambda_k y\sin\lambda_k\sin(\lambda_k y),\\
s_y(\lambda_k,y) &= \left(\frac{\nu-1}{2}\sin\lambda_k + \frac{1+\nu}{2}\lambda_k\cos\lambda_k\right)\cos(\lambda_k y) + \frac{1+\nu}{2}\lambda_k y\sin\lambda_k\sin(\lambda_k y),\\
t_{xy}(\lambda_k,y) &= \left(-\sin\lambda_k + \frac{1+\nu}{2}\lambda_k\cos\lambda_k\right)\sin(\lambda_k y) - \frac{1+\nu}{2}\lambda_k y\sin\lambda_k\cos(\lambda_k y).
\end{aligned} \tag{4}$$

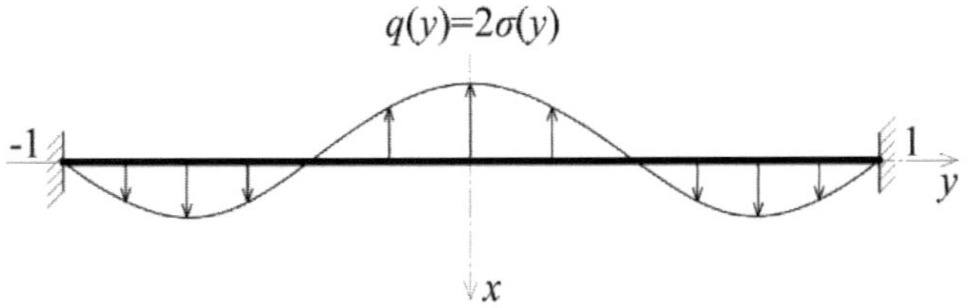

Figure 2. Clamped beam.

The Lagrange coefficients σ_k depend on the expanded function $\sigma(y)$ and are determined by the formulas

$$M_k = \frac{L'(\lambda_k)}{2}, \quad \sigma_k = \int_{-1}^{1} \sigma(y) x_k(y) dy, \quad x_k(y) = -\frac{\cos(\lambda_k y)}{4\lambda_k \sin \lambda_k}. \tag{5}$$

$x_k(y)$ are the finite parts of the functions biorthogonal to $s_x(\lambda_k, y)$. The expressions of the form M_k, $\lambda_k M_k$, $\lambda_k^2 M_k$ in formulas (2) are normalizing factors for the corresponding eigenfunctions.

Equation (3), in addition to complex roots, also has two real roots $\pm\lambda_1$. They correspond to the terms in front of the sums in (2). These terms are equal to

$$\begin{aligned}
U^1(x, y) &= \{-[\lambda_1 d \sinh(\lambda_1 d) + \cosh(\lambda_1 d)] \cosh(\lambda_1 x) + \lambda_1 x \sinh(\lambda_1 x) \cosh(\lambda_1 d)\} \, \sigma_1 \lambda_1 \frac{\xi(\lambda_1, y)}{\lambda_1 M_1 \Delta}, \\
V^1(x, y) &= \{-[\lambda_1 d \sinh(\lambda_1 d) + \cosh(\lambda_1 d)] \sinh(\lambda_1 x) + \lambda_1 x \cosh(\lambda_1 x) \cosh(\lambda_1 d)\} \, \sigma_1 \frac{\chi(\lambda_1, y)}{M_1 \Delta}, \\
\sigma_x^1(x, y) &= \{-[\lambda_1 d \sinh(\lambda_1 d) + \cosh(\lambda_1 d)] \sinh(\lambda_1 x) + \lambda_1 x \cosh(\lambda_1 x) \cosh(\lambda_1 d)\} \, \sigma_1 \frac{s_x(\lambda_1, y)}{M_1 \Delta}, \\
\sigma_y^1(x, y) &= \{-[\lambda_1 d \sinh(\lambda_1 d) - \cosh(\lambda_1 d)] \sinh(\lambda_1 x) + \lambda_1 x \cosh(\lambda_1 x) \cosh(\lambda_1 d)\} \, \sigma_1 \lambda_1^2 \frac{s_y(\lambda_1, y)}{\lambda_1^2 M_1 \Delta}, \\
\tau_{xy}^1(x, y) &= [-d \sinh(\lambda_1 d) \cosh(\lambda_1 x) + x \sinh(\lambda_1 x) \cosh(\lambda_1 d)] \, \sigma_1 \lambda_1^2 \frac{t_{xy}(\lambda_1, y)}{\lambda_1 M_1 \Delta}, \\
\Delta &= \lambda_1 d - \cosh(\lambda_1 d) \sinh(\lambda_1 d).
\end{aligned} \tag{6}$$

3. Solution for a narrow rectangle and its comparison with the beam solution

Consider a quite narrow rectangle, assuming $d = 0.1$. Let us take

$$\sigma(y) = \frac{1}{2}(5y^4 - 6y^2 + 1) \tag{7}$$

and find the solution to the problem, using formulas (2).

Let us compare this solution with the solution for the beam shown in figure 2, assuming

$$q(y) = 2\sigma(y) = 5y^4 - 6y^2 + 1. \tag{8}$$

The normal and tangential stresses in the beam are determined by the formulas

$$\sigma_b(x, y) = \frac{M(y)}{I} x, \quad \tau_b(x, y) = \frac{6Q(y)}{b(2d)^3}(d^2 - x^2), \quad I = \frac{b(2d)^3}{12}, \tag{9}$$

where I is the moment of inertia of the beam cross-section, $b = 1$ for the beam and for the plate.

The bending moment and shear force are, respectively, equal to

$$M(y) = \frac{y^6}{6} - \frac{y^4}{2} + \frac{y^2}{2} - \frac{19}{210} \quad \text{and} \quad Q(y) = y^5 - 2y^3 + y. \tag{10}$$

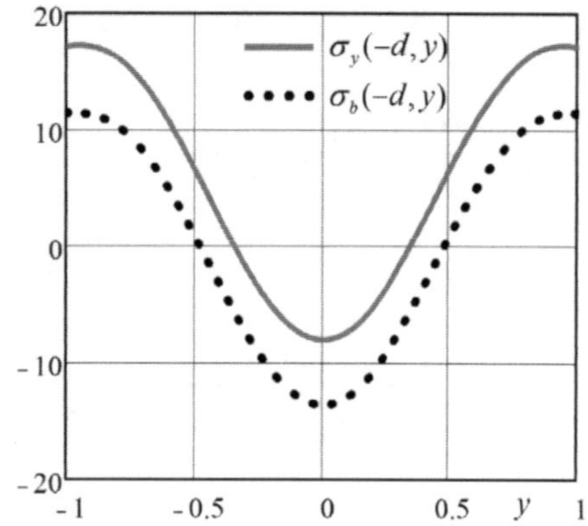

Figure 3. Distribution of the normal stresses $\sigma_y(-d, y)$ in the plate and stresses $\sigma_b(-d, y)$ in the beam.

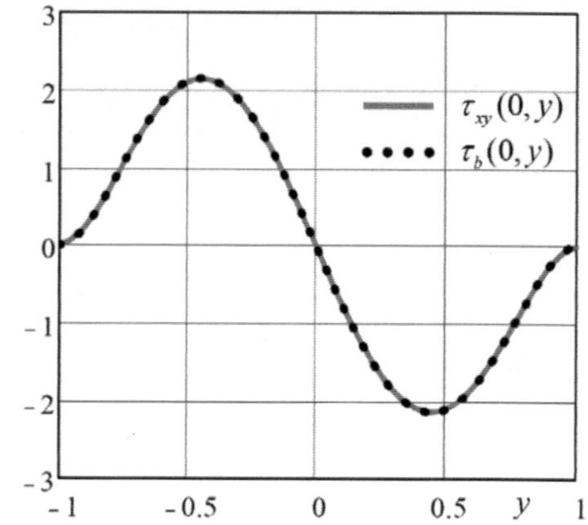

Figure 4. Distribution of the tangential stresses $\tau_{xy}(0, y)$ in the plate and stresses $\tau_b(0, y)$ in the beam.

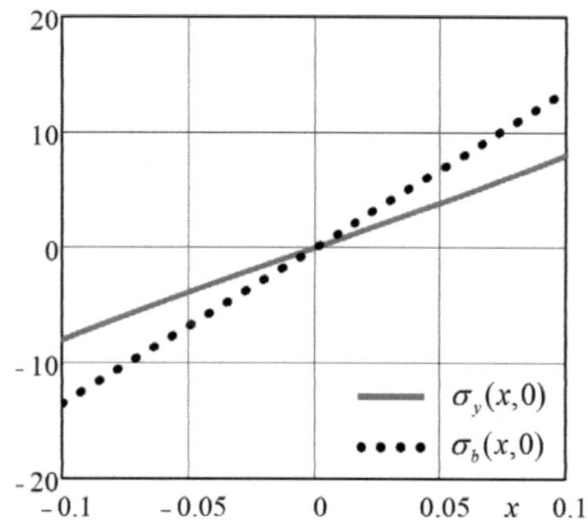

Figure 5. Distribution of the normal stresses $\sigma_y(x, 0)$ in the plate and stresses $\sigma_b(x, 0)$ in the beam.

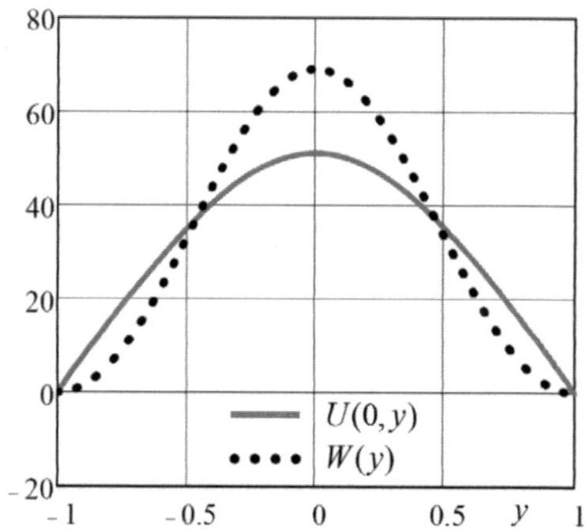

Figure 6. Distribution of the displacements $U(0, y)$ of the centreline $x = 0$ of the plate and the deflection $W(y)$ of the beam.

The beam deflection multiplied by the shear modulus is equal to

$$W(y) = \frac{(y^2 - 1)^2(5y^4 - 18y^2 + 29)}{840} \frac{1 + \nu}{I}. \tag{11}$$

Below in figures 3–6, the solid curves correspond to the solution for the plate, and the dotted curves correspond to the solution for the beam.

Figure 3 shows the changes in the normal stresses $\sigma_y(-d, y)$ in the plate and the stresses $\sigma_b(-d, y)$ in the beam. The curves differ in a constant. The distributions of the tangential stresses along the length of the beam at $x = 0$ (figure 4) are practically indistinguishable. Figure 5 shows the linear nature of the change in the normal stresses $\sigma_y(x, 0)$ along the width of the plate in comparison with the stresses $\sigma_b(x, 0)$ in the beam. The tangential stresses $\tau_{xy}(x, 0)$ in the plate and $\tau_b(x, 0)$ in the beam are practically indistinguishable. The last figure 6 shows the displacements $U(0,y)$ along the centreline $x=0$ of the plate and the deflection $W(y)$ of the beam. The curves differ significantly. This is due, first of all, to the fact that, in contrast to the beam, under no values of the variable x the derivative $\partial U(x,y)/\partial y \neq 0$ at $y=\pm 1$ since the functions $\partial \xi(\lambda_k, y)/\partial y \neq 0$ $(k=1,2,\ldots)$ at these points. For other loads, for example, for $\sigma(y)=1$, the nature of the differences remains unchanged.

4. Conclusions

1. Usually, solutions based on beam theory are compared with the corresponding solutions of the theory of bending of thin plates [2, 3], which is quite natural. The two solutions shown in the graphs make it possible to visually evaluate the contribution made by the normal stresses $\sigma_x(x,y)$ and displacements $V(x,y)$ that are not taken into account either in beam theory or in the theory of plate bending.

2. The exact solutions constructed in the paper, at the corner points, do not have a singularity characteristic of an infinite wedge, one edge of which is rigidly clamped and the other edge of which a load is applied to. The question of whether the solution for an infinite wedge can be considered as asymptotic in the corresponding solutions for a rectangle was the subject of numerous discussions in the 50–70 s of the last century. In the well-known solution for an infinite wedge [4] in the polar coordinate system, the type of boundary conditions changes when moving along one coordinate line (along the straight edges of the wedge), which becomes quite obvious if the opening angle of the wedge is 180 degrees. Therefore, the corner point in an infinite wedge is a geometric object. In any exact solutions for finite domains with singular boundary (corner points and points of change of the type of boundary conditions), it is an infinitesimal element similar to other points of the domain. This statement follows directly from the theory of constructing exact solutions in a half-strip or in a rectangle (for the first time, this feature of solutions for such domains was paid attention to by D. I. Sherman [5] and E. I. Shemyakin [6]). For greater clarity, the following example can be given. Imagine a half-strip to the vertical end of which two normal identical concentrated forces acting at the corner points are applied (even-symmetric deformation). If these forces are considered as tangential forces applied at the corner points on the long sides of the half-strip, then, obviously, we obtain a different solution to the problem. However, for an infinite wedge, these loads (normal in the first case and tangential in the second) are indistinguishable.

3. The solution to the problem under consideration will generally be irregular at the corner points [7]. In approximate numerical-analytical methods of solving boundary value problems for finite domains with corner points of the boundary, in which the type of boundary conditions changes, one must be careful not to consider an irregular solution to be a solution with a power singularity.

Acknowledgments

The research was funded by RFBR and NSFC according to the research project No. 20-51-53021.

References

[1] Kovalenko M D, Menshova I V, Kerzhaev A P, and Yu G 2020 A boundary value problem in the theory of elasticity for a rectangle: Exact solutions *Zeitschrift für angewandte Mathematik und Physik* **71** (6) 199 doi: 10.1007/s00033-020-01425-2
[2] Reddy J N 2007 *Theory and Analysis of Elastic Plates and Shells* (Boca Raton: CRS Press)
[3] Timoshenko S P and Woinowsky-Krieger S 1959 *Theory of Plates and Shells* (New-York: McGraw-Hill)
[4] Timoshenko S P and Goodier J N 1951 *Theory of Elasticity* (New-York: McGraw-Hill)
[5] Sherman D I 1940 On a problem of the theory of elasticity *Doklady Akademii Nauk SSSR* **27** (9) 907–10
[6] Shemyakin E I 1996 Boundary value problems in the elasticity theory: Domains with corner points (the planar case) *Doklady Physics* **41** (3) 135–8
[7] Muskhelishvili N I 1953 *Some Basic Problems of Mathematical Theory of Elasticity* (Groningen: Noordhoff)

On the influence of the pipe outer coating nonuniformity on the contact stresses arising under rigid bushes

K E Kazakov

Ishlinsky Institute for Problems in Mechanics of the Russian Academy of Sciences, Moscow, Russia

E-mail: kazakov-ke@yandex.ru

Abstract. The work is devoted to the formulation and construction of a solution in analytical form for the mechanical problem of contact interaction between a finite system of rigid cylindrical bushes and a pipe with an external coating, which has longitudinal nonuniformity. In industry and mechanical engineering, the presence of such additional coatings is required, for example, for protection the pipe main layer from external aggressive media, for its electrical or thermal insulation, etc. The nonuniformity of such a layer may be due to various additional requirements, as well as the obtaining process of this layer. When analyzing the stress-strain state of the pipe interacting with the bushes, it is necessary to take into account all the features. The paper shows that the mathematical model of the above-described interaction problem is a system of mixed integral equations containing integral operators of different types. Moreover, the resulting system includes functions describing the variable properties of the outer layer in the area of interaction with the bushings. The special approach described by the main theses in this paper makes it possible to obtain a solution in a form in which all complex functions are allocated as separate multipliers. This makes it possible to perform calculations even in cases where the coating is highly heterogeneous or consists of different materials.

1. Introduction

Pipes with several layers are often applied in different fields of industry, construction, and mechanical engineering. The main reasons for the use of such pipes are the need to ensure structural rigidity under the condition of a long service life, which can be reduced, for example, due to the action of external fields and media. Therefore, some of the layers are carriers, and some perform other functions. For example, the steel layer provides rigidity, and the polymer coating protects the steel from corrosion. Fixing of pipes and their connection is carried out with the help of various devices: bushings, inserts, couplings, clamps, etc. The pipe is deformed at the attachment points, therefore, when analyzing the service life of the pipe, it is necessary to take into account, among other things, the stresses arising in these areas. Increased stresses can cause the destruction of of polymer materials [1], which will lead, for example, to the formation of cracks and environmental effects on the base material. Some contact problems for coated pipes were formulated and solved in papers [2–5], however, in these works it was assumed that both the main carrier pipe and the outer layer are made of a homogeneous isotropic material. Studies of contact problems for pipes with external coatings and single bushes in the case when

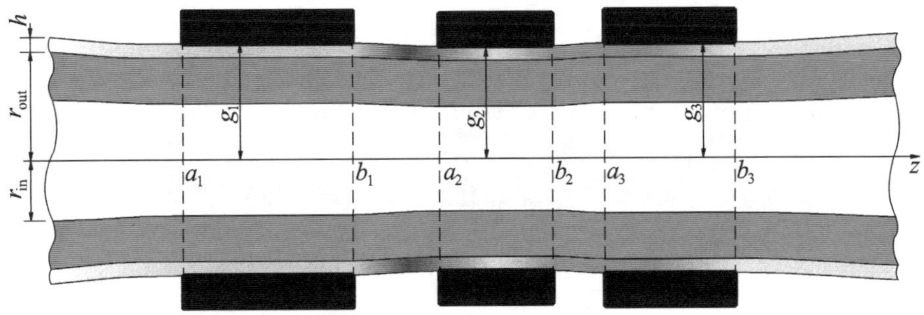

Figure 1. Interaction of the bushing system and coated pipe.

the external coating may have heterogeneity and the profiles of the contacting surfaces are described by complex (rapidly changing) functions, are presented in [6, 7]. Some analytical solutions for other contact problems, in which the interaction of bodies with complex properties and surfaces was considered, were obtained, for example, in [8–13]. In current article, we propose a solution for the problem for a viscoelastic aging pipe with an longitudinal nonuniform coating, on which several different rigid bushes are put on at the same time.

2. Mechanical and mathematical formulation of the problem

A coating made of elastic material is applied to the outer surface of a viscoelastic aging pipe. Its properties change along the longitudinal coordinate. It presupposed that thickness of such a coating is significantly less than the inner radius and thickness of internal pipe. At some point in time τ_0, n different rigid cylindrical bushes are placed on such a pipe with tension (see figure 1). The contact areas do not change over time and coincide with the inner surfaces of the bushings (i.e. ring contact occurs, and the lengths of the contact areas are equal to the widths of the bushings; there is no gap between the bushes and the pipe). The width of the contact area is also significantly greater than the coating thickness. For the sake of certainty, it is assumed that smooth contact is carried out everywhere. The problem is that it is necessary to obtain analytical expressions for contact stresses in the areas of interaction between the pipe and bushings.

Summarizing the results obtained in [2, 4, 5], it can be shown that the mathematical model of the problem is the following system of integral equations:

$$\frac{[1-\nu^2(z)]hq_i(z,t)}{E(z)} + \frac{2(1-\nu_{\text{in}}^2)}{\pi}\sum_{j=1}^{n}\left[\frac{1}{E_{\text{in}}(t-\tau_{\text{in}})}\int_{a_j}^{b_j} k_{\text{cyl}}\left(\frac{z-\xi}{r_{\text{in}}}\right)q_j(\xi,t)\,d\xi \right.$$
$$\left. - \int_{\tau_0}^{t}\frac{K_{\text{in}}(t-\tau_{\text{in}},\tau-\tau_{\text{in}})}{E_{\text{in}}(\tau-\tau_{\text{in}})}\int_{a_j}^{b_j} k_{\text{cyl}}\left(\frac{z-\xi}{r_{\text{in}}}\right)q_j(\xi,\tau)\,d\xi\,d\tau\right] = r_{\text{out}} + h - g_i = \delta_i, \tag{1}$$
$$z\in[a_i,b_i],\quad t\ge\tau_0,\quad i=1,2,\ldots,n,$$

where $q_i(z,t)$ is the contact pressure in the area of ith bush, $E(z)$ and $\nu(z)$ are coating elasticity modules, depending on the longitudinal coordinate z, $E_{\text{in}}(t)$ is elastic module of inner aging viscoelastic layer, ν_{in} is its Poisson's ratio, $K_{\text{in}}(t,\tau)$ is a creep kernel for the material of the inner layer ($K_{\text{in}}(t,\tau) = E_{\text{in}}(\tau)\partial[E_{\text{in}}^{-1}(\tau) + C_{\text{in}}(t,\tau)]/\partial\tau$, $C_{\text{in}}(t,\tau)$ is a creep function, see, e.g., [5, 13]), $\tau_{\text{in}} \le \tau_0$ is the time of pipe inner layer manufacture, a_i and b_i are the borders of the bushes (z-coordinates), matching with the boundaries of the areas of contact, r_{in} and r_{out} are the inner and outer radii of an aging pipe in undeformed state, h is the coating thickness, g_i is the inner radius of the ith bush, $k_{\text{cyl}}(s)$ is the kernel of the contact problem for cylindrical bodies, having the form (here we introduce Bessel functions $\mathrm{I}_0(u)$, $\mathrm{I}_1(u)$, $\mathrm{K}_0(u)$, and $\mathrm{K}_1(u)$)

$$k_{\text{cyl}}(s) = \int_0^\infty \frac{L(u)\cos(su)}{u}\,du, \quad L(u) = uS^{-1}(u)[u^2B_1^2(u) - f(1,u)D_1^2(u) - 1],$$
$$S(u) = A_1^2(u)k_r u^4 - B_1^2(u)u^2 f(k_r,u) - C_1^2(u)k_r u^2 f(1,u) + D_1^2(u)f(1,u)f(k_r,u)$$
$$+ k_r^{-1}f(1,u) + f(k_r,u), \quad f(r,u) = 2r^{-1}(1-\nu_{\text{in}}) + ru^2, \quad k_r = r_{\text{in}}^{-1}r_{\text{out}},$$
$$A_1(u) = \mathrm{I}_0(u)\mathrm{K}_0(uk_r) - \mathrm{I}_0(uk_r)\mathrm{K}_0(u), \quad B_1(u) = \mathrm{I}_0(u)\mathrm{K}_1(uk_r) + \mathrm{I}_1(uk_r)\mathrm{K}_0(u),$$
$$C_1(u) = \mathrm{I}_0(uk_r)\mathrm{K}_1(u) + \mathrm{I}_1(u)\mathrm{K}_0(uk_r), \quad D_1(u) = \mathrm{I}_1(u)\mathrm{K}_1(uk_r) - \mathrm{I}_1(uk_r)\mathrm{K}_1(u).$$

3. Solution of the problem

The mixed integral equation system (1) can be reduced to an operator form with dimensionless variables and functions

$$c^*(t^*)\mathbf{D}^*(z^*)\mathbf{q}^*(z^*,t^*) + (\mathbf{I} - \mathbf{V}^*)\mathbf{F}\mathbf{q}^*(z^*,t^*) = \boldsymbol{\delta}^*, \quad z^* \in [-1,1], \quad t^* \geq 1. \tag{2}$$

The following notation is introduced here $(i, j = 1, 2, \ldots, n)$

$$\mathbf{q}^*(z^*,t^*) = (q^{1*}(z^*,t^*) \;\; q^{2*}(z^*,t^*) \;\; \cdots \;\; q^{n*}(z^*,t^*))^T, \quad \boldsymbol{\delta}^* = (\delta^{1*} \;\; \delta^{2*} \;\; \cdots \;\; \delta^{n*})^T,$$
$$\mathbf{D}^*(z^*) = \begin{pmatrix} m^1(z^*) & 0 & \cdots & 0 \\ 0 & m^2(z^*) & \cdots & 0 \\ \vdots & \vdots & \ddots & \vdots \\ 0 & 0 & \cdots & m^n(z^*) \end{pmatrix}, \quad \mathbf{F} = \begin{pmatrix} \mathbf{F}^{11*} & \mathbf{F}^{12*} & \cdots & \mathbf{F}^{1n*} \\ \mathbf{F}^{21*} & \mathbf{F}^{22*} & \cdots & \mathbf{F}^{2n*} \\ \vdots & \vdots & \ddots & \vdots \\ \mathbf{F}^{n1*} & \mathbf{F}^{n2*} & \cdots & \mathbf{F}^{nn*} \end{pmatrix},$$
$$z^* = 2\bar{a}_i^{-1}(z - \eta_i), \quad \xi^* = 2\bar{a}_j^{-1}(\xi - \eta_j), \quad t^* = \tau_0^{-1}t, \quad \tau_{\text{in}}^* = \tau_0^{-1}\tau_{\text{in}},$$
$$\zeta^{i*} = \bar{a}^{-1}\bar{a}_i, \quad \eta^{i*} = 2\bar{a}^{-1}\eta_i, \quad \delta^{i*} = 2\bar{a}^{-1}\delta_i, \quad c^*(t^*) = E_\circ^{-1}E_{\text{in}}(t - \tau_{\text{in}}),$$
$$m^i(z^*) = \frac{h[1-\nu^2(z)]E_\circ}{\bar{a}_i(1-\nu_{\text{in}}^2)E(z)}, \quad q^{i*}(z^*,t^*) = \frac{2\zeta^{i*}(1-\nu_{\text{in}}^2)q_i(z,t)}{E_{\text{in}}(t-\tau_{\text{in}})},$$
$$\mathbf{F}^{ij*}f(z^*) = \int_{-1}^1 k^{ij}(z^*,\xi^*)f(\xi^*)d\xi^*, \quad \mathbf{V}^*f(t^*) = \int_1^{t^*} K^*(t^*,\tau^*)f(\tau^*)d\tau^*,$$
$$k^{ij}(z^*,\xi^*) = \pi^{-1}k_{\text{cyl}}[r_{\text{in}}^{-1}(z-\xi)], \quad K^*(t^*,\tau^*) = K_{\text{in}}(t-\tau_{\text{in}}, \tau-\tau_{\text{in}})\tau_0,$$

where $\eta_i = 0.5(a_i + b_i)$ is the midpoint of the i bush, $\bar{a}_i = b_i - a_i$ is its width, $\bar{a} = \min_{i=1,2,\ldots,n}\bar{a}_i$, and $E_\circ$ is the dimensional coefficient (for example, $E_\circ = E(0)$).

Obtained operator equation (2) practically coincides with equation (6) from paper [14]: In equation (6) from [14], it must be assumed that $\boldsymbol{\alpha}(t) \equiv 0$, $\mathbf{g}(x) \equiv 0$, and $\boldsymbol{\delta}(t) = \text{const}$. It can also be shown (see [7]) that the spatial kernels of these equations have similar properties. This means that the solutions to both problems are constructed identically, so in this paper we will use the results obtained in [14]. We will indicate further only the main steps of the approach used.

Step 1. By replacing variables using the formula $\mathbf{q}^*(z^*,t^*) = \mathbf{D}^{-1/2*}(z^*) \cdot \mathbf{Q}(z^*,t^*)$ and multiplying from the left by $\mathbf{D}^{-1/2*}(z^*)$ both sides of the equation (2), we obtain a new operator equation in which the matrix $\mathbf{D}^*(z^*)$ is missing before the first term, and the kernel of the new Fredholm operator $\tilde{\mathbf{F}}$ has the same properties as the kernel of the original operator $\mathbf{F}$:

$$c^*(t^*)\mathbf{Q}(z^*,t^*) + (\mathbf{I} - \mathbf{V}^*)\tilde{\mathbf{F}}\mathbf{Q}(z^*,t^*) = \mathbf{D}^{-1/2*}(z^*) \cdot \boldsymbol{\delta}^*, \quad z^* \in [-1,1], \quad t^* \geq 1. \tag{3}$$

Here $\mathbf{D}^{-1/2*}(z^*)$ is a diagonal matrix function, on the diagonal of which the functions $1/\sqrt{m^1(z^*)}$, $1/\sqrt{m^2(z^*)}$, ..., $1/\sqrt{m^n(z^*)}$ are located.

Step 2. Based on the orthogonalization of a system of linearly independent vector-functions

$$\mathbf{i}^1/\sqrt{m^1(z^*)}, z^*\mathbf{i}^1/\sqrt{m^1(z^*)}, (z^*)^2\mathbf{i}^1/\sqrt{m^1(z^*)}, \ldots, \mathbf{i}^2/\sqrt{m^2(z^*)}, z^*\mathbf{i}^2/\sqrt{m^2(z^*)},$$
$$(z^*)^2\mathbf{i}^2/\sqrt{m^2(z^*)}, \ldots\ldots, \mathbf{i}^n/\sqrt{m^n(z^*)}, z^*\mathbf{i}^n/\sqrt{m^n(z^*)}, (z^*)^2\mathbf{i}^n/\sqrt{m^n(z^*)}, \ldots,$$

we obtain a orthonormal basis that takes into account the presence in the equation of functions $m^i(z^*)$ associated with the replacement of variables with the function of coating inhomogeneity. Here $\mathbf{i}^i$ are unit vectors with the ith unit coordinate.

Step 3. After this, we find the eigenvectors of the $\varphi_k(z^*)$ $(k = 0, 1, 2, \ldots)$ of the Fredholm operator $\tilde{\mathbf{F}}$, which is included in the main integral equation (3). This makes it possible to bring the resulting system of equations for the functional coefficients of the decomposition of the solution to a diagonal form.

Step 4. By equating the corresponding coefficients for eigenfunctions in the left and right parts of the resulting equation, we finally obtain the following formulas for calculating the desired contact pressures $(i = 1, 2, \ldots, n,\ k = 0, 1, 2, \ldots,\ z^* \in [-1, 1],\ t^* \geq 1)$

$$q^{i*}(z^*, t^*) = \frac{1}{m^i(z^*)}\sum_{k=0}^{\infty} z_k(t^*) \sum_{m=0}^{\infty} \psi^i_{km} p^{i\circ}_m(z^*),$$

$$z_k(t^*) = (\mathbf{I}+\mathbf{W}_k)[c^*(t^*) + \gamma_k]^{-1} \sum_{i=1}^{n} \sqrt{J_{0,i}}\psi^i_{k0}\delta^{i*},$$

$$\mathbf{W}_k f(t^*) = \int_1^{t^*} R^*_k(t^*, \tau^*) f(\tau^*)\, d\tau^*, \quad J_{k,i} = \int_{-1}^{1} \frac{(\xi^*)^k}{m^i(\xi^*)}\, d\xi^*, \quad d_{-1,i} = 1, \tag{4}$$

$$p^{i\circ}_k(z^*) = \frac{1}{\sqrt{d_{k-1,i} d_{k,i}}} \begin{vmatrix} J_{0,i} & \cdots & J_{k,i} \\ \vdots & \ddots & \vdots \\ 1 & \cdots & (z^*)^k \end{vmatrix}, \quad d_{k,i} = \begin{vmatrix} J_{0,i} & \cdots & J_{k,i} \\ \vdots & \ddots & \vdots \\ J_{k,i} & \cdots & J_{2k,i} \end{vmatrix},$$

where $R^*_k(t^*, \tau^*)$ are resolvents of the kernels $[\gamma_k + c^*(t^*)]^{-1} K^*(t^*, \tau^*)\gamma_k$, ψ^i_{km} and γ_k can be determined from the solution of the spectral problem $\sum\limits_{l=0}^{\infty}\sum\limits_{j=1}^{n} \psi^j_{kl} K^{ij}_{ml} = \psi^i_{km}\gamma_k$ $(i = 1, 2, \ldots, n,$ $k, m = 0, 1, 2, \ldots)$, coefficients K^{ij}_{ml} are determined from the formula

$$K^{ij}_{ml} = \int_{-1}^{1}\int_{-1}^{1} \frac{k^{ij}(z^*, \xi^*) p^{i\circ}_m(z^*) p^{j\circ}_l(\xi^*)}{m^i(z^*) m^j(\xi^*)} dz^* d\xi^*, \quad i, j = 1, 2, \ldots, n, \quad m, l = 0, 1, 2, \ldots$$

The reverse substitution of variables using formulas (4) leads us to expressions for contact pressures in dimensional form

$$q_i(z, t) = \frac{E(z)}{1 - \nu(z)} \sum_{k=0}^{\infty} \tilde{z}_k(t) \sum_{m=0}^{\infty} \psi^i_{km} p^{i\circ}_m \left(\frac{2z - (a_i + b_i)}{b_i - a_i}\right), \quad i = 1, 2, \ldots, n,$$
$$\tilde{z}_k(t) = \frac{E_{\text{in}}(t - \tau_{\text{in}})\bar{a}}{2E_\circ h} z_k\left(\frac{t}{\tau_0}\right), \quad k = 0, 1, 2, \ldots, \quad z \in [a_i, b_i], \quad t \geq \tau_0. \tag{5}$$

Discussion and conclusions

In the expression (4) for dimensionless functions $q^{i*}(z^*, t^*)$ and in the expression (5) for contact stresses $q_i(z, t)$, the functions $m^i(z^*)$ and $[1 - \nu(z)]^{-1}E(z)$, respectively, are separate factors. This means that in applied calculations, in order to achieve acceptable accuracy, it is sufficient to limit ourselves to a relatively small number of terms in a series, even if the inhomogeneity of external layer is described by a discontinuous or rapidly changing function. Standard approaches (for example, the method of decomposition into a series of trigonometric or polynomial vector-functions) do not allow this. In these methods, in order to correctly describe complex functions, it is necessary to take a sufficiently large number of members of the series. However, a large number of members of the series leads to the accumulation of computational error. Moreover, even if these functions are constant (that is, if the coverage is homogeneous), the application of classical analytical methods will lead to the need to solve an infinite system of Volterra integral equations, which causes significant difficulties.

Thus, a solution has been constructed in the work that allows taking into account the influence of inhomogeneity, rheological properties of the pipe and the proximity of the bushings on the distribution of contact pressures, which makes it possible to make an accurate analysis of such a coupling when assessing the durability and reliability of the structure as a whole.

Acknowledgments

The present work was supported in part by the Ministry of Science and Higher Education of the Russian Federation in the framework of a Federal Target Program (contract No. AAAA-A20-120011690132-4) and supported in part by the Russian Foundation for Basic Research (project No. 19-51-60001).

References

[1] Arutyunyan A R and Arutyunyan R A 2017 Aging and fracture of compressible elastic-viscous media *Vestnik of Saint Petersburg University. Mathematics. Mechanics. Astronomy* **4(62)** (2) 258–65 doi: 10.21638/11701/spbu01.2017.209
[2] Manzhirov A V and Chernysh V A 1988 On the interaction of a rigid reinforcing sleeve and inhomogeneous aging high-pressure pipes *Mechanics of Solids* **23** (6) 104–10
[3] Manzhirov A V and Chernysh V A 1990 Contact problem for a layered inhomogeneous aging cylinder reinforced by a rigid ring *Journal of Applied Mechanics and Technical Physics* **31** (6) 894–900 doi: 10.1007/BF00854204
[4] Manzhirov A V and Chernysh V A 1991 The contact problem of the discrete fitting of an inhomogeneous viscoelastic ageing cylinder with a system of rigid collars *Journal of Applied Mathematics and Mechanics* **55** (6) 901–9 doi: 10.1016/0021-8928(91)90143-I
[5] Arutyunyan N Kh and Manzhirov A V 1999 *Contact Problems of the Theory of Creep* (Erevan: Izdat. NAN RA) [in Russian]
[6] Kazakov K E and Kurdina S P 2020 Wear of a high pressure layered pipe by a rough rigid bush *IOP Conference Series: Materials Science and Engineering* **904** 012010 doi: 10.1088/1757-899X/904/1/012010
[7] Kazakov K E 2020 On the interaction of rigid bush and pipe with nonuniform coating *AIP Conference Proceedings* **2293** 210012 doi: 10.1063/5.0026560
[8] Arutyunyan N Kh and Kolmanovskii V B 1983 *Creep Theory of Inhomogeneous Solids* (Moscow: Nauka) [in Russian]
[9] Krejčí P and Petrov A 2017 A mathematical model for the third-body concept *Mathematics and Mechanics of Solids* **23** (3) 420–32 doi: 10.1177/1081286517732827
[10] Argatov I I and Chai Y S 2020 A self-similar model for fretting wear contact with the third body in gross slip *Wear* **466467** 203562 doi: 0.1016/j.wear.2020.203562
[11] Lyubicheva A N, Kudritskii V G, and Goryacheva I G 2018 Viscoelastic contact of textured surfaces *Journal of Friction and Wear* **39** (6) 517–21 doi: 10.3103/S1068366618060090
[12] Goryacheva I G 2006 Mechanics of discrete contact *Tribology International* **39** (5) 381–6 doi: 10.1016/j.triboint.2005.04.020
[13] Goryacheva I G and Torskaya E V 2019 Contact of multi-level periodic system of indenters with coated elastic half-space *Facta Universitatis, Series: Mechanical Engineering* **17** (2) 149–59 doi: 10.22190/FUME190511022G
[14] Kazakov K E and Parshin D A 2019 On the indentation of a system of punches into a layered foundation *Engineering Letters* **27** (1) 251–5

Development of a mathematical model of a force actuator for the opening of a space structure with transformable configuration

V N Zimin*, **A V Krylov**, **G N Kuvyrkin**, **and A O Shakhverdov**

E-mail: *zimin@bmstu.ru

Abstract. The prospects for the development of radio astronomy, solar energy, space communications, exploration of the Earth's surface and other planets from space are currently associated with the creation of fundamentally new large-sized space structures. The inevitable complication of the design schemes of promising large-sized systems due to the increase in their operational functionality requires the development of mathematical models that adequately describe the mechanical properties of structures. When designing large-sized systems, the scheme of which allows for automatic configuration change of the structure, it is necessary to take into account shock loads. They inevitably arise when the working state of the structure fixes in orbit upon completion of the process of its opening. To ensure smooth, reliable and shock-free opening of large-sized space structures, it is proposed to use force actuators with active elements made of titanium nickelide material with a shape memory effect. During the tests of the active elements, the main parameters of the force actuator were determined: the generated force, the actuation time and the length of the working stroke. The length of the working stroke was determined by the change in the relative elongation of the active element of the force actuator during its heating. The conducted experimental and theoretical studies are aimed at developing a mathematical model of the functioning of an active element made of titanium nickelide with a shape memory effect for the opening of a space structure with transformable configuration.

1. Introduction

To date, many variants of space structures that have different configurations in transport and operating positions have been developed [1]. The inevitable complication of the design schemes of promising space systems due to the increase in their operational functionality leads to the necessity of creating large space structures [2–5]. At the same time, large-sized structures with changeable configuration (transformable) become key components of the technology of modern space systems, figure 1.

The ground experiment plays a significant role in the creation of space structures. The difficulties that arise when modeling operating conditions of space in a ground experiment require the creation of expensive stands for simulating weightlessness of large transformable structures and vacuum chambers unique in size. Therefore, the design of such systems is directly related to the development of a mathematical model of the structure that adequately describes its mechanical properties [6].

The operability of space structures with transformable configuration is mainly determined by how great are the forces that arising in them during deployment [7], therefore, ensuring reliable

a) b)
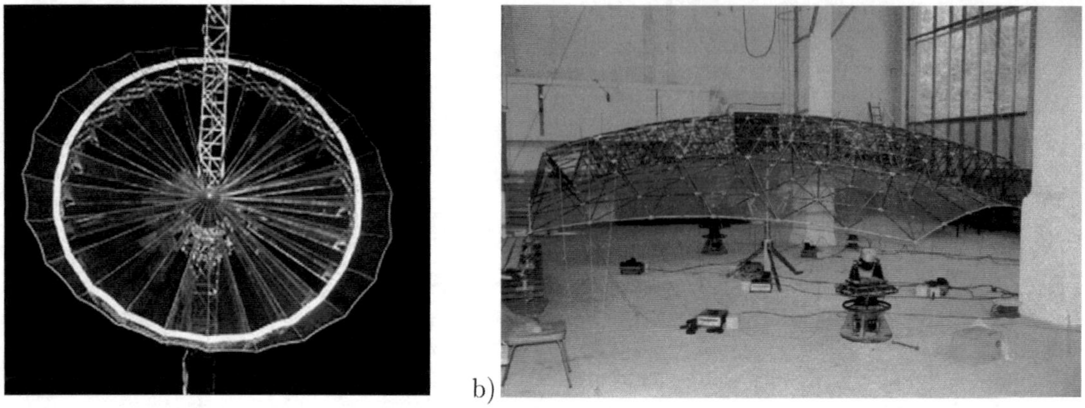

Figure 1. Large space structures: umbrella type reflector (*a*); truss-type reflector (*b*).

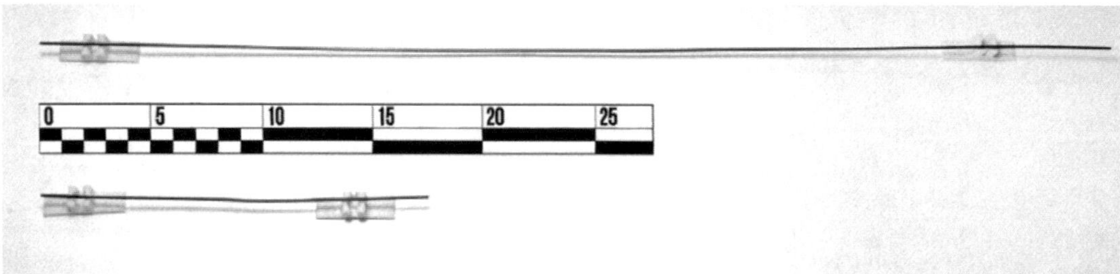

Figure 2. Active elements of the force actuator with SME made of titanium nickelide material.

opening is associated with solving complex problems of mechanics. The process of changing the configuration occurs under the influence of force actuators, which, generally, use various kinds of springs and electric motors.

Research in the field of the use of materials with shape memory effect (SME) for the creation of thermomechanical force actuators for space systems is currently of increased interest to scientists, engineers and designers [8–11]. The material with SME can be used to create force actuators due to its ability to restore large deformations and create significant force [12–14]. The active element of such a force actuator can be made in the form of a wire made of a material with SME such as titanium nickelide [8, 11, 15]. For the successful application of such actuators for the opening of a space structure with a transformable configuration, it is necessary to determine both the shape of the active element and their functionality [10]. Theoretical and experimental studies were carried out to determine the main characteristics of the active element: the developed operating force, the actuation time, the working stroke (maximum displacement).

2. Experimental studies

The mathematical difficulties of describing the behavior of materials with SME and the need to use a sufficiently large amount of actual experimental data in mathematical models have not yet allowed to develop engineering methods for calculating force actuators for the opening of space structures with transformable configuration [16, 17]. Active elements made of titanium nickelide wire with a diameter of 1.5 mm and lengths of 160 mm and 460 mm are used in the force actuators under consideration (figure 2). They are heated during operation by passing an electric current through them.

The total deformation of a wire made of a material with SME consists of elastic, temperature

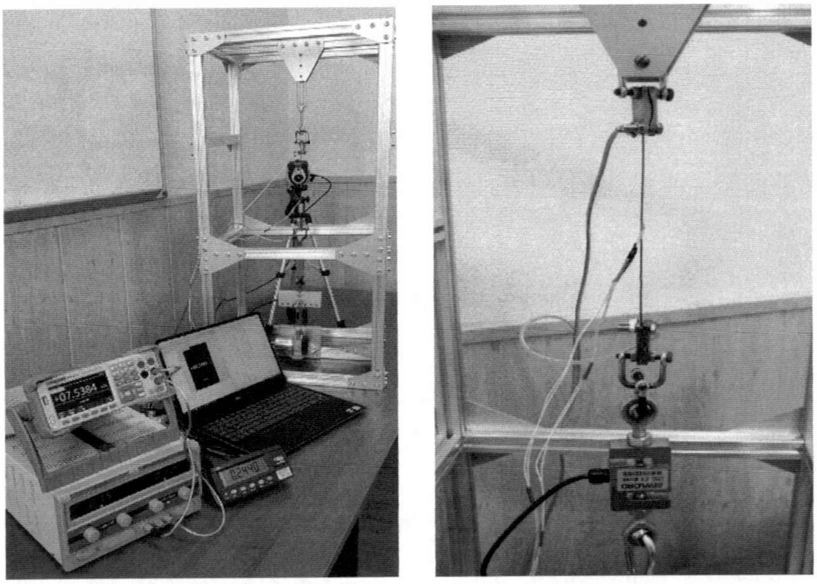

Figure 3. Experimental setup for determining the main characteristics of the active element.

and phase deformations:

$$\varepsilon = \varepsilon_{\mathrm{e}} + \varepsilon_{\mathrm{T}} + \varepsilon_{\mathrm{tr}},$$

where $\varepsilon_{\mathrm{e}} = \sigma/E$ is elastic deformation, σ is tension at the current cross-sectional point, E is elasticity modulus, $\varepsilon_{\mathrm{T}} = \alpha_{\mathrm{T}}\Delta T$ is temperature deformation, α_{T} is coefficient of thermal expansion, $\varepsilon_{\mathrm{tr}}$ is phase deformation.

If a tensile force is applied to the wire (deformed), then when a certain tension is reached, it is deformed plastically and this deformation is called phase deformation $\varepsilon_{\mathrm{tr}}$. After the tensile force is removed, the phase deformation $\varepsilon_{\mathrm{tr}}$ remains in the wire. With subsequent heating of the wire, as a result of the reverse martensitic transformation, the phase deformation is partially restored (the material "remembers" its shape). The restored deformation is called shape memory deformation $\varepsilon_{\mathrm{rec}}$. The phase deformation can reach very large values for the titanium nickelide material 6–10%. Shape memory deformation is always smaller.

When designing active elements from a material with SME using a mathematical apparatus, dependencies that establish a relationship between the amount of phase deformation and the operational loads of the force actuator are required. These dependences are called force-deformation dependences and can be obtained as a result of an experiment.

During the preparatory tests, the active elements were subjected to various types of heat treatment. At this stage, the heating temperature of the furnace varied from 450°C to 850°C. The heat treatment time of the active elements varied from 30 to 60 minutes, with subsequent cooling taking place in the furnace. Heat treatment was carried out in order to eliminate possible internal stresses in the active elements. Further, a tensile force was applied to the active element, which created a tensile stress in its cross-sections. Then the active element was heated. The tests were carried out on the setup shown in figure 3.

The temperature of the active element at the time it began to reduce its length was 70°C. To simulate the operational force developed by the active element during experimental studies, a spring element with a stiffness of 17 N/mm was used. The tension created by the spring element before the start of the heating process of the active element was 0.04 kN. Intensive shortening of its length by the active element stopped at temperature 110°C.

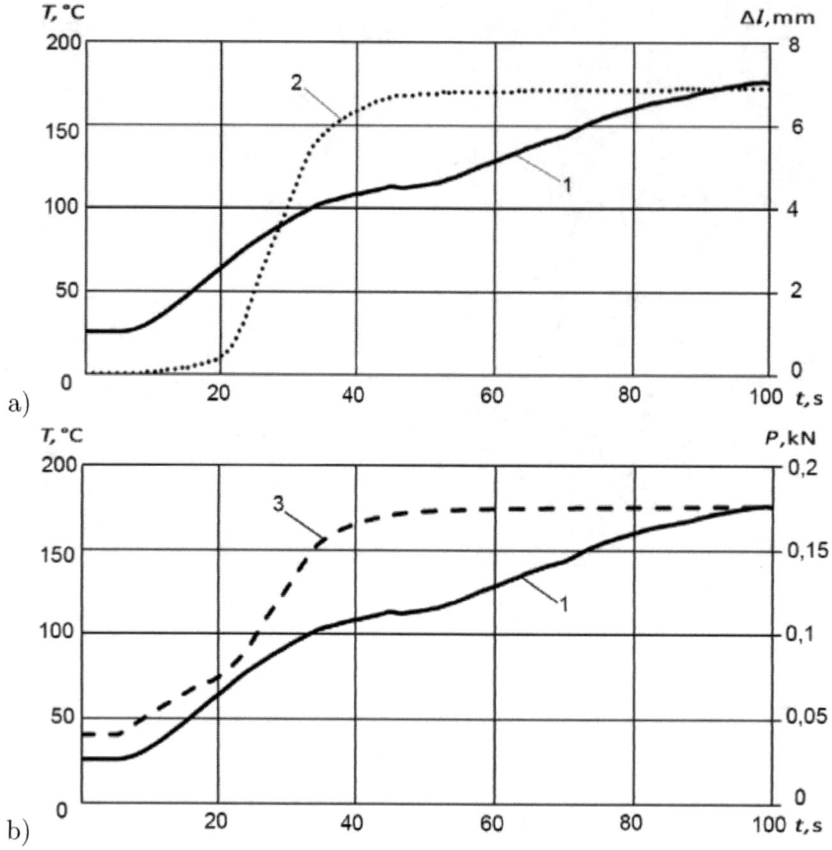

Figure 4. Test results: change in temperature and elongation of the active element (*a*); change in temperature and the force developed by the active element (*b*). *1* — temperature; *2* — displacement; *3* — force.

Figure 4 shows the test results. The active element with a length of 160 mm was deformed to a length of 176 mm due to the application of a tensile force. After removing the load, the length of the active element decreased to 174 mm. When heated, the length of the active element length decreased to 167 mm. The working stroke was 4% of the length of the active element before heating. Therefore, the maximum displacement was 7 mm, and the maximum force was 0.177 kN.

In the course of further experimental studies of the active elements of the force actuator with a length of 160 mm and 460 mm operating under uniaxial compression, their force-deformation characteristics were obtained. Active elements made of materials with SME can have only a temperature field as an operating influence. At the same time, along with the problem of deformation, there is a necessity to solve the problem of thermal conductivity. For materials with SME, both of these problems are nonlinear and related.

3. Functional model of the active element

In the course of ground-based experimental studies, it can be assumed that the change in the thermal energy of the active elements of the force actuator is equal to the amount of heat received due to electrical energy minus heat losses from natural convection

$$cm\frac{dT}{dt} = RI_1^2 - \alpha S(T - T_\mathrm{a}), \tag{1}$$

where c is specific heat capacity, m is mass of the active element, T is temperature of the active element, t is time; R is resistance of the active element, I_1 is current in the active element during testing, α is heat exchange coefficient, S is heat exchange surface area, T_a is ambient temperature.

Following [11], the solution of equation (1) will have the form:

$$T(t) = \bar{T} - (\bar{T} - T_0)\exp(-t/\beta),$$

where $T_0 = T(0)$; $\beta = cm/(\alpha S)$, $\bar{T}$ is equilibrium temperature.

For calculating the temperature change of the active elements, the selection of the heat exchange coefficient was carried out. Also, the calculations took into account the change in the resistance and heat capacity of the active elements during heating. During the numerical experiment, the obtained results correlated well with the test results.

When operating in orbit, it can be assumed that the change in the thermal energy of the active element of the force actuator is equal to the amount of heat received due to electrical energy minus thermal losses from radiation in orbit

$$cm\frac{dT}{dt} = RI_2^2 - \varepsilon\sigma_0 T^4 S, \tag{2}$$

where I_2 is current in the active element in orbit, ε is emissivity, σ_0 is Stefan-Boltzmann constant. The equilibrium temperatures from (1) and (2) under the condition $dT/dt = 0$

$$\bar{T}_1 = \frac{RI_1^2}{\alpha S} + T_\mathrm{a}, \quad \bar{T}_2 = \sqrt[4]{\frac{RI_2^2}{\varepsilon\sigma_0 S}}.$$

To ensure an approximate similarity of processes during ground tests and in orbit, it is necessary that

$$\bar{T}_1 = \bar{T}_2. \tag{3}$$

Then from (3) we obtain

$$\frac{RI_1^2}{\alpha S} + T_\mathrm{a} = \sqrt[4]{\frac{RI_2^2}{\varepsilon\sigma_0 S}}. \tag{4}$$

Convert (4) to the form

$$I_2 = \sqrt{\frac{\varepsilon\sigma_0 S}{R}\left(\frac{RI_1^2}{\alpha S} + T_\mathrm{a}\right)^2}. \tag{5}$$

In the course of conducting experimental studies in ground conditions, the initial length of the active element of the force actuator was equal to 460 mm, the current strength was 7.7 A. The resistance of the active element of the force actuator, obtained during the tests, was 0.22 Ω. Substituting these values in formula (5), we obtain an approximate value of the current strength I_2 in orbit, which is necessary to fulfill condition (3),

$$I_2 \approx 3.3\,\mathrm{A}.$$

The following values included in relation (5) were used in the calculation:

$$\alpha = 50\,\frac{\mathrm{W}}{\mathrm{m^2K}}, \quad S = 21.7\times 10^{-4}\,\mathrm{m}^2, \quad T_\mathrm{a} = 298\,\mathrm{K}, \quad \varepsilon = 0.66, \quad \sigma_0 = 5.67\times 10^{-8}\,\frac{\mathrm{W}}{\mathrm{m^2K^4}}.$$

One of the important characteristics of the actuator for the opening is the actuation time. To set the required actuation time of the active element of the actuator, we need to select the appropriate voltage supplied to the active element of the actuator. When changing the voltage supplied to the active element during the tests, it was found that its actuation time changed significantly, table 1. It should also be noted that despite the significant variation in the actuation time of the active element, its working stroke remained constant, and the process was smooth and shock-free.

Table 1. Dependence of the actuation time on the heating power.

No.	Working length (mm)	Voltage (V)	Current (A)	Average actuation time (s)	Working stroke (%)
1	175	1.9	7.8–8.7	230	5.7
2	173	2.0	8.2–9.1	110	5.7
3	175	2.1	8.6–9.5	65	5.7
4	175	2.2	9.1–10.1	40	5.7
5	175	2.3	9.4–10.4	35	5.7
6	175	2.4	9.8–10.8	30	5.7
7	174	2.5	10.2–11.1	25	5.7
8	175	2.6	10.6–11.5	20	5.7
9	175	2.7	11.1–12.0	15	5.7
10	175	2.8	11.5–12.5	10	5.7
11	174	2.9	11.9–12.8	10	5.7

Active elements, for which temperature change is used to control deformations, can provide slow (quasi-static) smooth shock-free movements of space structures with transformable configuration. This is due to the fact that the heating process of active elements is physically unrealistic to make fast.

4. Conclusion

In the course of quantitative and qualitative studies of the active elements of a force actuator with SME, it was found that it can provide a smooth opening of a large-sized space structure. This eliminates dynamic shock loads on its structural elements when fixing the opened operating state of the system in orbit with locking devices. Experimental studies have confirmed that force actuators operating in compression mode have an optimal combination of the created force and the amount of working stroke. Therefore, a wire made of titanium nickelide material with SME is most appropriate to use as an active element of force actuators for opening space structures with transformable configuration.

Acknowledgments

The study was supported by the Ministry of Science and Higher Education of Russia (project 0705-2020-0047).

References

[1] Lopatin A V and Rutkovskaya M A 2007 The review of designs of modern transformed space antennas (Part 1) *Vestnik Sibirskogo Gosudarstvennogo Aerokosmicheskogo universiteta imeni akademika M.F. Reshetneva* No. 2(18) 51–7

[2] Hasanzade V, Sedighy S H, and Shahravi M 2017 Compact deployable umbrella antenna design with optimum communication properties *Journal of Spacecraft and Rockets* **54** (3) 782–8 doi: 10.2514/1.A33710

[3] Zheng F and Chen M 2015 Affordable system conceptual structure design of new deployable spaceborne antenna *33rd AIAA International Communications Satellite Systems Conference and Exhibition* Art. No. 4343 doi: 10.2514/6.2015-4343
[4] Ponomarev S V 2011 Transformable reflectors of spacecraft antennas *Vestnik Tomskogo Gosudarstvennogo Universiteta, Matematika i Mekhanika* No. 4(16) 110–9
[5] Im E, Thomson M, Fang H, et al. 2007 Prospects of large deployable reflector antennas for a new generation of geostationary Doppler weather radar satellites *AIAA SPACE 2007 Conference & Exposition* Art. No. 9917 doi: 10.2514/6.2007-9917
[6] Banichuk N V, Karpov I I, Klimov D M, et al. 1997 *Mechanics of Large Space Structures* (Moscow: Izdat. "Faktorial") [in Russian]
[7] Zimin V N, Zikun Z, Krylov A V, and Churilin S A 2019 Mathematical modelling of the deployment of a large transformable space structure *AIP Conference Proceedings* **2171** 030002 doi: 10.1063/1.5133168
[8] Likhachev V A, Razov A I, Cherniavsky A G, et al. 1994 Truss mounting in space by shape memory alloys *Proceedings of the First International Conference on Shape Memory and Superelastic Technologies, California, USA* pp. 245–8
[9] Karla S, Bhattacharya B, and Munjal B S 2017 Design of shape memory alloy actuated intelligent parabolic antenna for space applications *Smart Materials and Structures* **26** (9) 095015 doi: 10.1088/1361-665X/aa7468
[10] Barvinok V A, Bogdanovich V I, Groshev A A, et al. 2013 Design method of power drives from material shape memory effect for rocket and space technology *Izvestiya Samarskogo Nauchnogo Tsentra RAN* **15** (6) 272–7
[11] Zimin V N, Krylov F V, and Shakhverdov A O 2021 Development of the mathematical model of the force actuator for deployment of large-sized space structures *Journal of Physics: Conference Series* **1902** 012115 doi: 10.1088/1742-6596/1902/1/012115
[12] Riad A, Ainamany A, and Benzohra M 2017 The shape memory alloy actuator controlled by the Sun's radiation *Materials Research Express* **4** (7) 075701 doi: 10.1088/2053-1591/aa75bb
[13] Liang C and Rogers C A 1997 Design of shape memory alloy actuators *Journal of Intelligent Material Systems and Structures* **8** (4) 303–13 doi: 10.1177/1045389X9700800403
[14] De Laurentis K J, Fisch A, Nikitczuk J, and Mavroidis C 2002 Optimal design of shape memory alloy wire bundle actuators *Proceedings 2002 IEEE International Conference on Robotics and Automation (Cat. No. 02CH37292)* Vol. 3, pp. 2363–8 doi: 10.1109/ROBOT.2002.1013585
[15] Schiedeck F, Hemsel T, and Wallaschek J 2006 The use of shape memory alloy wires in actuators *Solid State Phenomena* **113** 195–8 doi: 10.4028/www.scientific.net/SSP.113.195
[16] Mishustin I V 2012 Three-level model of shape memory alloy with application to reversing shape memory effect simulation *Mekhanika Kompozitsionnykh Materialov i Konstruktsii* **18** (4) 463–74
[17] V. A. Barvinok, V. I. Bogdanovich, O. V. Lomovsky, etc. 2011 Development of reversible power drives from materials with the effect of memory of form for devices, applied in the knots of check off of space vehicles *Izvestiya Samarskogo Nauchnogo Tsentra RAN* **13** (4(2)) 301–6

Supersonic divergence of a panel with a free edge initially loaded in two directions tensile and compressive forces

S R Martirosyan

Institute of Mechanics of the National Academy of Sciences of Armenia, Yerevan, Republic of Armenia

E-mail: mechinsstella@mail.ru

Abstract. In the article, in a linear formulation, the problem of the static stability of a rectangular plate in a supersonic gas is investigated, initially loaded in two directions, respectively, stretched along the gas flow and compressed in the direction perpendicular to the gas flow velocity, under the assumption that the gas flow runs on its free edge. An analytical solution to the problem is obtained.

The space of parameters of the "plate-flow" system was partitioned into areas of stability and static instability in the form of panel divergence and in the form of localized divergence. The corresponding critical velocities are found. The ratio of the stress coefficients characterizing the tensile and compressive forces, leading to the dominance of the stabilization effect, has been established. It is shown that compressive forces have an insignificant effect on the stability of sufficiently elongated plates, in comparison with tensile forces. The boundary value of the tensile stress coefficient is found, starting from which the elongated plates are stable at the beginning of the interval of supersonic velocities.

1. Introduction

In many aircraft and ship constructions, the plates experience different types of forces at the same time [1, 2, 11]. In particular, tensile and compressive forces that are located in the middle surface of the plate. In this case, since the width or length of the plate, which is the panel of an aircraft wing, the deck of a ship, etc., is usually much small compared to the dimensions of the constructions, in many cases the forces can be considered uniformly distributed along the corresponding side of the plate.

Consideration of the problems of aero elastic stability of plates and shells under combined loading is of great applied and theoretical importance. The question of the elastic stability of aircraft skin panels, which are flat plates or shallow shells, inevitably arises at the design and construction stage of any aircraft to ensure flight safety.

Theoretical studies of these problems make it possible to identify various types of buckling caused by the nature of deformations, as well as to assess the effect of combined loads on the stability threshold, with the aim of further analyzing the possibility of managing it. Therefore, in modern research, it is especially important to study the dynamic behavior of the "plate-flow" system, if possible, by analytical methods, along with numerical ones. A huge number of works are devoted to the study of the static and dynamic instability of plates and shells, a review of

Journal of Physics: Conference Series 2231 (2022) 012030 doi:10.1088/1742-6596/2231/1/012030

which is mainly contained in monographs and in articles [1–5]. However, in these works, with the exception of the works of A. A. Movchan [5], approximate solutions have been constructed, and an effective estimate of the accuracy of these approximations has not been given.

In the proposed article, an analytical solution is obtained for the static stability problem for a linear "plate-flow" system in a supersonic gas flow, under the assumption that a rectangular plate with one free edge and three hinged edges is initially loaded in two directions by tensile and compressive forces. And also, it was assumed that the supersonic gas flow runs on the free edge of the plate. An analytical solution to the problem was obtained using a method whose algorithm is described in detail in work [14]. This method is used to obtain an analytical solution to a number of problems of aero elastic stability [15–19].

The space of the parameters of the system "plate-flow" was partitioned into areas of stability and static instability: the area of panel divergence and the area of localized divergence. The boundary of the stability area is investigated. The critical velocities of panel divergence and localized divergence, the boundary values of the tensile and compressive stress coefficients, as well as the boundary value of the parameter of the ratio of the sides of the rectangular plate, delimiting the area of stability and the areas of instability in the form of panel divergence and localized divergence, are found, under the assumption that at the moment of "buckling" only bending stresses arise in it. The estimation of the influence on the stability of the "plate-flow" system of the stress coefficients characterizing the initial combined loaded state is determined.

2. Statement of the problem

Consider a thin elastic rectangular plate occupying, in a Cartesian coordinate system $Oxyz$, a domain $0 \le x \le a$, $0 \le y \le b$, $-h \le z \le h$. The Cartesian coordinate system $Oxyz$ is chosen so that the Ox- and Oy-axes lie in the plane of the undisturbed plate, and the Oz-axis is perpendicular to the plate and directed to the side of the supersonic gas flow streamlining it on one side in the direction of the Ox-axis with an undisturbed velocity V. It is assumed that the gas flow is plane and potential. Suppose that the edge $x = 0$ of the plate is free, whereas the edges $x = a$, $y = 0$ and $y = b$ are hinged. The hinges are assumed consider to be perfect.

We also assume that the plate initially, even before the gas flow around it, is loaded in two directions, respectively, by tensile $N_x = 2h\sigma_x$ and $N_y = 2h\sigma_y$ forces, which are the result of heating, or some other reason; tensile σ_x and compressive σ_y stresses are assumed to be uniformly distributed over the entire middle surface of the panel, to be constant and unchanged with a change in the deflection $w = w(x, y)$ of the plate [1, 2, 10, 11].

The deflection of the plate $w = w(x, y)$ will cause excessive pressure Δp on the upper streamlined surface of the plate from the side of the flowing gas flow, which is taken into account by the approximate formula of the "piston theory" [1, 2, 6, 7]: $\Delta p = -a_0\rho_0 V \frac{\partial w}{\partial x}$, where a_0 is the sound velocity in an undisturbed gas medium, and ρ_0 is the density of the undisturbed gas flow. The deflections $w = w(x, y)$ are small as compared with the plate thickness $2h$.

Let us find out the conditions under which, along with the unperturbed form of equilibrium (unbent plate), a curved form of equilibrium (curved plate) is possible in the case in which the bending of the plate is caused by the corresponding aerodynamic loads Δp as well as the tensile σ_x and compressive σ_y stresses in the middle surface of the plate. We assume that the tensile stresses σ_x are small than the lower limit of fluidity $(\sigma_x)_{\text{pr.}}$ [1, 11, 16] and the compressive stresses σ_y are small as compared to the critical stresses $(\sigma_y)_{\text{cr.}}$ that can produce buckling of the plate in the absence of the gas flow ($V = 0$) [1, 2, 8, 10, 11, 15, 17–19]. The values of $(\sigma_x)_{\text{pr.}}$ and $(\sigma_y)_{\text{cr.}}$ are within the elastic range.

Then, under the assumption of the validity of the Kirchhoff hypothesis and the "piston theory" [6, 7], the differential equation of bending of a thin rectangular plate is described by the relation [1, 2]:

$$D\Delta^2 w - N_x \frac{\partial^2 w}{\partial x^2} + N_y \frac{\partial^2 w}{\partial y^2} + a_0 \rho_0 V \frac{\partial w}{\partial x} = 0, \quad w = w(x, y); \tag{1}$$

and the corresponding boundary conditions:

$$\frac{\partial^2 w}{\partial x^2} + \nu \frac{\partial^2 w}{\partial y^2} = 0, \quad D \frac{\partial}{\partial x} \left(\frac{\partial^2 w}{\partial x^2} + (2 - \nu) \frac{\partial^2 w}{\partial y^2} \right) - N_x \frac{\partial w}{\partial x} = 0, \quad x = 0, \tag{2}$$

$$w = 0, \quad \frac{\partial^2 w}{\partial x^2} = 0, \quad x = a; \tag{3}$$

$$w = 0, \quad \frac{\partial^2 w}{\partial y^2} = 0, \quad y = 0 \text{ and } y = b; \tag{4}$$

where $\Delta^2 w = \Delta(\Delta w)$, Δ is the Laplace's differential operator; D is the cylindrical flexural stiffness; ν is Poisson's ratio.

It is required to find the critical velocity V_{cr} or the lowest velocity value in the interval of supersonic and hypersonic velocities [1, 2]:

$$V_{\text{cr.}} \in (a_0 M_0, a_0 M_{2\cos m.}), \quad M_0 = \sqrt{2}, \quad M_{2\cos m.} \approx 33.85; \tag{5}$$

starting from which the "plate-flow" system loses its static stability at velocities $V \geq V_{\text{cr.}}$ under the assumption

$$\sigma_x < (\sigma_x)_{\text{pr.}}, \quad \sigma_y < (\sigma_y)_{\text{cr.}}. \tag{6}$$

Here, M_0 and $M_{2\text{cosm.}}$ are the Mach numbers corresponding to the boundary values of the interval of supersonic and hypersonic velocities [1, 2]; $(\sigma_x)_{\text{pr.}}$ is the lower limit of fluidity; $(\sigma_y)_{\text{cr.}}$ is the critical stresses, leading to buckling of the plate in the absence of flow ($V = 0$) and tensile forces ($\sigma_x = 0$). Its values are found in [15–19].

Thus, analysis of the stability of the "plate–flow" system (1)–(4) is reduced to the study of the differential equation (1) with the corresponding boundary conditions (2)–(4) for the deflection $w = w(x, y)$ under the assumption (5) and (6).

Let us denote by γ the ratio of the width a of the panel (downstream side) to its length b

$$\gamma = ab^{-1} \in (0, \infty). \tag{7}$$

Obviously, the limiting value of $\gamma = 0$ corresponds to an infinitely elongated plate, and the value of $\gamma = \infty$ corresponds to a semi-infinite plate-strip.

3. General solution of the stability problem

The general solution of equation (1), satisfying the boundary conditions (2)–(4), will be sought in the form of harmonic vibrations

$$w(x, y) = \sum_{n=1}^{\infty} C_n \exp(\mu_n r x) \sin(\mu_n y), \quad \mu_n = \pi n b^{-1}, \quad \text{for all} \quad \gamma \in (0, \infty), \tag{8}$$

where C_n are the arbitraries constants: $\sum_{n=1}^{\infty} C_n^2 \neq 0$, n is the number of half-waves along the side b of the plate.

Substituting the solution (8) into the differential equation (1), we obtain the characteristic equation, which is an algebraic equation of the fourth degree

$$r^4 - 2(1 + \beta_x^2) r^2 + \alpha_n^3 r + (1 - \beta_y^2) = 0, \tag{9}$$

$$\alpha_n^3 = a_0 \rho_0 V D^{-1} \mu_n^{-3}, \quad \beta_x^2 = \tfrac{1}{2} N_x D^{-1} \mu_n^{-2} = h \sigma_x D^{-1} \mu_n^{-2}, \quad \beta_y^2 = N_y D^{-1} \mu_n^{-2} = 2h \sigma_y D^{-1} \mu_n^{-2}. \tag{10}$$

In accordance with constraint (5) and condition (6), we have

$$\alpha_n^3 \in a_0^2\rho_0 D^{-1}\mu_n^{-3}(M_0, M_{2\text{cosm.}}), \tag{11}$$

$$\beta_x^2 = \beta_x^2(n,\gamma,\nu) < (\beta_x^2)_{\text{pr.}} = h(\sigma_x)_{\text{pr.}} D^{-1}\mu_n^{-2}, \quad \beta_y^2 = \beta_y^2(n,\gamma,\nu) < (\beta_y^2)_{\text{cr.}} = 2h(\sigma_y)_{\text{cr.}} D^{-1}\mu_n^{-2}. \tag{12}$$

It is clear that parameter α_n^3 of the gas flow velocity V characterizes the non–conservative component of the load, and, accordingly, the stress coefficients β_x^2 and β_y^2 characterize the conservative component of the load.

Equation (9) after simple transformations can be reduced to the form

$$(r^2 - 1 - \beta_x^2)^2 + \alpha_n^3 r = \beta_x^4 + 2\beta_x^2 + \beta_y^2, \quad \alpha_n^3 > 0, \quad \beta_x^2 > 0, \quad \beta_y^2 > 0. \tag{13}$$

It obviously follows from this that in the space of "essential" parameters of system (1)–(4), characteristic equation (9) has two negative roots $r_{1,2} \in R$ and a pair of complex conjugate roots $r_{3,4} \in W$ with a positive real part, or two pairs of complex conjugate roots $r_{1,2} \in W$, $r_{3,4} \in W$.

Similarly to what was shown in articles [13, 14] for $\alpha_n^3 \neq 0$, $\beta_x^2 = \beta_y^2 = 0$, [15] and [16] respectively for $\alpha_n^3 \neq 0$, $\beta_x^2 = 0$, $\beta_y^2 \neq 0$ and for $\alpha_n^3 \neq 0$, $\beta_x^2 \neq 0$, $\beta_y^2 = 0$, here also for $\alpha_n^3 \neq 0$, $\beta_x^2 \neq 0$, $\beta_y^2 \neq 0$ the characteristic equation (9) has either two negative roots and a pair of complex conjugate roots, or two pairs of complex conjugate roots. However, in the latter case, the corresponding characteristic determinant at $\alpha_n^3 > 0$ ($V > 0$) has no real roots in its domain.

Thus, it can be argued that the original problem (1)–(4) at $\alpha_n^3 \neq 0$ has a solution if and only if two roots the characteristic equation (9) are two negative roots: $r_1 < 0$, $r_2 < 0$, and the other two roots are complex conjugate ($r_{3,4} \in W$) with a positive real part.

Let us investigate the characteristic equation (9) using the algorithm of the Ferrari solution method [12–14]. According to the algorithm of the Ferrari method, characteristic equation (9) can be represented as the multiplication of two square trinomials:

$$\left[r^2 + \sqrt{2(q+1+\beta_x^2)}\,r + q - \sqrt{q^2 - 1 + \beta_y^2}\right] \times \left[r^2 - \sqrt{2(q+1+\beta_x^2)}\,r + q + \sqrt{q^2 - 1 + \beta_y^2}\right] = 0, \tag{14}$$

where q is the only real root of the cubic equation [12]

$$8(q+1+\beta_x^2)(q^2 - 1 + \beta_y^2) - \alpha_n^6 = 0, \quad q \in (q_0, \infty) \subset R. \tag{15}$$

It is easy to show that the determinant of the cubic equation (15) is positive, which implies the uniqueness of its real root.

Using graphical-analytical research methods, it can be shown that the characteristic equation (9) in the form (14) for all $\gamma \in (0,\infty)$ and $\gamma = \infty$ has two negative roots $r_{1,2}$ and a pair complex conjugate roots $r_{3,4}$ with a positive real part in the interval of values

$$q \in (q_0, \infty) \subset R, \tag{16}$$

$$q_0 = \begin{cases} \frac{1}{3}\left[-(1+\beta_x^2) + 2\sqrt{(1+\beta_x^2)^2 + 3(1-\beta_y^2)}\right], & \beta_y^2 \leq 1 + \frac{1}{3}(1+\beta_x^2)^2 < (\beta_y^2)_{\text{cr.}}, \\ 1, & 1 + \frac{1}{3}(1+\beta_x^2)^2 < \beta_y^2 < (\beta_y^2)_{\text{cr.}}, \end{cases} \tag{17}$$

$$\beta_x^2 < (\beta_x^2)_{\text{pr.}}.$$

The expressions of roots r_i are easily determined from the solution of the corresponding quadratic equations (14):

$$r_{1,2} = -0.5\sqrt{2(q+1+\beta_x^2)} \pm \sqrt{\sqrt{q^2 - 1 + \beta_y^2} - 0.5(q - 1 - \beta_x^2)}, \quad r_1 < 0, r_2 < 0, \tag{18}$$

$$r_{3,4} = 0.5\sqrt{2(q+1+\beta_x^2)} \pm i\sqrt{\sqrt{q^2 - 1 + \beta_y^2} + 0.5(q - 1 - \beta_x^2)}. \tag{19}$$

Note that expression (18) is obviously a necessary condition for the loss of stability of a sufficiently wide plates ($\gamma \geq \gamma_*$) in the form of a divergence localized in the vicinity of edge $x = 0$, like a semi-infinite plate-strip ($\gamma = \infty$) for all $\beta_y^2 < (\beta_y^2)_{\text{loc.}} < 1$, $\beta_x^2 < (\beta_x^2)_{\text{pr.}}$ [3, 6, 16, 17, 19]. Here, $(\beta_y^2)_{\text{loc.}}$ is the critical value of the stress coefficient β_y^2 at which the unperturbed form of equilibrium of a sufficiently wide panel $\gamma \geq \gamma_*$ loses its stability in the form of localized instability at the its free edge $x = 0$ in the absence of flow ($V = 0$) [8, 9, 10, 13–16].

Unlike rather wide plates ($\gamma \geq \gamma_*$), rather elongated plates and plates of moderate dimensions ($\gamma < \gamma_*$) lose their static stability in the form of panel divergence [14–16].

Thus, $\gamma_* = \gamma_*(\nu, \beta_x^2, \beta_y^2)$ is the limiting value of the parameter $\gamma = ab^{-1}$ delimiting the domains of panel divergence ($\gamma < \gamma_*$) and localized divergence ($\gamma \geq \gamma_*$) in the space of "essential" parameters of the stability problem (1)–(4) [13, 14–19].

In accordance with expressions (18) and (19), the general solution (8) will be rewritten as a double series in the form:

$$w(x,y) = \sum_{n=1}^{\infty}\sum_{k=1}^{4} C_{nk}\exp(\mu_n r_k x)\sin(\mu_n y), \quad \text{when } \gamma \in (0, \gamma_*) \tag{20}$$

in the case of sufficiently wide plates ($\gamma \geq \gamma_*$), as well as a semi-infinite plate-strip ($\gamma = \infty$), the general solution (8) will be written in the form [8, 9, 10, 13]:

$$w(x,y) = \sum_{n=1}^{\infty}\sum_{k=1}^{2} C_{nk}\exp(\mu_n r_k x)\sin(\mu_n y), \quad \text{when } \gamma \in [\gamma_*, \infty] \tag{21}$$

taking into account the condition of damping of vibrations at the edge $x = a$.

Substituting expression (11) into relation (15), after simple transformations, we obtain an explicit form of the dependence of the gas flow velocity V on the parameters of the system (1)–(4), namely: in the case of the system with an elongated plate or with a plate of moderate dimensions ($\gamma \in (0, \gamma_*)$), we obtain

$$V(q, n, \gamma, \nu, \beta_x^2, \beta_y^2) = 2\sqrt{2(q + 1 + \beta_x^2)(q^2 - 1 + \beta_y^2)\pi^3 n^3 \gamma^3 D (a_0 \rho_0 a^3)^{-1}}. \tag{22}$$

And, in the case of the system with sufficiently wide plates ($\gamma \in [\gamma_*, \infty)$), as with as a semi-infinite plate-strip ($\gamma = \infty$), we obtain

$$V(q, n, \nu, \beta_x^2, \beta_y^2) = 2\sqrt{2(q + 1 + \beta_x^2)(q^2 - 1 + \beta_y^2)\pi^3 n^3 D (a_0 \rho_0 b^3)^{-1}}. \tag{23}$$

In this case, holds for all "essential" parameters of the stability problem (1)–(4), as applied to the interval of supersonic velocities (5), it is obvious that

$$V(q) \in (V(q_0), a_0 M_{2\text{cosm.}}) \subseteq (a_0 M_0, a_0 M_{2\text{cosm.}}). \tag{24}$$

4. Sufficient sign of loss of static stability of the system "plate-flow". Characteristic determinants

4.1. Dispersion equation for the rectangular plates

Substituting the general solution (20) of the differential equation (1) into the boundary conditions (2)–(4), we obtain a homogeneous system of fourth–order algebraic equations with respect to constants C_{nk}, $k = \overline{1,4}$. The determinant of this system equated to zero leads to the dispersion equation of the panel divergence for all $\gamma \in (0, \infty)$ in the form:

$$\begin{aligned}
F_{\text{div.pan.}} &= F(q, n, \gamma, \nu, \beta_x^2, \beta_y^2) \\
&= \sqrt{2(q+1+\beta_x^2)}\Big[\Big(q+1-\sqrt{q^2-1+\beta_y^2}\Big)^2-2(q+1)\nu-(1-\nu)^2+2\beta_x^2\Big(q-\sqrt{q^2-1+\beta_y^2}\Big)\Big]B_1B_2 \\
&-\sqrt{2(q+1+\beta_x^2)}\Big[\Big(q+1+\sqrt{q^2-1+\beta_y^2}\Big)^2-2(q+1)\nu-(1-\nu)^2+2\beta_x^2\Big(q+\sqrt{q^2-1+\beta_y^2}\Big)\Big]B_1B_2 \\
&\times \exp\big[-2\sqrt{2(q+1+\beta_x^2)}\pi n\gamma\big] +2\Big(\Big\{(4q^2+2q-1)\sqrt{q^2-1+\beta_y^2}-(2q^2-4q+1)(q+1) \\
&-\Big(q-1-\sqrt{q^2-1+\beta_y^2}\Big)\beta_y^2+\Big(2q^2+4q-1+2q\sqrt{q^2-1+\beta_y^2}+\beta_y^2+2q\beta_x^2\Big)\beta_x^2 \\
&-2\Big[(2q-1)(q+1)-q\sqrt{q^2-1+\beta_y^2}+\beta_y^2+q\beta_x^2\Big]\nu+\Big(q+1+\sqrt{q^2-1+\beta_y^2}+\beta_x^2\Big)\nu^2\Big\}\sinh(\pi n\gamma B_1) \\
&+2\sqrt{2(q+1+\beta_x^2)(q^2-1+\beta_y^2)(q+1+\beta_x^2)}B_1\cosh(\pi n\gamma B_1)\Big)B_2\cos(\pi n\gamma B_2) \\
&\times \exp\big[-\sqrt{2(q+1+\beta_x^2)}\pi n\gamma\big]+2\Big(-B_1\Big\{(4q^2+2q-1)\sqrt{q^2-1+\beta_y^2}+(2q^2-4q+1)(q+1) \\
&+\Big(q-1+\sqrt{q^2-1+\beta_y^2}\Big)\beta_y^2-\Big(2q^2+4q-1-2q\sqrt{q^2-1+\beta_y^2}+\beta_y^2+2q\beta_x^2\Big)\beta_x^2 \\
&+2\Big[(2q-1)(q+1)+q\sqrt{q^2-1+\beta_y^2}+\beta_y^2+q\beta_x^2\Big]\nu-\Big(q+1-\sqrt{q^2-1+\beta_y^2}+\beta_x^2\Big)\nu^2\Big\}\cosh(\pi n\gamma B_1) \\
&-\sqrt{2(q+1+\beta_x^2)(q^2-1+\beta_y^2)}[3(q^2-1)+2\beta_y^2-(2+\beta_x^2)\beta_x^2]\sinh(\pi n\gamma B_1)\Big)\sin(\pi n\gamma B_2) \\
&\times \exp[-\sqrt{2(q+1+\beta_x^2)}\pi n\gamma]=0,
\end{aligned} \tag{25}$$

$$B_1=\sqrt{\sqrt{q^2-1+\beta_y^2}-0.5(q-1-\beta_x^2)},\quad B_2=\sqrt{\sqrt{q^2-1+\beta_y^2}+0.5(q-1-\beta_x^2)},$$
$$B_1>0,\quad B_2>0.$$

The hyper surface (25) with a value of $q = q_{\text{crdiv}}$, where q_{crdiv} is the real solution to the equation (25), delimits the domain of stability ($F_{\text{div.pan.}} > 0$) and the domain of divergence ($F_{\text{div.pan.}} < 0$).

4.2. The dispersion equation of the localized divergence for semi-infinite plate-strip

Substituting the general solution (21) of the differential equation (1) into the boundary conditions (2) and (3), we obtain a homogeneous system of second-order algebraic equations with respect to constants C_{nk}, $k = \overline{1,2}$. The determinant of this system equated to zero leads to the dispersion equation of the localized divergence in the form [15–19]:

$$\begin{aligned}
F_{\text{loc.div.}} = F(q,\nu,\beta_x^2,\beta_y^2) &= \Big(q+1-\sqrt{q^2-1+\beta_y^2}\Big)^2-2(q+1)\nu- \\
&-(1-\nu)^2+2\beta_x^2\Big(q-\sqrt{q^2-1+\beta_y^2}\Big)=0,\quad \beta_x^2<(\beta_x^2)_{\text{pr.}},\quad \beta_y^2<(\beta_y^2)_{\text{loc.}}<1.
\end{aligned} \tag{26}$$

Likewise, the hyper surface (26) with a value of $q = q_{\text{loc.div}}$, where $q_{\text{loc.div}}$ is the only real solution to the equation (26), delimits the domain of stability ($F_{\text{loc.div.}} > 0$) and the domain of localized divergence ($F_{\text{loc.div.}} < 0$).

Using analytical research methods, it is easy to show that for all n, $\beta_x^2 < (\beta_x^2)_{\text{pr.}}$, $\beta_y^2 < (\beta_y^2)_{\text{loc.}} < 1$, it is true

$$\lim_{\gamma\to\infty}(F_{\text{div.pan.}} = F(q,n,\gamma,\nu,\beta_x^2,\beta_y^2)=0)=(F_{\text{loc.div.}}(q,\nu,\beta_x^2,\beta_y^2)=0). \tag{27}$$

Table 1. The values of function γ_*.

β_x^2	0	1.5	1.8	3.0	3.6
γ_*	1.96	1.96	1.8	1.6	1.4

Numerical research methods have shown the fulfillment of condition (27) not only in the case of a semi-infinite plate-strip ($\gamma = \infty$), but also in the case of sufficiently wide plates, starting from the value $\gamma = \gamma_*$, where $\gamma_* = \gamma_*(\nu, \beta_x^2, \beta_y^2)$ (table 1).

So, the boundary value

$$\gamma = \gamma_* = \gamma_*(\nu, \beta_x^2, \beta_y^2) \tag{28}$$

separates the domains of the static instability: the domain of the panel divergence and the domain of localized divergence.

Substituting the values $q_{\text{cr.div.}}$ and $q_{\text{loc.div.}}$ in expressions (22) and (23) we obtain the critical panel divergence velocity $V_{cr.div.}$ and localized divergence velocity $V_{loc.div.}$ respectively. At gas flow velocities of $V \geq V_{cr.div.}$ the rectangular plate ($\gamma < \gamma_*$) loses stability in the form of panel divergence, and at velocities of $V \geq V_{\text{loc.div.}}$ the sufficiently wide plate ($\gamma \geq \gamma_*$) and the semi–infinite plate-strip ($\gamma = \infty$) loses stability in the form of localized divergence in the vicinity of the free edge $x = 0$ of the plate. In both cases, the transitions from the domain of stability to the domain of instability are "soft".

Accordingly, during the panel divergence, the additional stresses arising during the flow lead to the "buckling" of the entire surface of the plate at a limited velocity. In the case of localized divergence, additional stresses lead to "buckling" of only a narrow strip of the plate surface along its free edge $x = 0$. Due to the limited velocity of change in the surface of the plate, these deformations are considered static.

In accordance with expressions (22) and (23), by virtue of condition (5), it is obvious that

$$V_{\text{cr.div.}} = V(q_{\text{cr.div.}}, n, \gamma, \beta_x^2, \beta_y^2) \in (V(q_0, n, \gamma, \beta_x^2, \beta_y^2), a_0 M_{2\text{cosm.}}) \subseteq (a_0 M_0, a_0 M_{2\text{cosm.}}),$$
$$V_{\text{loc.div.}} = V(q_{\text{loc.div.}}, n, \beta_x^2, \beta_y^2) \in (V(q_0, n, \beta_x^2, \beta_y^2), a_0 M_{2\text{cosm.}}) \subseteq (a_0 M_0, a_0 M_{2\text{cosm.}}).$$

4.3. Dispersion equation for an infinite elongated plate

Tending $b \to \infty$, the differential equation (1) and boundary conditions (2)–(4) of the stability problem, take the form:

$$D\frac{d^4w}{dx^4} - N_x\frac{d^2w}{dx^2} + a_0\rho_0 V\frac{dw}{dx} = 0, \tag{29}$$

$$\frac{d^2w}{dx^2} = 0, \quad D\frac{d^3w}{dx^3} - N_x\frac{dw}{dx} = 0, \quad x = 0, \qquad w = 0, \quad \frac{d^2w}{dx^2} = 0, \quad x = a. \tag{30}$$

Hence, it is obvious that in the case of an infinitely elongated panel ($b \to \infty$), the description of the stability problem (1)–(4) for $N_x \neq 0$, $N_y \neq 0$ and $N_x \neq 0$, $N_y = 0$ is the same.

The stability problem (1)–(4) for $N_x \neq 0$, $N_y = 0$ is studied in detail in article [16].

That's why, according to the results of article [19], the dispersion equation in the case of an infinitely elongated plate $b \to \infty$ ($\gamma = 0$) has the following description:

$$F_\infty(\tilde{q}, \tilde{\beta}_x^2) = \Big\{ - \exp\Big[- 1.5\sqrt{2(\tilde{q} + \tilde{\beta}_x^2)}\Big] \cos\sqrt{0.5(3\tilde{q} - \tilde{\beta}_x^2)}\Big\}\sqrt{(3\tilde{q} - \tilde{\beta}_x^2)(\tilde{q} + \tilde{\beta}_x^2)^3}$$
$$- (3\tilde{q}^2 - \tilde{\beta}_x^4) \sin\sqrt{0.5(3\tilde{q} - \tilde{\beta}_x^2)} = 0. \tag{31}$$

Here $\tilde{q} \in (\tilde{q}_0, \infty)$ is the only real root of the cubic equation

$$8\tilde{q}^2(\tilde{q}+\tilde{\beta}_x^2)-S^6=0, \quad \tilde{\beta}_x^2=h\sigma_x D^{-1}a^2, \quad S^3=a_0\rho_0 VD^{-1}a^3 \quad (\tilde{q}_0 > \tilde{\beta}_x^2 \cdot \tfrac{1}{3} \gg 1). \tag{32}$$

It obviously follows from this that

$$V(\tilde{q}, \tilde{\beta}_x^2) = 2\sqrt{2(\tilde{q} + \tilde{\beta}_x^2)\tilde{q}D(a_0\rho_0 a^3)^{-1}}, \quad \tilde{q} \in (\tilde{q}_0, \infty), \quad \tilde{\beta}_x^2 = h\sigma_x D^{-1}a^2. \tag{33}$$

As applied to the interval of supersonic velocities (5), we obtain

$$\tilde{q} \in (\tilde{q}_0, \tilde{q}(a_0 M_{2\text{cosm.}})) \subseteq (\tilde{q}(a_0 M_0), \tilde{q}(a_0 M_{2\text{cosm.}})). \tag{34}$$

In this case, also, the hyper surface (31) is the boundary of the stability domain ($F_\infty(\tilde{q}, \tilde{\beta}_x^2) > 0$) and the instability domain ($F_\infty(\tilde{q}, \tilde{\beta}_x^2) < 0$) in the form of the panel divergence. Substituting the first real root $\tilde{q}^{(1)}_{\text{cr.div.}} = \tilde{q}_{\text{cr.div.}}(\tilde{\beta}_x^2)$ of equation (31) into formula (33), we obtain the reduced critical panel divergence velocity $\tilde{V}_{\text{cr.div.}}D^{-1}(a_0\rho_0 a^3)$ of an infinitely elongated plate ($\gamma = 0$).

4.4. Dispersion equation for the sufficiently elongated plates
As is known [14–17], for sufficiently elongated plates ($\gamma \ll 1$) we have

$$q \in (q_0, \infty), \quad q_0 \gg 1. \tag{35}$$

In this case, it is true

$$\lim_{\gamma \to \gamma_{**}} (F_{\text{div.pan.}} = F(q, n, \gamma, \nu, \beta_x^2, \beta_y^2) = 0) = (F_{\text{elong.}}(q, n, \gamma_{**}, \beta_x^2) = 0) \tag{36}$$

for all n, $\beta_x^2 < (\beta_x^2)_{\text{pr.}}$, $\beta_y^2 < (\beta_y^2)_{\text{cr.}}$, when $\gamma_{**} \le 10^{-2}$;

$$F_{\text{elong.}}(q, n, \gamma, \beta_x^2) = \Big\{ - \exp\Big[- 1.5\sqrt{2(q + \beta_x^2)}\pi n\gamma\Big] + \cos\Big[\sqrt{0.5(3q - \beta_x^2)}\pi n\gamma\Big]\Big\}$$
$$\times \sqrt{(3q - \beta_x^2)(q + \beta_x^2)^3} - (3q^2 - \beta_x^4)\sin[\sqrt{0.5(3q - \beta_x^2)}\pi n\gamma] = 0. \tag{37}$$

Equation (37) is the dispersion equation for the panel divergence of a sufficiently elongated plate for all $\gamma \in (0, \gamma_{**}]$, where [14, 16]

$$\gamma_{**} \approx 10^{-2}. \tag{38}$$

The hyper surface (37) with the value $q = q_{\text{div.elong.}}$ delimits the domain of stability ($F_{\text{elong.}} > 0$) and the domain of panel divergence ($F_{\text{elong.}} < 0$). Here $q_{\text{div.elong.}}$ is the real solution to the equation (37).

Then, in accordance with condition (34), expression (22) for all $\gamma \le \gamma_{**} \approx 10^{-2}$ can be written in the form:

$$V(q, n, \gamma, \beta_x^2) = 2\sqrt{2(q + \beta_x^2)q\pi^3 n^3\gamma^3 D(a_0\rho_0 a^3)^{-1}}. \tag{39}$$

Substituting the first real root $q_{\text{div.elong.}} = q_{\text{div.elong.}}(n, \gamma, \beta_x^2)$ of equation (37) into formula (39), we obtain the reduced critical panel divergence velocity $V_{\text{cr.div.}}D^{-1}(a_0\rho_0 a^3)$ of a sufficiently elongated plate ($\gamma \in (0, \gamma_{**}]$).

Taking into account condition (5), it follows that

$$V(q, n, \gamma, \beta_x^2) \in (V(q_0, n, \gamma, \beta_x^2), a_0 M_{2\text{cosm.}}) \subseteq (a_0 M_0, a_0 M_{2\text{cosm.}}), \quad \gamma \in (0, \gamma_{**}].$$

In accordance with expressions (37) and (39), we can say that the reduced critical divergence velocity $VD^{-1}(a_0\rho_0 a^3)$ of a sufficiently elongated panel ($\gamma \le \gamma_{**} \approx 10^{-2}$) does not depend on the stress coefficient β_y^2 of compressive forces and Poisson's ratio ν, up to an accuracy of the order of a high degree of smallness, as in the case infinitely elongated panel ($\gamma = 0$).

Note that the boundary value (38), obtained using analytical methods for analyzing estimates of the terms of equation (25), was also confirmed by numerical methods of analysis.

Table 2. The values of function $(\beta_x^2)_{\min}$.

$2ha^{-1}$	0.006	0.010	0.012	0.015
$(\beta_x^2)_{\min.}$ at $\gamma = 0$	7.90	2.30	1.50	0.82
$(\beta_x^2)_{\min.}$ at $\gamma = 0.01$	7900.0	2300.0	1500.0	827.0
$(\beta_x^2)_{\min.}$ at $\gamma = 0.1$	79.0	23.0	15.0	8.27
$(\beta_x^2)_{\min.}$ at $\gamma = 0.3$	9.60	1.98	1.17	0.62

5. Numerical results

Using the numerical analysis of dispersion equations (25), (26), (31), and (37) families of curves $\{q(n,\gamma,\nu,\beta_x^2,\beta_y^2)\}$ and $\{q(n,\nu,\beta_x^2,\beta_y^2)\}$ were constructed, appropriately parameterized: the most representative of these families of curves were selected and investigated.

The static instability domain was divided into the domain of panel divergence and localized divergence. The values of function $\gamma_* = \gamma_*(\nu,\beta_x^2,\beta_y^2)$, delimiting these domains, are found (table 1).

As it turned out, γ_* is a decreasing function of the stress coefficient of tensile forces β_x^2: as β_x^2 increases from 0 to 12, it decreases by about 1.6 times. And vanishingly little depends on the stress coefficient of compressive forces β_y^2 and Poisson's ratio ν. This means that with an increase of β_x^2, the domain of panel divergence narrows, and the domain of localized divergence, on the contrary, increases.

It is shown that elongated plates ($\gamma \in (0, 0.33)$), like an infinitely elongated plate ($\gamma = 0$), of relative thickness $2ha^{-1} \in [0.006, 0.015]$, being unstable near $a_0\sqrt{2}$, which is the beginning of the interval of supersonic velocities (5) [1, 2, 14–19], become stable at $\beta_x^2 \in [(\beta_x^2)_{\min.}, 15.76)$ for all ν, β_y^2. The values of function $(\beta_x^2)_{\min} = (\beta_x^2)_{\min}(\gamma,\nu,\beta_y^2)$ are found. Function $(\beta_x^2)_{\min}(\gamma,\nu,\beta_y^2)$ is monotonically increasing from ν and β_y^2, but with increasing γ it decreases (table 2). It is clear that at values of $\beta_x^2 > (\beta_x^2)_{\min.}$ the influence of tensile forces on the stability of the system dominates: tensile forces lead to stabilization of the state of the system, in contrast to compressive forces.

Shown is the dependence of $(\beta_x^2)_{\min}$ on the relative thickness $2ha^{-1}$ of the plate. For steel plates, the values of the function $(\beta_x^2)_{\min.}$ of $2ha^{-1}$ are given in table 2.

Thus, in the case of elongated plates ($\gamma \in (0, 0.33)$), the effect of destabilization of compressive forces σ_y on the stability of the system (1)–(4) is vanishingly small in comparison with the effect of stabilization of tensile forces σ_x, which is quite consistent with the results of [15, 16].

The critical velocities of the panel divergence $V_{\text{cr.div.}}$ and localized divergence $V_{\text{loc.div.}}$ for all values of the parameters of the system (1)–(4) are found. With $n = 1$ and all $\gamma \in (0, \infty]$, critical velocities $V_{\text{cr.div.}}$ and $V_{\text{loc.div.}}$ take the lowest value with fixed values of the other parameters. And also, they are smaller in plates made of materials with a high Poisson's ratio ν. The critical velocities $V_{\text{cr.div.}}$ and $V_{\text{loc.div.}}$ are an increasing function of β_x^2 and decreasing of β_y^2. In the case of plates of moderate sizes $\gamma \in [0.33, \gamma_*)$ (table 1) at $\beta_x^2 \in [0, 5.2)$ and $\beta_y^2 \in [0, 1)$, the critical panel divergence velocity $V_{\text{cr.div.}}$ increases by about 6–8 times; in the case of sufficiently wide plates $\gamma \in [\gamma_*, \infty)$, the critical velocity of localized divergence $V_{\text{loc.div.}}$ at $\beta_x^2 \in [0, 3.7)$ and $\beta_y^2 \in [0, 0.73)$ increases by about 3 times.

At a certain ratio of forces σ_x and σ_y in the middle surface of the plate, respectively, stretching and compressing, the effect of their "mutual balancing" is observed, most clearly manifested in the case of plates at $\gamma \in [0.33, \infty)$ (table 3).

As follows from the data in table 3, in the case of sufficiently wide plates at $\gamma \in [\gamma_*, \infty)$ under condition $\varphi = \beta_x^2(\beta_y^2)^{-1} = 1$ for all permissible values of $\beta_x^2 \in [0, 3.7)$, $\beta_y^2 \in [0, 0.73)$, the critical velocities of localized divergence are exactly equal to the critical velocities of localized

Table 3. The values of functions β_x^2 and β_y^2.

γ	0.5	0.8	1.0	$\geq \gamma_*$
$(\beta_x^2;\beta_y^2)$	$(0.1;0.27)$	$(0.1;0.12)$	$(0.1;0.105)$	$\beta_x^2=\beta_y^2\in[0,0.73)$
	$(0.3;0.8)$	$(0.3;0.35)$	$(0.3;0.312)$	
		$(0.5;0.59)$	$(0.5;0.515)$	

divergence of a panel with unloaded edges ($\beta_x^2=0$, $\beta_y^2=0$) [13, 14].

At $\gamma\in[0.33,\gamma_*)$, we have $\varphi=\beta_x^2(\beta_y^2)^{-1}<1$ which implies the dominance of the stabilizing effect of tensile forces σ_x on the stability of the panel, in comparison with the destabilizing effect of compressive forces σ_y. At the same time, under condition $\varphi=\beta_x^2(\beta_y^2)^{-1}<1$, the critical panel divergence velocities are exactly equal to the critical velocities $V_{\text{cr.div.}}$ in the absence of the initial stress state of the plate ($\beta_x^2=0$, $\beta_y^2=0$) [14].

Thus, thanks to the proposed analytical method, it became possible to partition the parametric space of the problem stability (1)–(4) into domains of dominance of the influence of tensile forces, dominance of the influence of compressive forces, as well as on the domain in which their actions are balanced.

6. Conclusions
The graphical-analytical research method presented in this paper can be used to obtain an analytical solution for a wide class of stability problems for rectangular plates in a supersonic gas flow under initial combined loading.

References
[1] Volmir A S 1963 *The Stability of the Elastic System* (Moscow: Fizmatgiz) [in Russian]
[2] Bolotin V V 1961 *Nonconservative Problems of the Theory of Elastic Stability* (Moscow: Fizmatlit) [in Russian]
[3] Novichkov Yu N 1978 A flutter of plates and shells In *Results of Science and Technology.* Vol. 11. *Mechanics of Deformable Solids* (Moscow: VINITI) pp. 67–122
[4] Algazin S D and Kijko I A 2006 *Flutter of Plates and Shells* (Moscow: Nauka) [in Russian]
[5] Movchan 1956 On of Oscillations of a Plate Moving in a Gas *Prikl. Mat. Mekh.* **20** (2) 211–22.
[6] Il'yushin A A 1956 Law of Flat Sections at the Big Supersonic Velocity *Prikl. Mat. Mekh.* **20** (6) 733–55.
[7] Ashley G H and Zartarian G 1956 Piston theory – a new aerodynamic tool for the aeroelastician *Journal of the Aeronautical Sciences* **23** (12) 1109–18 doi: 10.2514/8.3740
[8] Ishlinskii A Yu 1954 About one limiting transition in the theory of stability of elastic rectangular plates *Doklady AN SSSR* **95** (3) 38–46
[9] Konenkov Yu K 1960 A Rayleigh–type flexural wave *Soviet Physics Acoustics* **6** (1) 122–23
[10] Tovstik P E 1995 *Stability of the Thin Shells: Asymptotic Methods* (Moscow: Nauka, Fizmatlit) [in Russian]
[11] Birger I A and Panovko Ya G (Editors) 1968 *Strength. Stability. Vibrations* (Moscow: Mashinostroenie) [in Russian]
[12] Korn G A and Korn T M 1968 *Mathematical Handbook* (McGrow–Hill Book Company)
[13] Belubekyan M V and Martirosyan S R 2012 The localized instability of the elastic plate-strip, streamlined by supersonic gas flow *Mechanics — Proceedings of National Academy of Sciences of Armenia* **65** (1) 29–34 doi: 10.33018/65.1.3
[14] Belubekyan M V and Martirosyan S R 2014 On the problem of the flutter of an elastic rectangular plate, when the supersonic gas flow is in a direction perpendicular to the free edge *Mechanics — Proceedings of National Academy of Sciences of Armenia* **67** (2) 14–44 doi: 10.33018/67.2.2
[15] Belubekyan M V and Martirosyan S R 2017 On divergence of compressed panel in supersonic gas flow, an accumulating on its free edge *Mechanics — Proceedings of National Academy of Sciences of Armenia* **70** (4) 12–34 doi: 10.33018/70.4.2
[16] Belubekyan M V and Martirosyan S R 2019 On divergence of the panel, stretched on the direction of

supersonic gas flow, an accumulating on its free edge *Mechanics — Proceedings of National Academy of Sciences of Armenia* **72** (2) 24–41 doi: 10.33018/72.2.2

[17] Belubekyan M V and Martirosyan S R 2020 Supersonic flutter of a compressed elongated plate in the presence of concetrated inertial masses and moments *Mechanics — Proceedings of National Academy of Sciences of Armenia* **73** (4) 58–74 doi: 10.33018/73.4.5

[18] Belubekyan M V and Martirosyan S R 2021 Supersonic flutter panel with one free edge, compressed in the direction, perpendicular to the gas flow velocity, in presence of concentrated inertial masses and moments *Mechanics — Proceedings of National Academy of Sciences of Armenia* **74** (2) 33–59 doi: 10.33018/74.2.3

[19] Belubekyan M V and Martirosyan S R 2021 On the stability of a panel with a free edge compressed in a direction perpendicular to the velocity of a supersonic gas flow in the presence of concentrated inertial masses and moments *Mechanics — Proceedings of National Academy of Sciences of Armenia* **74** (3) 19–36 doi: 10.33018/74.3.2

AUTHOR INDEX